Solid–Liquid Electrochemical Interfaces

ACS SYMPOSIUM SERIES **656**

Solid–Liquid Electrochemical Interfaces

Gregory Jerkiewicz, EDITOR
Université de Sherbrooke

Manuel P. Soriaga, EDITOR
Texas A&M University

Kohei Uosaki, EDITOR
Hokkaido University

Andrzej Wieckowski, EDITOR
University of Illinois at Urbana-Champaign

Developed from a symposium sponsored by the
International Chemical Congress of Pacific Basin Societies at the
1995 International Chemical Congress of Pacific Basin Societies

American Chemical Society, Washington, DC

Library of Congress Cataloging-in-Publication Data

Solid—liquid electrochemical interfaces / Gregory Jerkiewicz, editor . . . [et al.].

p.   cm.—(ACS symposium series, ISSN 0097–6156; 656)

"Developed from a symposium sponsored by the International Chemical Congress of Pacific Basin Societies at the 1995 International Chemical Congress of Pacific Basin Societies, Honolulu, Hawaii, December 17–22, 1995."

Includes bibliographical references and indexes.

ISBN 0–8412–3480–9

1. Solid—liquid interfaces—Congresses. 2. Electrochemistry—Congresses.

I. Jerkiewicz, Gregory. II. International Chemical Congress of Pacific Basin Societies (1995: Honolulu, Hawaii). III. Series

QD509.S65S64   1996
541.3'724—dc21

96–49774
CIP

This book is printed on acid-free, recycled paper.

Foreword

THE ACS SYMPOSIUM SERIES was first published in 1974 to provide a mechanism for publishing symposia quickly in book form. The purpose of this series is to publish comprehensive books developed from symposia, which are usually "snapshots in time" of the current research being done on a topic, plus some review material on the topic. For this reason, it is necessary that the papers be published as quickly as possible.

Before a symposium-based book is put under contract, the proposed table of contents is reviewed for appropriateness to the topic and for comprehensiveness of the collection. Some papers are excluded at this point, and others are added to round out the scope of the volume. In addition, a draft of each paper is peer-reviewed prior to final acceptance or rejection. This anonymous review process is supervised by the organizer(s) of the symposium, who become the editor(s) of the book. The authors then revise their papers according to the recommendations of both the reviewers and the editors, prepare camera-ready copy, and submit the final papers to the editors, who check that all necessary revisions have been made.

As a rule, only original research papers and original review papers are included in the volumes. Verbatim reproductions of previously published papers are not accepted.

ACS BOOKS DEPARTMENT

Contents

Preface xi

1. From Electrochemistry to Molecular-Level Research
   on the Solid–Liquid Electrochemical Interface: An Overview 1
   Gregory Jerkiewicz

2. Molecular Dynamics Simulation of Interfacial Electrochemical
   Processes: Electric Double Layer Screening 13
   Michael R. Philpott and James N. Glosli

3. Computer Simulation of the Structure and Dynamics of Water
   Near Metal Surfaces.. 31
   E. Spohr

4. Fundamental Thermodynamic Aspects of the Underpotential
   Deposition of Hydrogen, Semiconductors, and Metals 45
   Alireza Zolfaghari and Gregory Jerkiewicz

5. Micrometer-Scale Imaging of Native Oxide on Silicon Wafers
   by Using Scanning Auger Electron Spectroscopy 61
   Mikio Furuya

6. Growth Kinetics of Phosphate Films on Metal Oxide Surfaces 71
   T. S. Murrell, M. G. Nooney, L. R. Hossner,
   and D. W. Goodman

7. The Effects of Bromide Adsorption on the Underpotential
   Deposition of Copper at the Pt(111)–Solution Interface 87
   Nenad M. Marković, C. A. Lucas, Hubert A. Gasteiger,
   and Philip N. Ross, Jr.

8. Electron Spectroscopy Studies of Acidified Water Surfaces 106
   D. Howard Fairbrother, H. S. Johnston, and G. A. Somorjai

9. Electrochemical Digital Etching: Atomic Level Studies
   of CdTe(100).. 115
   T. A. Sorenson, B. K. Wilmer, and J. L. Stickney

10. **Adsorption of Bisulfate Anion on the Au(111), Pt(111), and Rh(111) Surfaces: A Comparative Study**................ 126
S. Thomas, Y.-E. Sung, and Andrzej Wieckowski

11. **Ex Situ Low-Energy Electron Diffraction and Auger Electron Spectroscopy and Electrochemical Studies of the Underpotential Deposition of Lead on Cu(100) and Cu(111)** 142
Gessie M. Brisard, Entissar Zenati, Hubert A. Gasteiger, Nenad M. Markovic, and Philip N. Ross, Jr.

12. **Anion Adsorption and Charge Transfer on Single-Crystal Electrodes** 156
Roberto Gómez, José M. Orts, and Juan M. Feliu

13. **In Situ Scanning Tuneling Microscopy of Organic Molecules Adsorbed on Iodine-Modified Au(111), Ag(111), and Pt(111) Electrodes** 171
K. Itaya, N. Batina, M. Kunitake, K. Ogaki, Y.-G. Kim, L.-J. Wan, and T. Yamada

14. **Structure of the GaAs(100) Surface During Electrochemical Reactions Determined by Electrochemical Atomic Force Microscopy**.. 189
Kohei Uosaki, Michio Koinuma, Namiki Sekine, and Shen Ye

15. **Microfabrication and Characterization of Solid Surfaces Patterned with Enzymes or Antigen–Antibodies by Scanning Electrochemical Microscopy**.. 202
H. Shiku, Y. Hara, T. Takeda, T. Matsue, and I. Uchida

16. **Electroactive Polymers: An Electrochemical and In Situ Scanning Probe Microscopy Study** 210
P. Forrer, G. Repphun, E. Schmidt, and H. Siegenthaler

17. **X-ray Photoelectron Spectroscopy and Scanning Tunneling Microscopy Studies of Thin Anodic Oxide Overlayers on Metal and Alloy Single-Crystal Surfaces**.................................... 236
P. Marcus and V. Maurice

18. **Structure and Catalysis of Rh–Pt(100), Rh–Pt(110), Pt–Rh(100), and Pt–Rh(110) Surfaces Prepared by Electrochemical Metal Deposition** 245
K. Tanaka, Y. Okawa, A. Sasahara, and Y. Matsumoto

19. **Electron Spectroscopy and Electrochemical Scanning Tunneling Microscopy of the Solid–Liquid Interface: Iodine-Catalyzed Dissolution of Pd(110)**............ 274
Manuel P. Soriaga, W. F. Temesghen, J. B. Abreu, K. Sashikata, and K. Itaya

20. **Electrocatalysis of Formic Acid and Carbon Monoxide with Probe Adlayers of Carbon and Ethylidyne on Pt(111)** 283
D. E. Sauer, R. L. Borup, and E. M. Stuve

21. **Photooxidation Reaction of Water on an n-TiO$_2$ Electrode: Investigation of a Previously Proposed New Mechanism by Addition of Alcohols to the Electrolyte**.................................. 297
Y. Magari, H. Ochi, S. Yae, and Y. Nakato

22. **The Behavior of Pyrazine and Monoprotonated Pyrazine Adsorbed on Silver Electrodes: A Surface-Enhanced Raman Scattering Study** 310
A. G. Brolo and D. E. Irish

23. **Temperature Dependence of Growth of Surface Oxide Films on Rhodium Electrodes**... 323
Francis Villiard and Gregory Jerkiewicz

INDEXES

Author Index.. 345

Affiliation Index... 345

Subject Index... 346

Preface

ELECTROCHEMICAL SURFACE SCIENCE has undergone rapid development in recent years due to the adaptation of ultrahigh-vacuum- (UHV-) based experimental techniques such as low-energy electron diffraction (LEED), Auger electron spectroscopy (AES), X-ray photoelectron spectroscopy (XPS), and high-resolution electron energy loss spectroscopy (HREELS). These techniques have allowed the establishment of a direct correlation between the composition and structure of the electrode surface and the mechanism of electrode processes. The adaptation of these techniques has buttressed a variety of classical electrochemical techniques that demand theoretical models in order to arrive at a semblance of mechanistic features at the atomic level.

In addition to utilization of combined UHV and electrochemistry (UHV–EC), the enormous growth witnessed in the past ten years can be attributed to the inclusion of scanning tunneling microscopy (STM) and atomic force microscopy (AFM) in the analytical arsenal. STM–AFM-based data were employed to support or dispute earlier results from purely electrochemical or UHV–EC techniques; stimulating scientific discussions ensued. The scientific debate about the mechanism of electrode processes vis-à-vis the structure and composition of the electrochemical interface subsequently spurred theorists to wade in and join the debate.

The motivation behind the "Symposium on Electron Spectroscopy and STM–AFM Analysis of the Solid–Liquid Electrochemical Interface" was to assemble in one place some major players in electrochemical surface science. The obvious rationale was that such a gathering would help distill and focus future work to issues deemed most critical to further progress in the area. The processes that were discussed at the symposium included electrodeposition and electrocrystallization, passivation of metals and alloys, anodic dissolution of metals and semiconductors, oxidation of small molecules, assembly of semiconducting layers, hydrogen adsorption, and charge transfer at surface-modified electrodes.

The wide scope of topics covered makes this book of particular importance to new and established researchers in physical chemistry, electrochemistry, interfacial science, and materials research. The subjects treated also make the volume substantive reading for researchers in other more applied sciences.

The volume contains 23 chapters. The first chapter is an overview of the field. The three chapters that follow deal with molecular dynamic simulations, phase transitions, and thermodynamics of the electrochemical interface. The next seven chapters describe UHV–EC applications. The next five chapters focus on strategies based on STM and AFM. The three chapters that follow are unique in that they describe work that employed tandem UHV–EC–STM–AFM. The closing four chapters provide examples of relevant data that can be obtained from spectroscopic and classical electrochemical techniques.

Acknowledgments

This book was developed from the symposium organized under the auspices of the 1995 International Chemical Congress of Pacific Basin Societies (Pacifichem) which took place in Honolulu, Hawaii, in December 1995. The symposium was made possible by the support of the Scientific Program Committee directed by Jean Lessard. The editors gratefully acknowledge generous financial support from Digital Instruments, Inc., and Physical Electronics. The editors thank the session chairpersons for their help during the symposium. They express their gratitude to all the speakers and attendees who provided the presentations and interesting discussions at the symposium, and contributed the chapters included in this volume. Finally, the editors thank Maureen W. Matkovich, Laura J. Manicone, and Charlotte McNaughton of the Books Department of the American Chemical Society for their precious guidance and help in organizing, assembling, and preparing this volume for publication.

GREGORY JERKIEWICZ
Département de chimie
Université de Sherbrooke
Sherbrooke
Québec J1K 2R1, Canada

MANUEL P. SORIAGA
Department of Chemistry
Texas A&M University
College Station, TX 77843–3255

KOHEI UOSAKI
Division of Chemistry
Graduate School of Science
Hokkaido University
Sapporo 060, Japan

ANDRZEJ WIECKOWSKI
School of Chemical Sciences
University of Illinois
  at Urbana-Champaign
600 South Mathews Avenue
Urbana, IL 61801

September 25, 1996

Chapter 1

From Electrochemistry to Molecular-Level Research on the Solid–Liquid Electrochemical Interface

An Overview

Gregory Jerkiewicz

Département de chimie, Université de Sherbrooke, Sherbrooke, Québec J1K 2R1, Canada

Electrochemistry is often regarded as a classical domain of physical chemistry. Indeed, a historical survey reveals that Volta invented the first battery already in 1800, known as the Volta pile, to produce electricity. Concurrently, Nicholson and Carlisle constructed and applied Voltaic piles to conduct water electrolysis at platinum and gold electrodes. These experiments performed in Italy and England mark the origin of electrochemistry (*1*). It is apparent that right from the very beginning, electrochemistry was becoming an empirical area of physical chemistry with little understanding of the fundamental processes occurring at what became known later on as the anode and the cathode. It was only in 1834 that Faraday proposed two fundamental laws, discovered empirically, that defined the relation between the amount of the deposited material, or evolved gas, and the quantity of electricity passed through the electrolyte (*2,3*). The experimental work of the 1800's led to the conclusion that the solution conducts electricity in a manner different than the solid and that the region where the solution and the solid are in contact, *the interface*, must have its particular properties. In 1806, Grotthuss postulated that "galvanic action" polarizes water to such a degree that the elements can pass through the solution in opposite directions, always remaining bonded to a partner until released at the poles (the electrodes). This concept did not take into account the necessity of charge transfer at the electrodes and was proven incorrect. Nevertheless, it inspired Daniell and Miller to propose the first visual representation of the electrified solid-liquid interface in 1844, Figure 1, which illustrates ideas about the mechanism of conductivity in the electrolyte (*1*).

In 1879, Helmholtz recognized the importance of the electrical nature of the solid surface being in contact with the electrolyte, related it to electrode processes, and subsequently proposed a model of the electrical double layer, Figure 2a. According to his concept, a negatively charged surface will attract positive charges from the solution and vice versa, thus giving rise to a notion of *a fixed (rigid) double layer*. Between 1910 and 1913, this idea was modified by Gouy and Chapman who pointed out that the Helmholtz model neglected the Boltzmann distribution of the ions present in the electrolyte. Their concept involved *a diffuse double layer* comprising ions and solvent molecules and extending some distance from the solid surface,

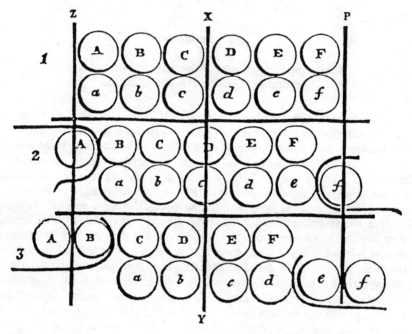

Figure 1. Diagram given by Daniell and Miller in 1844 illustrating the ideas of Grotthuss about the conductivity mechanism in the electrolyte and "the galvanic action". In this model elements pass through the solution in opposite directions, always remaining bonded to a partner until released at the poles. In 1, **A** is bonded to **B** and **e** is bonded to **f**. In 2, **A** is being separated from **B** and **e** is being separated from **f**. In 3, **A** is separated from **B** and **e** is separated from **f**. This diagram is the first visual representation of the solid-liquid interface.

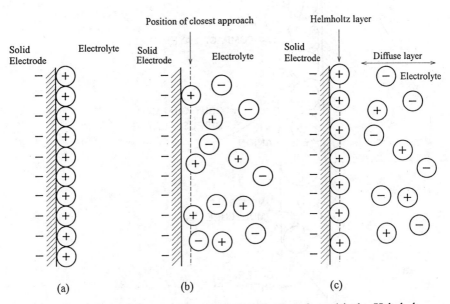

Figure 2. Three models of the electrochemical interface: **(a)** the Helmholtz fixed (rigid) double layer, 1879; **(b)** the Gouy-Chapman diffuse double layer 1910-1913; **(c)** the Stern double layer, 1924, being a combination of the Helmholtz and the Gouy-Chapman concepts.

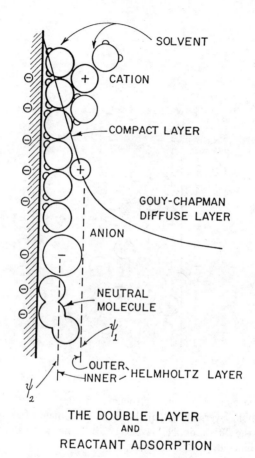

THE DOUBLE LAYER
AND
REACTANT ADSORPTION

Figure 3. General representation of the double layer taking into account presence of oriented solvent molecules, adsorbed neutral species, cations in the outer (diffuse) layer and specifically adsorbed anions, 1965 (Reproduced with permission from reference 4).

Figure 2b. The two distinct models were finally combined into a new model by Stern in 1924 who recognized that the electrified solid-liquid interface comprised both the fixed Helmholtz layer and the diffuse one of Gouy and Chapman, Figure 2c (*4-7*).

The invention of polarography by Heyrovsky in 1920's as a new experimental method followed by detail theoretical treatment in 1930's which led to quantitative interpretation of the polarographic current-potential transients marked a milestone in electrochemistry (*8,9*). This technique permitted the electrochemist to interpret mass transfer electrode processes in terms of diffusion or convection; it was subsequently applied in studies of reaction kinetics. Although the technique was limited to dropping mercury electrodes, its impact on electrochemistry and development of novel electroanalytical techniques was so tremendous that the inventor was awarded a Nobel Prize in Chemistry in 1959.

A new model of the double layer was developed by Parsons (*10*) in 1954 who took into account distribution of ions and solvent molecules in the double-layer. In this model, cations are regarded as remaining outside a layer of strongly oriented and adsorbed solvent molecules. The theory had the advantage of taking into account the presence of adsorbed solvent molecules that had been neglected in previous theories of the double layer. It took into account in a realistic way the presence of adsorbed solvent molecules that had been neglected in the previous models. It also recognized changes of the dielectric constant of the solvent, approaching the limiting Maxwell value in the inner region (in the inner layer) and a higher value in the outer region (the outer layer), Figure 3 (*4,10-12*).

In order to understand the interfacial behavior of adsorbed species at the electrode surface, work was carried out on development of adsorption isotherms. This approach although of pure thermodynamic nature led to evaluation of thermodynamic state functions, the Gibbs free energy of adsorption, the entropy of adsorption and the enthalpy of adsorption, which governed the electro-adsorption process. Based on an analysis of the behavior of the Gibbs free energy of adsorption versus the coverage by the adsorbed species one was able to assess the attractive or repulsive nature of lateral interactions between the adsorbed species (*13,14*). Following the research on electrochemical adsorption isotherms, advancement was made in understanding of the mechanism of interfacial electron transfer and the role of the electrode potential, the double layer, temperature, the solvent effect, the reactant-surface interaction in the process (*15-17*).

In pursuit of comprehension of the relation between the electrode reactivity, surface-chemical composition and structure, the electrochemist had to search new experimental techniques which would provide molecular-level data inaccessible by application of electrochemical techniques. It was realized that chemistry and physics of the solid-gas interface had been enjoying a tremendous progress due to development of ultra-high vacuum, UHV, based experimental techniques such as low-energy electron diffraction, LEED, Auger electron spectroscopy, AES, X-Ray photoelectron spectroscopy, XPS, and high resolution electron energy loss spectroscopy, HREELS (*18-20*). Employment of these powerful techniques to electrochemical research by combining a UHV surface analysis chamber with an electrochemical chamber provided first insights into the abundance of atomic-level data (*21-28*). Doubts were raised with regard to the validity of the results on the ground that the surface characterization was conducted on emersed electrodes, thus

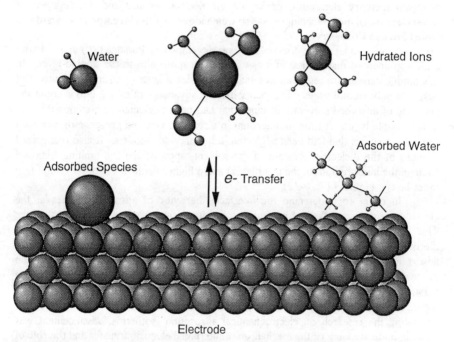

Water

Hydrated Ions

Adsorbed Water

Adsorbed Species

e- Transfer

Electrode

Figure 4. Contemporary model of the electrified solid-liquid interface taking into account the electrode structure, specifically adsorbed anions in the inner layer, structure of the solvent (here water) molecules at the electrode surface, hydrated ions in the diffuse layer and the interfacial electron transfer, 1995 (Provided by and reproduced with permission of K. Itaya).

in absence of the electrochemical environment employing presence of the electric field and the electrolyte bulk. Complementary experiments engaging other techniques such as IR and Raman spectroscopies (*29-31*), radiochemistry (*32*), chronocoulometry (*33*) or X-Ray-based spectroscopies (*34*) dispelled most of the doubts and UHV-based techniques were recognized as a powerful tool in studies of the solid-liquid electrified interface.

Concurrently with employment of UHV-base techniques to electrochemical studies on single-crystal surfaces, Clavilier (*35*) and Hamelin (*36*) developed and mastered a new methodology which allowed to conduct experiments on well-defined noble-metal electrode surfaces without expensive UHV apparatus. This novel experimental approach was quickly applied in various research laboratories and led to rapid growth of electrochemistry at well-defined surfaces (*37*).

Development of scanning tunneling microscopy, STM (*38,39*), and atomic force microscopy, AFM (*40,41*), became a milestone in surface science. It was immediately realized in the electrochemical community that one could conduct simultaneously electrochemical and STM experiments taking advantage of a bipotentiostat, a well-established device in the arsenal of electrochemical instrumentation. Moreover, it was recognized that one could conduct coupled electrochemical and AFM experiments without the need of a bipotentiostat normally required in electrochemical STM, EC STM. These new techniques were originally applied for verification of the atomic-level order of single-crystal electrodes during cyclic-voltammetry, CV, measurements or the structure of under-potential deposited metallic layers (*42-46*). Moreover, it became apparent that one could observe and image surface-morphology changes upon polarization of semiconducting species (*47,48*) or initial formation followed by 3D growth of metallic layers (*49,50*), thus to visualize the hitherto not-well understood under-potential deposition, electrodeposition and electrocrystallization. The versatility of these techniques was so enormous that they quickly were put to use in studies of electrodissolution of metals (corrosion), mechanism of growth of surface oxide films and morphology of self-assembled monolayers (*51*).

In recent years, much attention was given to the role of neutral species and specifically adsorbed anions on the structure of the electrochemical interface, the electric field distribution in their vicinity and their role in electrocatalysis. EC STM became a key experimental technique in providing insight into the structure of the adsorbed neutral species or the specifically adsorbed anions copresent with under-potential deposited metallic layers (*52-54*) and in supporting data on the anion surface coverage based on chronocoulometry experiments (*55-57*). These structural results derived from various experimental approaches have led to a contemporary model of the electrified solid-liquid interface which is presented in Figure 4.

One of the most fundamental problems in electrochemical surface science is distribution of the electric potential and the particles at the interface. The classical model which prevailed until about 1980 treated the electrode surface as a perfect and structureless conductor and did not take into account the surface electronic structure. The electrolyte was considered as an ensemble of hard, point ions immersed in a dielectric continuum. This approximation neglected the fine structure of the solvent molecules and the solute as well as their discrete interactions. In recent years, much progress has been made in providing a more realistic model of the solid-liquid electrochemical interface by applying quantum mechanical theories to model the metal surfaces and statistical mechanics to describe the behavior of the electrolyte (*58-61*). The solid electrode can be treated in term of the jellium model which treats the positive charges of the metal ions as smeared out into a positive background charge interacting with an inhomogeneous electron gas. This approximation does not take into account any information on the surface lattice structure and it is applicable to polycrystalline surfaces only (*62*). This approach can be further modified to describe behavior of single-crystal surfaces by incorporating a lattice pseudopotential by introducing the Ashcroft potential (*58*).

A good model of the electrolyte must describe the ions, solvent molecules and their orientation at the molecular level. Molecular dynamics simulations that are performed to visualize the orientation of the electrolyte molecules in the vicinity of the electrode surface are based on a set of parameters that can be varied in order to best described the properties of the system under investigation. The most reasonable models for solvent-solvent and ion-solvent interactions consider distribution of point charges on solvent molecules and take into account Lennard-Jones-type potentials that are strongly repulsive at short distances. Molecular dynamics simulations are typically performed on a system confined between two metal electrodes and the number of confined ions and solvent molecules is often limited by the computing power of modern computers. Some representative examples of results of such calculations are given in ref. *60,63-68*.

In summary, electrochemistry, to the large extent due to development of electrochemical surface science, is undergoing renewed interest and application of modern surface-science techniques, theoretical research and powerful molecular dynamics simulations performed using state-of-the-art computing facilities make it one of the most active areas of physical chemistry as accurately pointed out in ref. *69*. Improvement and modification of the existing theoretical models or development of new ones will lead not only to more detail comprehension of the structure of the electrochemical double layer but also to evaluation of such important parameters as cohesive forces acting between the under-potential deposited species, or specifically adsorbed anions, and the metal electrode, forces that are responsible for adhesion and mechanical stability of this metal and semiconductor deposits (*70,71*), an aspect of vital importance to modern electrochemical technology. Finally, the role of hydrogen as the fuel of the future should not be underestimated in the light of the ever-increasing pollution. Thus research on development of rechargeable metal-hydride batteries or hydrogen-based fuel cells seems to be a prospective direction in electrochemistry (*72,73*). Yet, despite enormous experimental and theoretical efforts spent on understanding of the interfacial behavior of hydrogen little is know about the mechanism of its electrochemical interfacial transfer or about the role of the site blocking elements (*74*), often referred to as surface poisons or surface promoters, in inhibiting or enhancing H absorption into the host metal or alloy (*75,76*). Comprehension of surface electrochemical processes at the molecular level is of vital importance to advancement in design of non-polluting high-energy output batteries and fuel cells which are substantial energy creating devices (*77,78*). Therefore, electrochemical surface science still remains a discipline of physical chemistry where a lot of pioneering research still can be done and discoveries can be made.

Acknowledgments

Acknowledgment is made to the NSERC of Canada and the FCAR du Québec for research support. The author is indebted to Prof. A. Wieckowski of The University of Illinois at Urbana-Champaign for stimulating discussion on the most recent developments in electrochemical surface science and his comments on the Overview Chapter. The author wishes to express his thanks to Kathleen for her love, patience, encouragement and support throughout this work.

References

1. Laidler, K. J. *The World of Physical Chemistry*; Oxford University Press: New York, 1993.
2. Williams, L. P. *Michael Faraday: A Biography*; Chapman and Hall: London, 1965.
3. *Electrochemistry, Past and Present*; Stock, J. T.; Orna, M. V., Eds.; ACS Symposium Series 390; ACS: Washington, D.C., 1989.
4. Conway, B. E. *Theory and Principles of Electrode Processes*; Ronald Press: London, 1965.
5. Conway, B. E. *Ionic Hydration in Chemistry and Physics*; Elsevier: New York, 1981.
6. Gileadi, E.; Kirowa-Eisner, E.; Penciner, J. *Interfacial Electrochemistry*; Addison-Wesley: New York, 1975.
7. Brett, C. M. A.; Brett, A. M. O. *Electrochemistry. Principles, Methods, and Applications*; Oxford University Press: New York, 1993.
8. Heyrovsky, J. *Chem. Listy* **1922**, *16*, 256.
9. Heyrovsky, J.; Kuta, J. *Principles of Polarography*; Academic Press: New York, 1966.
10. Parsons, R. In *Modern Aspects of Electrochemistry*; Bockris, J. O'M.; Conway, B. E., Eds.; Academic Press: New York, 1954; Vol. 1, pp 103-179.
11. Bockris, J. O'M.; Devanathan, M. A. V.; Müller, K. *Proc. Roy. Soc. London* **1963**, *A274*, 55-79.
12. Bard, A. J.; Faulkner, L. R. *Electrochemical Methods*; John Wiley & Sons: New York, 1980.
13. Frumkin, A. N. In *Advances in Electrochemistry and Electrochemical Engineering*; Delahay, P., Ed.; Interscience Publishers: New York, 1984; Vol. 3, pp 287-391.
14. Frumkin, A. N. *Electrode Processes*; Nauka: Moscow, 1987 (in Russian).
15. Marcus, R. A. *J. Phys. Chem.* **1963**, *67*, 853-857; Marcus, R. A. *J. Chem. Phys.* **1965**, *43*, 679-701.
16. Marcus, R. A. In *Annual Review of Physical Chemistry*; Eyring, H.; Christensen, C. J.; Johnston, H. S., Eds.; Annual Reviews: Palo Alto, California, 1964; Vol. 15, pp 155-196.
17. Weaver, M. J. In *Comprehensive Chemical Kinetics*; Compton, R. G., Ed.; Elsevier: New York, 1987; Vol. 27, pp 1-60.
18. Somorjai, G. A. *Principles of Surface Chemistry*; Prince-Hall: Englewood Cliffs, New Jersey, 1972.

19. Somorjai, G. A. *Chemistry in Two Dimensions: Surfaces*; Cornell University Press: Ithaca, New York, 1981.
20. Ertl, G; Küppers, J. *Low Energy Electrons abd Surface Chemistry*; VCH: Deerfield Beach, Florida, 1985.
21. Hubbard, A. T. *Accts. Chem. Res.* **1980**, *13*, 177-184.
22. Hubbard, A. T. *Chem. Rev.* **1988**, *88*, 633-656.
23. Sherwood, P. M. A. In *Practical Surface Analysis*; Briggs, D.; Seah, M. P., Eds.; John Wiley & Sons: Chichester, New York, 1983, pp 445-476.
24. Kolb, D. M. *Z. Phys. Chem. Bd.* **1987**, *154*, 179-199.
25. *Electrochemical Surface Science*; Soriaga, M. P., Ed.; ACS Symposium Series 378; ACS: Washington, D.C., 1988.
26. Ross, P. N.; Wagner, F. T. In *Advances in Electrochemistry and Electrochemical Engineering*; Gerischer, H., Ed.; John Wiley & Sons: New York, 1984; Vol. 13, pp 69-112.
27. Kötz, R. In *Advances in Electrochemical Science and Engineering*; Gerischer, H.; Tobias, C. W., Eds.; VCH: New York, 1990; Vol. 1, pp 75-126.
28. Soriaga, M. P.; Harrington, D. A.; Stickney, J. L.; Wieckowski, A. In *Modern Aspects of Electrochemistry*; Conway, B. E.; Bockris, J. O'M.; White, R. E., Eds.; Academic Press: New York, 1996; Vol. 28, pp 103-179.
29. Nichols, J. N. In *Adsorption of Molecules at Metal Electrodes*; Lipkowski, J.; Ross, P. N., Eds.; VCH: New York, 1992, pp 347-389.
30. Pettinger, B. In *Adsorption of Molecules at Metal Electrodes*; Lipkowski, J.; Ross, P. N., Eds.; VCH: New York, 1992, pp 285-345.
31. Iwasita, T.; Nart, F. C. In *Advances in Electrochemical Science and Engineering*; Gerischer, H.; Tobias, C. W., Eds.; VCH: New York, 1995; Vol. 4, pp 123-216.
32. Krauskopf, E. K.; Wieckowski, A. In *Adsorption of Molecules at Metal Electrodes*; Lipkowski, J.; Ross, P. N., Eds.; VCH: New York, 1992, pp 119-169.
33. Lipkowski, J.; Stolberg, L. In *Adsorption of Molecules at Metal Electrodes*; Lipkowski, J.; Ross, P. N., Eds.; VCH: New York, 1992, pp 171-238.
34. *Electrochemical Interfaces*; Abruña, H. D., Ed.; VCH: New York, 1991.
35. Clavilier, J.; Faure, R.; Guinet, G.; Durand, R. *J. Electroanal. Chem.* **1980**, *107*, 205-209; see also Clavilier, J. *J. Electroanal. Chem.* **1980**, *107*, 211-216.
36. Hamelin, A. In *Modern Aspects of Electrochemistry*; Bockris, J. O'M.; Conway, B. E.; White, R. E., Eds.; Plenum Press: New York, 1985; Vol. 16, pp 1-101.
37. Markovic, N.; Marinkovic, N.; Adzic, R. R. *J. Electroanal. Chem.* **1988**, *241*, 309-328; see also Markovic, N. M.; Marinkovic, N. S.; Adzic, R. R. *J. Electroanal. Chem.* **1991**, *314*, 289-306.
38. Binnig, G.; Rohrer, H.; Gerber, Ch.; Weibel, E. *Phys. Rev. Lett.* **1982**, *49*, 57-61.
39. Chen, J. *Introduction to Scanning Tunneling Microscopy*; Oxford University Press: New York, 1993.
40. Binnig, G.; Quate, C. F.; Gerber, Ch.; Weibel, E. *Phys. Rev. Lett.* **1986**, *56*, 930-933.
41. Sarid, D. *Scanning Force Microscopy*; Oxford University Press: New York, 1994.
42. Schardt, B. C.; Yau, S.-L.; Rinaldi, F. *Science* **1989**, *243*, 1050-1053.

43. Itaya, K.; Sugawara, S.; Sashikata, K.; Furuya, N. *J. Vac. Sci. Technol. A* **1990**, *8*, 515-519.

44. Hondo, H.; Sugawara, S.; Itaya, K. *Anal. Chem.*, **1990**, *62*, 2424-2429.

45. Manne, S.; Hansma, P. K.; Massie, J.; Elings, V. B.; Gewirth, A. A. *Science*, **1991**, *251*, 183-186.

46. Nichols, R. J.; Kolb, D. M.; Behm, R. J. *J. Electroanal. Chem.* **1991**, *313*, 109-119.

47. Erriksson, S.; Carlsson, P.; Holmström, B.; Uosaki, K. *J. Electroanal. Chem.* **1991**, *313*, 121-128.

48. Allongue, P. In *Advances in Electrochemical Science and Engineering*; Gerischer, H.; Tobias, C. W., Eds.; VCH: New York, 1995; Vol. 4, pp 1-66.

49. Budevski, E. B.; Staikov, G. T.; Lorenz, W. J. *Electrochemical Phase Formation and Growth*; VCH: New York, 1996.

50. Schneir, J.; Elings, V.; Hansma, P. K. *J. Electrochem. Soc.* **1988**, *135*, 2774-2777.

51. Wilbur, J. L.; Biebuyck, H. A.; MacDonald, J. C.; Whitesides, G. M. *Langmuir*, **1995**, *11*, 825-831.

52. Wan, L.-J.; Yau, S.-L.; Itaya, K. *J. Phys. Chem.* **1995**, *99*, 9507-9513.

53. Sawaguchi, T.; Yamada, T.; Okinaka, Y.; Itaya, K. *J. Phys. Chem.* **1995**, *99*, 14149-14155.

54. Kunikate, M.; Batina, N.; Itaya, K. *Langmuir*, *1995*, *11*, 2337-2340.

55. Shi. Z.; Lipkowski, J. *J. Phys. Chem.* **1995**, *99*, 4170-4175;

56. Savich, W.; Sun, S.-G.; Lipkowski, J.; Wieckowski, J. *J. Electroanal. Chem.* **1995**, *388*, 233-237.

57. Shi, Z.; Wu, S.; Lipkowski, J. *Electrochim. Acta* **1995**, *40*, 9-15.

58. Schmickler, W. In *Structure of Electrified Interfaces*; Lipkowski, J.; Ross, P. N., Eds.; VCH: New York, 1993, pp 201-238.

59. Schmickler, W. *Interfacial Electrochemistry*; Oxford University Press: New York, 1996.

60. Heinzinger, K. In *Structure of Electrified Interfaces*; Lipkowski, J.; Ross, P. N., Eds.; VCH: New York, 1993, pp 239-275.

61. Borkowska, Z.; Stimming, U. In *Structure of Electrified Interfaces*; Lipkowski, J.; Ross, P. N., Eds.; VCH: New York, 1993, pp 277-307.

62. *Theory of the Inhomogeneous Electron Gas*; Lundquist, S.; March, N. H., Eds.; Plenum: New York, 1983.

63. Allen, M. P.; Tildsley, D. J. *Computer Simulations of Liquids*; Oxford University Press: New York, 1987.

64. Spohr, E. *J. Phys. Chem.* **1989**, *93*, 6171-6180.

65. Glosli, J. N.; Philpott, M. R. *J. Chem. Phys.* **1992**, *96*, 6962-6969.

66. Blum, L. *J. Phys. Chem.* **1977**, *81*, 136-147; see also Huckaby, D. A.; Blum, L. *J. Electroanal. Chem.* **1991**, *315*, 255-261.

67. Attard, P.; Wei, D.; Patey, G. N. *J. Chem. Phys.* **1991**, *96*, 3767-3771.

68. Torrie, G. M.; Patey, G. N. *Electrochim. Acta* **1991**, *36*, 1677-1684.

69. Bard, A. J.; Abruña, H. D.; Chidsey, C. E.; Faulkner, L. R.; Feldberg, S. W.; Itaya, K.; Majda, M.; Melroy, O.; Murray, R. W.; Porter, M. D.; Soriaga. M. P.; White, H. S. *J. Phys. Chem.* **1993**, *97*, 7147-7173.

70. Jerkiewicz, G.; Zolfaghari, A. *J. Electrochem. Soc.* **1996**, *143*, 1240-1248.
71. Jerkiewicz, G.; Zolfaghari, A. *J. Phys. Chem.* **1996**, *100*, 8454-8461.
72. *Hydrogen in Metals*; Parts I and II; Alefeld, G.; Volkl, J., Eds.; Springer-Verlag: New York, 1978.
73. *Hydrogen in Intermetalic Compounds*; Schlapbach, L., Ed.; Springer-Verlag: New York, Part I, 1988; Part II, 1992.
74. Protopopoff, E.; Marcus, P. *J. Chim. Phys.* **1991**, *88*, 1423-1452.
75. Conway, B. E.; Jerkiewicz, G. *J. Electroanal. Chem.* **1993**, *357*, 47-66.
76. Jerkiewicz, G.; Borodzinski, J. J.; Chrzanowski, W.; Conway, B. E. *J. Electrochem. Soc.* **1995**, *142*, 3755-3763.
77. Vincent, C. A.; Bonino, F.; Lazzari, M.; Scrosati, B. *Modern Batteries*; Edward Arnold: London, 1984.
78. *Handbook of Batteries and Fuel Cells*; Linden, D., Ed.; McGraw Hill: New York, 1989.

Chapter 2

Molecular Dynamics Simulation of Interfacial Electrochemical Processes: Electric Double Layer Screening

Michael R. Philpott[1] and James N. Glosli[2]

[1]IBM Almaden Research Center, 650 Harry Road,
San Jose, CA 95120–6099
[2]Lawrence Livermore National Laboratory, University of California,
Livermore, CA 94551–9900

The status of computer simulations of electric double layers is briefly summarized and a road map for solving the important problems in the atomic scale simulation of interfacial electrochemical processes is proposed. As examples efforts to simulate screening in electric double layers are described. Molecular dynamics simulations on systems about 4 nm thick, containing up to 1600 water molecules and NaCl at 1M to 3M concentration, displayed the main features of double layers at charged metal surfaces including: bulk electrolyte zone, diffuse ionic layer that screens the charge on the electrode and a layer of oriented water next to the surface.

This paper describes the application of molecular dynamics to chemical processes at the interface between a charged metal electrode and aqueous electrolyte. The long range goal is a scheme for the dynamics of chemical reactions on surfaces important in the electrochemical technology of power sources, electroplating, and corrosion control. The paper begins with a summary of our view of the current state of computer simulation applied to interfacial electrochemistry. The status is accompanied with a commentary on problems. To illustrate progress in the field we describe our simulations of screening of charged electrodes by aqueous electrolytes, including previously unpublished work. Double layers are some of the basic organizations found in electrochemical and biological systems that shield fields of layers and arrays of electric charge in contiguous structures. It is important to understand their properties using models that can be solved without making additional approximations. In this paper the structure of the aqueous part of the double layer is given in terms of time independent water and ion probability distribution functions averaged parallel to the metal surface. Electric fields and potentials are calculated from the microscopic charge density profile. These calculations provide a consistent microscopic picture of ions and water in a double layer including the species next to the charged surface (inner layer), in the 'diffuse layer' (also called the screening layer) and in the bulk zone. The effect of finite sized ions and water are clearly evident, as is the effect of the electric field on the orientation of surface water molecules.

Status of Molecular Dynamics Simulations

Figure 1 (top) is a sketch of the electric potential using Gouy-Chapman-Stern theory[1] in which the diffuse layer ions are treated as point charges, the water as a a dielectric continuum, and OHP and IHP (outer and inner Helmholtz planes) are introduced to mark the distance of closest approach of strongly hydrated ions and contact adsorbing ions respectively. Figure 1 (bottom) shows a molecular scale cartoon of ions and water near a flat charged metal surface. Ideas embodied by pictures like this together with Gouy-Chapman-Stern theory and the thermodynamic theory of surface excess quantities have been used to analyze and interpret experimental electrochemical data[2-5]. The electric potential shown is not consistent with the cartoon because an oriented layer of water would result in a strong oscillation in the potential. The simulations described here and by others show such oscillations near the electrode. They are not predicted by Gouy-Chapman theory. The advent of risc based work stations allows the testing of atomic scale models with thousands of molecules. Monte-Carlo and molecular dynamics computer simulations of ions and water molecules interacting by simple potentials are routinely performed for a few

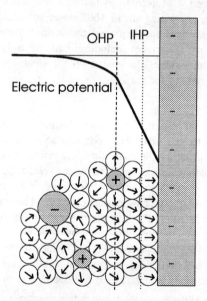

Figure 1. Schematic diagram (top) of electric potential across the double layer based on Gouy-Chapman-Stern model in which the solvent is a continuum dielectric. The cartoon (bottom) depicts a hypothetical arrangement of solvent and ions near a charged surface. Similar pictures are found in electrochemical texts. The labels IHP and OHP mark the inner and outer Helmholtz planes.

thousand water molecules for times up to several nanoseconds. We comment first on electrostatics and potential energy surfaces before other issues.

Electrostatics. In bodies with large but finite numbers of charges, three dimensional sums of electrostatic interactions (eg., ion-ion, dipole-dipole) can be decomposed into separate bulk-like and surface-boundary parts. The latter part is responsible for the phenomenon of conditional convergence when the size of the system is taken to infinity. For electrified interfaces it is essential that the long range part of the electrostatic interaction be computed without truncation in a manner consistent with the boundary conditions. For some geometries (eg., planar) the electric field of surface charge can be calculated by the method of images. There is some evidence from simulations on ions in polar solvents that truncated long range electrostatics result in correlations that cause like charged ions to attract each other. The crystal optics based methods of Ewald and Kornfeld are the simplest for calculating electrostatic fields[6]. The algorithm works for all space group symmetries. At best this method is order $N^{1.5}$ in the number of charges N. There are 2D summation methods that are faster[7, 8]. In our simulations we use the order N fast multipole method (fmm) developed by Greengard and Rokhlin [9-12]. This is a useful method for electrochemical simulations where a variety of boundary conditions (periodic, Dirichlet, Neumann or mixed boundaries) are encountered. It can also be adapted so that regions of low charge density are not subdivided when the charge count falls below a specified integer[11]. It is restricted to cubic simulation cells. The fmm is faster than Ewald for systems exceeding a few hundred charges[13]. Particle-mesh methods have been extensively used for long range r^{-1} potential problems[14]. For most systems P^3M is faster than fmm and can be used with orthorhombic simulation cells[15]. Parallel metal surfaces have an infinite set of multiple electrostatic images that have to be summed in plane-wise fashion[8, 16]. Recently we calculated distributions for ions between parallel metal plates held at a constant potential difference[17, 18]. These distributions defined charge distributions that were used to compute time independent electric fields and potentials across the cell. The effect of additional averaging over space volumes comparable to water molecules was also explored[19, 20]. Though most experiments are done at constant potential, there are very few simulations at constant potential[17, 18].

Molecule-Molecule and Molecule-Metal Potentials. There are continuing improvements in molecule-molecule potentials. High quality efforts are directed at improving the interaction, including electronic polarizability and other tensor properties[21, 22]. There are also potentials that include three body terms explicitly[23]. Possibly the best atom-metal potential is due to Barker[24]. The Barker potential for Xe/Pt(111) is an excellent fit to a large body of experimental data. There have been numerous quantum chemistry studies of ions and water on metal clusters some with applied fields and others on charged clusters to imitate the electrochemical environment (for water references see Zhu[25], for ions on clusters see Pacchioni[26]). So far only a few have been parameterized into forms that can be easily used in an MD code[25, 27-29]. Electric double layers on Ag and Au are thermodynamically stable over wide potential range[30]. It is unfortunate that simulations have focussed almost exclusively on Pt for which there is little experimental data. The reason is there is an easy to use set of potentials derived from quantum chemistry calculations of small Pt clusters with adsorbed water and ions. In principle cluster calculations performed with Cu, Ag or Au would be more reliable because relativistic orbital contractions and spin orbit effects. are not so important in these sp metals and because the d shell is more tightly bound. Recently several publications have reported cluster calculations for water and ions on Hg[31, 32] including a parameterization to give a pair potential for MD studies.

Dynamics. There are calculations in which the metal is modeled as an Einstein solid with harmonic vibrations[33]. When surface molecules and ions are strongly adsorbed molecular dynamics becomes an inefficient way to study surface processes due to the slow exchange between surface and solution. In this case it is possible to use umbrella sampling to compute distribution profiles[34, 35]. Recently the idea underlying Car-Parrinello was used for macroion dynamics[36, 37] in which the solvent surrounding charged macroions is treated as a continuum in a self consistent scheme for the potential controlling ion dynamics. Dynamical corrections from the solvent can be added. There is a need to develop statistical methods to treat the dynamics of complex objects that evolve on several different time scales.

Interfacial Electron Transfer. There have been several studies of electron transfer reactions[17, 38]and the connection with Marcus's theory[39]. It may be possible to use a Car-Parrinello like scheme on that part of the system directly affecting the electron transfer. There has also been very interesting studies of the ferro-ferri redox couple in solution[40, 41] that address many issues related to electron transfer from an electrode to a hydrated ion. Slow processes can be treated by transition state methods like the ones used in solid state ionic conductivity[42].

Ensembles. One goal of simulations is the calculation of experimental quantities. The most challenging are the Gibbs free energies of adsorption. Currently there is no proven scheme for constant chemical potential simulations of electrolyte adsorption on metals. It seems possible to develop Andersen's method[43] for (N,P,T) ensembles. Another possibility is an analog of the Gibbs ensemble[44] for electrolytes between plates with a bulk sample forming the second phase. In an interesting development the grand canonical Monte Carlo method was used for atomic fluid mixtures in a slit pore[45, 46].

Road Map

The intention is modest though the title of this section sounds pretentious. Given the current state of theory and simulation can we identify a path to useful computation of reactions important to electrochemical technologies? The science problems cover a broad range: deposition and dissolution of metal, formation of oxide layers, electron transfer from electrodes to ions, and charge migration in complex fluid phases. We comment first on the possibility of developing two existing methods, ab initio molecular dynamics and potential energy surface dynamics, to electrochemistry.

Ab Initio Molecular Dynamics. Chemical reactions involve the reorganization of electrons about the nucleii involved in the bond changes. The ab initio molecular dynamics scheme developed by Car and Parrinello[36] permits an accurate description of both electronic and nuclear rearrangements that occur during a reaction. The penalty for including electronic coordinates explicitly in an electrochemical simulation is the restriction to relatively few water molecules. Liquid water and proton transfer have been studied[47, 48]. The computational problem is immense so that at present the study of hundreds of water molecules takes too long. This number is quite enough for studying H-bonding and dissociation and the dynamics of the hydration of ions but is insufficient to deal with double layer structure or reactions of hydrated ions with charged metals.

Potential Energy Surfaces. For systems where some or all of the dynamics can be described by a potential energy surface (PES) it is possible to avoid solving the electronic Schrodinger equation, and use instead a PES parameterized with experimental data. Several cases already exist. First is the well known S_N2 reaction

$$Cl^- + H_3C - Cl \rightarrow Cl - CH_3 + Cl^-. \qquad [1]$$

The molecular dynamics of this reaction has been studied in water[49]. Another quite different example is the Brenner bond order potential[50] used to describe the dynamics of homolytic bond fission and formation in carbon and hydrogen containing systems[51]. These two examples involve the making or breaking of two electron bonds between low Z atoms. Explicit degrees of freedom for the electrons are avoided though the use of a PES. In gas-surface adsorbate theory considerable progress has been made understanding chemical dissociation reactions and physical sputtering using a combination of LEPS potential for the diatomic on the metal with a many body embedded atom-like potential for the metal[52]. The electrochemical problems are harder. They involve charged metal surfaces, charge transfer in polar environment and dynamics on vastly different time scales.

Road Map. Consider the deposition and dissolution of metal ions on charged surfaces. If systems can be identified where adiabatic potentials describe the nuclear motion then by analogy with the experience gained with the dissociation of diatomics on metal surfaces[52] and with carbon-hydrogen chemistry[53] it should be possible to describe how metal ions adsorb on metals and how metals dissolve. A key step is the calculation of surface charge on the metal and how it fluctuates at adatoms sites. Solvated counter charge amplifies the fluctuations and increases the probability of charge transfer at the site to a degree that metal deposition or dissolution begins[54]. Continuing this avenue it should also be possible to formulate schemes that describe the creation of insoluble metal oxide layers. This would require a model for water decomposition on metals. The central force model which permits water dissociation[55, 56], could be developed for a surface environment. Consider the process where an electron crosses a bridging ligand to an ionic center or organic moiety to a initiate chemical reaction. Some of these reactions could be studied by treating the electron as a quantum particle moving in a potential defined by the classical motion of molecules and ions in the double layer. Solvated electrons in liquids have been studied this way. If the electron cannot be so easily decoupled then a hybrid Car-Parrinello π-orbital scheme might be developed. Finally we comment on ion mobilities in complex fluids in lithium batteries. Progress here may require ideas borrowed from small polaron motion in solids, using transition state dynamic methods of Bennett[42], or using recent theories of macroion motions[37]. Though we are presently a long way from providing technologically useful information, being able to model aspects of these processes would provide atomic scale insight as to how these technologies work.

Details of the Model

Screening is treated using the immersed electrode model[13, 57, 58]. A layer of electrolyte with an excess of positive ions is confined by a semi-infinite metal on the rhs and a restraining non-metallic wall on the lhs. There is no external electric field. The metal is grounded to zero electric potential. The charge on the electrode equals the image charge of the excess positive ions (-4e or -8e) so that the metal charge is completely screened by the ions in solution. The difference between these and earlier simulations [13, 57-60] is size, here there are enough ions and water to form a bulk electrolyte region where the solution is locally neutral. Because there is only one metal surface (shown on the rhs in all the figures) there are no multiple electrostatic images in the direction perpendicular to the metal[8, 16]. The electric field and potentials are calculated using previously developed methods[19, 20].

This work uses the SPCE water model[61] (three charged mass points, q_H = 0.4238e, bond angle 109.5°, OH bond length 0.1 nm, Lennard-Jones sphere with radius σ = 0.317 nm and well depth ε= 0.650 kJ/mole) and associated parameters for NaCl[62]. The coordinate origin is at the center of the cell, and the axes are perpendicular (z) and parallel (x,y) to the metal surface. The simulation cell has edge length L=3.724 nm, the flat metal plane is at z = 1.862 nm and the flat restraining wall at z = -1.862 nm. The cell contains upto 1600 water molecules and the ion concentrations are approximately 1M, 2M or 3M NaCl. The metal charge density is either -0.046 Cm^{-2} (-4eL^{-2}) or -0.092 Cm^{-2} (-8eL^{-2}). Contact adsorption of ions is minimized. The cation has a smaller radius than the anion, so its hydration shell is strongly bound making it difficult for it to contact adsorb. The negative metal charge makes it energetically unfavorable for the anions Cl$^-$ to contact adsorb.

Two potentials are used to describe the interaction of water and ions with the metal. A 9-3 potential is used to for the Pauli repulsion and the attractive dispersive interactions between molecules or ions and the metal. The interaction between a charge on an ion or water with the conduction electrons of the metal is modeled with a classical electrostatic image potential. The position of image plane and origin plane (same as the plane through nuclei of the surface) of the 9-3 potential was taken to be coincident. In real materials the image and nuclear planes are not coincident. This is not important in the simulations because the thickness of the repulsive wall is large (ca. 0.247 nm). The 9-3 atom-surface wall parameters describing interaction with nonconduction electrons were chosen to be the same as that used by Lee etal[63], A=17.447x10^{-6} kJ(nm)^6/mole and B=76.144x10^{-3} kJ(nm)^3/mole for the O atom. The A and B parameters for H were set to zero. The potential corresponding to these parameters describe a graphite-like surface. A useful reference point in the wall potential is at z_w=1.615 nm where the 9-3 wall potential changes sign. Each simulation was run for about a nanosecond, and the instantaneous positions of all the atoms recorded every picosecond. The first 100 ps were discarded as anneal time. The density probability functions $\rho(z)$ were constructed by binning the configurations in bins (width L/800) parallel to the metal surface to give functions of z only.

Screening of Charged Metal Electrodes in SPCE Electrolyte

We begin with remarks on the screening of charged surfaces by aqueous electrolytes. At high salt concentrations the region with excess ionic charge is microscopically small. For dilute solutions (<0.1M) a crude estimate of the screening zone thickness is given by the Debye-Hückel screening length[64]

$$d = \sqrt{\left(\epsilon kT/(8\pi e^2 n_b v^2)\right)} \; . \qquad [2]$$

Here ϵ is the macroscopic dielectric constant of the solvent (ca. 80 for water), v the valence of the ion (one in this paper), and n_b is the bulk concentration of the ions. Typical values of d are: 3 nm in 0.01M, 1 nm in 0.1M, 0.3 nm in 1.0M, and 0.2 nm in 3M NaCl solutions. In high salt concentrations the screening should be more efficient and the screening length smaller, but since the length for 1M NaCl is the diameter of a water molecule, and the value for 3M NaCl is even smaller, these values mean nothing without the additional insight provided by MD simulations. At high concentrations there are many problems with simple Gouy-Chapman theory[65-67] and many modifications have been proposed[64]. There are three main problems: the dielectric constant of water in a high surface field, the lower length scale due to the finite molecular size, and correlated motion amongst ions and water. For example there is no reason to believe that the value of ϵ is 80 near a surface or in a field high enough for dielectric saturation to occur. This aspect has been discussed many times in the electrochemical literature[68]. Though we take the Debye length for concentrated solutions as an rough measure of double layer thickness, we should be very cautious when d approaches the size of a water molecule.

Figure 2 shows the distribution of the ions for three separate calculations with concentrations 1M, 2M and 3M NaCl. The charge on the electrode was -4e for 1M and 2M and -8e for 3M. The temperature was 30°C for 1M and 2M, and 100°C for the 3M NaCl solution. Note that there is no significant contact adsorption. All the peaks in the ion distributions occur away from the position of closest approach z_w=1.612 nm to the metal. This is precisely the situation we contrived for the reasons given before. The ion concentrations are approximately the same for $|z| < 1.0$ nm. This identifies the region of the system with bulk electrolyte properties. For $|z| < 0.5$ nm the two ion profiles are the same within 10 to 20%, for all the simulations. The bulk region is smaller for 1M NaCl because the screening layer is thicker. The integrated ion densities are monotonically increasing curves in Figure 2. They provide a rough measure of overall charge neutrality from the restraining wall to the metal. The vertical arrows indicate roughly where the ion charge densities start to diverge because of screening. Also shown in Figure 2 are the results of a calculation of ion densities using simple Gouy-Chapman theory. For optimal comparison the electrode was assume to start at z=1.612 nm. These superimposed curves show how lack of atomic scale structure limits the application of Gouy-Chapman theory. This is graphic evidence that there are important details in MD simulations due to hydration, molecular size, and water layering near charged surfaces. In each case the width of the screening layer is too small to justify the description as 'diffuse' (though we will continue to use this term). To estimate the diffuse layer width we measure from the z=1.615 nm (where wall potential changes sign) to the point where the difference in integrated ion densities is e^{-1} of the metal charge (8e for 3M, 4e for 1M and 2M).

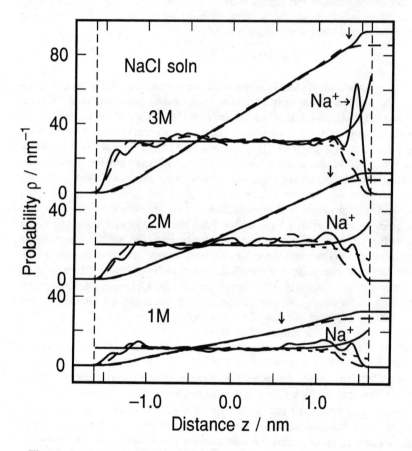

Figure 2. Ion probability distribution profiles for 1M to 3M NaCl solutions. Solid line Na^+ ion, broken line Cl^-. 1M and 2M solution at 30°C and -4e electrode charge, 3M solution at 100°C and -8e electrode charge. Vertical dash lines at |z|=1.615 nm mark where the 9-3 wall potential passes through zero, and beyound which water and ions are repelled. Inclined curves rising monotonically from left to right are the integrals of the corresponding ion distributions. Gouy-Chapman theory ion profiles shown as 'flat' curves for |z| < 1.615nm that rise (Na^+) or fall (Cl^-) for z > 1.0 nm.

For the solutions we find: 0.5 nm (0.31 nm) for 1M, 0.4 nm (0.24 nm) for 2M, 0.2 nm (0.18 nm) for 3M. The Debye shielding lengths calculated using Eqn(2) are in parentheses. Made with these favorable assumptions the agreement is remarkably close. It would change if we used $\varepsilon = 6$, a value more appropriate for a zone in which the dielectric properties of water are at saturation values[68].

Looking at the fine structure in the density profiles we see that on the metal side all the chloride distributions have weak features at ca. 1.2 nm and 1.4 nm. Both appear to be associated with peaks in the cation distribution and may therefore be due to contact pairs or solvent separated pairs. Of course the SPCE model for water was not designed with high salt concentrations in mind, so the ion pairs may be more a feature of the model and not nature. Correlation between ions at high salt concentrations alter the distribution near the charged surface. Note that there is no association of oppositely charged ion peaks at the left wall in Figure 2.

1M NaCl Solution

Figures 3, 4 and 5 show the results of an MD simulation using 1M NaCl solution. The simulation cell contains 32 Na^+ ions, 28 Cl^- ions, and 1576 water molecules at 30°C. The electrostatic charge on the electrode surface due to the difference in number of positive over negative ions is $-4|e|$ or $-0.046 \, Cm^{-2}$. The top panel in Figure 3 shows the probability density profiles for the water proton, water mass center, Na^+ ions and Cl^- ions. Both ion distributions have been smoothed to permit clearer identification of variations in density with position. We saw in Figure 2 the near coincidence of the integrals of the ion density for $z < 0.7$ nm which shows that the electrolyte is approximately charge neutral before this point. For $z > 0.7$ nm the integrated densities systematically diverge as expected for a transition from the locally neutral 'bulk' electrolyte into the 'diffuse' part of the electric double layer. The Na^+ ion distribution shows well defined structure in the form of a broad peak at ca. 1.1 nm, and a sharp peak at 1.4 nm. The water and proton distributions appear flat for $|z| < 0.8$ nm. On the metal side the water probability distribution has peaks at 0.9 nm, 1.2 nm, and a strong asymmetric peak at 1.6 nm. This latter feature is a superposition of a broad feature at 1.5 nm and a narrow peak at 1.6 nm. The peaks in the H distribution are at 0.9 nm, 1.2 nm, 1.55 nm, and ca. 1.7 nm. The last peak at 1.7 nm comes from protons in water OH bonds pointing at the metal.

The bottom panel in Figure 3 shows the potential calculated using the atom method, and the components of the potential calculated by the molecule method[19, 20]. The potential at the metal surface is about -2V. The potential in the 'bulk-like' zone comes mainly from the water quadrupole. This is a sensitive feature of the water model. Most water models are parameterized on bulk properties and not sensitive to changes in quadrupole. The electronic polarizability has been included in a average way. Tuning the properties of the water model to correctly account for these subtle effects is an important goal of future research. The monopole curve is from the ionic charge. If the system were truly neutral then the monopole curve would be flat and zero all the way to the beginning of the diffuse layer. The transition to monotonic decrease starting near $z = 0.7$ nm is another indicator of where the diffuse layer begins in this simulation. Note that combined monopole and dipole

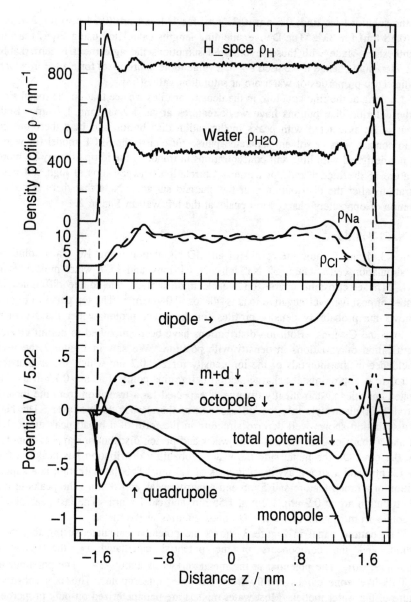

Figure 3. 1M NaCl solution at 30°C and -4e electrode charge. Top. Probability distribution profiles across the cell for H atom, water, and ions N⁺ and Cl⁻. Bottom. Total electric potential and component (monopole, dipole, combined monopole plus dipole, quadrupole and octopole) electric multipole potentials. The total potential was calculated from the total electric charge distribution. Note that though the monopole and dipole components go off scale their sum is finite and weaker than the quadrupole component of the potential for the SPCE water.

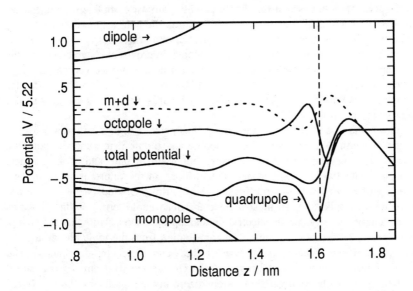

Figure 4. 1M NaCl solution at 30°C and -4e electrode charge. Detail of individual electric multipole component potentials shown in Figure 3 (bottom) for z > 0.8 nm.

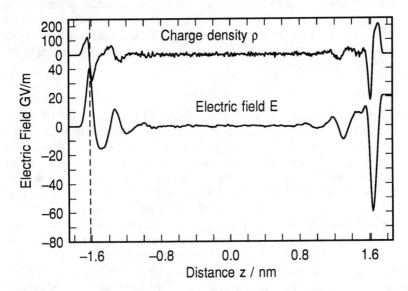

Figure 5. 1M NaCl solution at 30°C and -4e electrode charge. Total microscopic electric charge density and total electric field obtained by integrating the former.

potential curve m+d shows that the dipole potential almost completely compensates the monopole. Adding the quadrupole to m+d shifts the core region downwards by 3V and brings the molecular calculation of potential into fair correspondence with the atom method of calculation. Adding the octopole improves the agreement at the walls. The molecular method has larger extrema near the surface compared to the atom method. The reason for this is the need to include many high order multipoles in the molecule method. However the contact potential is the same in each case since it depends only on m and d[69].

Figure 4 shows a detail of the the potentials shown in Figure 3 (bottom) for the region z > 0.8 nm. We note that the monopole and dipole determine the potential at contact. The quadrupole potential is very important in bulk and at the surface. The octopole potential is important near the surface. In the future as water models improve it will be very important to include these terms. Obviously the surface potential relative to bulk solution must include the quadrupole term. Figure 5 shows the atomic charge density and the electric field along the z axis. Note that the charge density appears flat for |z| < 0.8 nm. The contribution from the ionic charge for z > 0.8 nm is not evident because the charge on the water molecules dominates. The electric field was obtained by integration of the charge density from -∞ to position z. The field is flat with small variations around zero in the region |z| < 0.8 nm. Near the metal the electric field undergoes a series of rapid oscillations due to the water packing structure at the interface. The small overall rise in field from bulk to surface is due to the excess Na^+ charge in the screening layer. The oscillation near z = -1.615 nm is due to layering at the restraining wall. This is an unwanted artifact of the immersed electrode model. It does not occur if the water density is decreased to the point that a vapor-liquid interface opens up at the restraining wall. Space prevents discussion of emersed electrodes[70, 71] which have a vapor-liquid interface[72-74] and a liquid-solid interface[75].

2M NaCl Solution

Figure 6 shows the results for a 2M NaCl solution at 30°C. There are 62 Na ions, 58 Cl ions and 1516 water molecules, the image charge on the electrode is -4|e|. The cation and anion concentrations are approximately the same for |z| < 1.0 nm. The bulk region appears larger than in 1M NaCl solution consistent with narrower screening zone. There are many similarities between Figures 3 and 6. The detailed proton and water profiles for z > 1.4 nm look the same. The bottom panel in Figure 6 shows the total potential calculated using the atom method, and the components of the potential calculated by the molecule method. The contact potential is about -2V, just the same as in the 1M case. The difference between 1M and 2M come mainly from the ion distributions and their direct interaction with the waters. Thus the monopole and dipole potentials are different, but since as already seen for 1M, they cancel each other, the main contribution to the total potential in the bulk region comes from the quadrupole. The water distributions for 1M and 2M are also similar and so are the potentials which come from water. Again if the system were truly neutral the monopole curve would be flat all the way to the edge of the diffuse layer. It is flatter than the 1M case and oscillates about zero before diving down to large

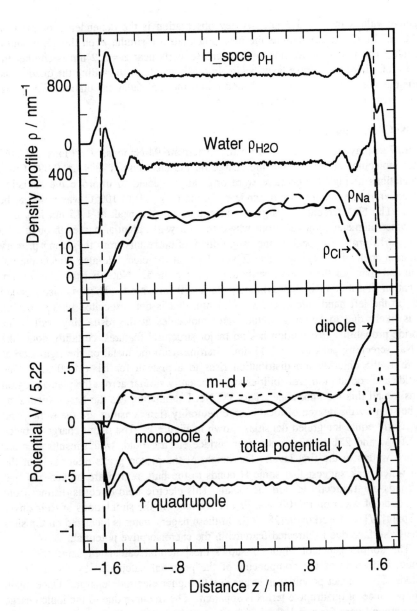

Figure 6. 2M NaCl solution at 30°C and -4e electrode charge. Top. Probability density distribution profiles across the cell for H atom, water, Na⁺ and Cl⁻ ions. Bottom. Total electric potential and component multipole potentials. Total potential calculated by atom method (see text). Note that the combined monopole and dipole components give a flat contribution across most of the cell. Octopole potential not displayed for clarity.

negative values for z > 1.2 nm. A key observation is the dependence of m+d on position. In all the simulations the m+d potentials are similar even though m and d are not. The transition to monotonic decrease starts near z = 1.2 nm is another indicator of where the diffuse layer begins in this simulation. Including the quadrupole shifts the 'core' region down by -3V and brings the molecular calculation of potential into correspondence with the atom potential.

3M NaCl Solution

Figure 7 displays the results for 3M NaCl. There are 94 Na ions, 86 Cl ions and 1463 water molecules at 100°C, and the charge on the electrode is -8|e| or -0.092 Cm^{-2}. Calculations at high temperature were originally selected to improve the statistics. Subsequently the temperature dependence in the range 30 to 100°C was found to be weak. The Na$^+$ screening charge is concentrated in the peak at 1.45 nm less than one water diameter removed from where the 9-3 wall potential passes through zero (z_w = 1.615 nm). The peak in the cation distribution results from the strong primary hydration shell of the cation. The solvent layer at the electrode also effects the position and shape of the cation distribution near the metal. Note that for |z| < 1.0 nm the cation distribution is approximately flat at 30 ions nm^{-1}. There is also a small peak on the left hand side at ca. z = -1.4 nm, that is not associated with screening but is likely due to layering of the water molecules at the restraining wall. The chloride probability distribution has no major structural features, certainly none like the Na$^+$ screening peak at z = 1.45 nm. Starting from the metal on the right side of Figure 1, the chloride ion distribution rises to a plateau for |z| < 1.00 nm. The chloride and sodium ion probabilities are sufficiently similar across the plateau region for us to call this the bulk zone. This 3M NaCl simulation has the best statistics, as can be seen by the degree of local charge neutrality (ion densities are the same), and very nearly equal integrated densities shown in Figure 2. The metal charge is twice that of 1M and 2M creating a stronger surface electric field, which results in more oriented waters in the first layer. The height of the proton peaks either side of the main water peak suggest that some H bonds to the bulk region are broken and that OH bonds point directly toward the metal. This electric field effect is distinct from localization of water on Pt(100) and Pt(111) surfaces in the simulations of Heinzinger and Spohr[27], and Berkowitz[28, 29]. In these papers water is localized on top sites of the Pt surface due to directed features in the chemisorptive potential.

The bottom panel of Figure 7 shows the potential calculated using the atom method, and some of the components of the potential calculated by the molecule method. The contact potential is larger due to higher electrode charge. Once again the importance of quadrupole terms is apparant. The m curve due to the ionic charge hovers around zero for z < 1.4 nm. The monopole potential is quite flat outside the screening layer. The transition to monotonic decrease starting near z = 1.4 nm is an indication of where the diffuse layer begins. Again m+d shows that the dipole potential completely compensates the monopole outside the screening zone. Including the quadrupole shifts the core region by -3V and brings the m+d potential into closer correspondence with the total potential calculated by the atom method.

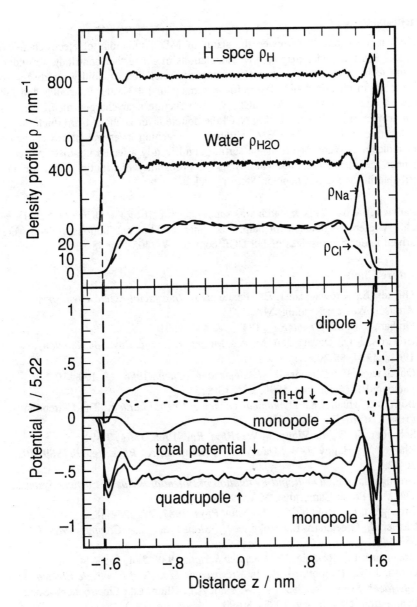

Figure 7. 3M NaCl solution at 30°C and -4e electrode charge. Top. Probability density distribution profiles across the cell for H atoms, water, Na⁺ and Cl⁻ ions. Bottom. Total electric potential and component multipole potentials. Total potential was calculated by atom method (see text). Note that monopole and dipole potentials are flat across most of the cell because the screening zone is narrow for 3M NaCl solution. Octopole potential not displayed for clarity.

Conclusions

In this paper we briefly reviewed the status of MD simulations of electrochemical interfaces and outlined a map of how calculations might further contribute to understanding technology. As an example of current simulation capability we discussed the structure of electric double layers for systems with 1500 water molecules and salt concentrations from 1M to 3M NaCl. Water structure near charged metal walls is not an artifact of small size. In 1M NaCl the double layer is about 1 nm thick (about three layers of water) while in 3M solution the screening layer was narrower than a water molecule. Water layers at the surface significantly affected the distribution of ions near the metal, creating features in the probability distributions that are not describable in the Gouy-Chapman-Stern model.

Acknowledgments. This research was supported in part by the Office of Naval Research. A referee is thanked for helpful comments. The contributions of JNG were performed under the auspices of US DOE contract W-7405-Eng-48.

Literature Cited

1. Gileadi, E., Kirowa-Eisner, E., Penciner, J. *Interfacial Electrochemistry*; Addison-Wesley: Reading, MA, 1975.
2. Grahame., D. C. *Chem. Rev.* **1947**, *41*, 441 - 501.
3. Brockris, J. O., Devanathan, M. A., Müller, K. *Proc. Roy. Soc.(London)* **1963.**, *A274*, 55-79.
4. Dutkiewicz, E., Parsons, R. *J. Electroanal. Chem.* **1966**, *11*, 100-110.
5. Parsons, R. *Chem. Rev.* **1990**, *90*, 813-826.
6. Born, M., Huang, K. *Dynamical Theory of Crystal Lattices*; The Clarendon Press: Oxford, England, 1954.
7. Schacher, G. E., de Wette, F. W. *Phys. Rev.* **1965**, *136A*, 78-91.
8. Rhee, Y. J., Halley, J. W., Hautman, J., Rahman, A. *Phys. Rev. B* **1989**, *40*, 36-42.
9. Greengard, L. F. *The Rapid Evaluation of Potential Fields in Particle Systems*; The MIT Press: Cambridge, MA, 1987.
10. Greengard, L., Rokhlin, V. *J. Comp. Phys.* **1987**, *73*, 325-348.
11. Carrier, J., Greengard, L., Rokhlin, V. *Siam J. Sci. Stat. Comput.* **1988**, *9*, 669-686.
12. Greengard, L., Rokhlin, V. *Chemica Scripta* **1989**, *29A*, 139-144.
13. Glosli, J. N., Philpott, M. R. In *Microscopic Models of Electrode-Electrolyte Interfaces. Symp. Proc. 93-5*; J. W. Halley L. Blum, Ed.; Electrochem. Soc.: Pennington, New Jersey, 1993. 80-90.
14. Hockney, R. W., Eastwood, J. W. *Computer Simulation using Particles*; McGraw-Hill: New York, 1981.
15. Pollock, E. L., Glosli, J. N. *Comput. Phys. Commun.* **1996**, *95*, 93 - 110.
16. Hautman, J., Halley, J. W., Rhee, Y. *J. Chem. Phys.* **1989**, *91*, 467-472.
17. Halley, J. W., Hauptman, J. *Phys. Rev. B* **1988**, *38*, 11704-11710.
18. Philpott, M. R., Glosli, J. N., Roberts, J. *unpublished* **1996**.
19. Glosli, J. N., Philpott, M. R. *Electrochimica Acta* **1996**, *41*, 2145 - 2158.

20. Philpott, M. R., Glosli, J. N. *J. Electroanal. Chem.* **1996**, *409*, 65 - 72.
21. Stone, A. J. *Molec. Phys.* **1985**, *56*, 1065 - 1082.
22. Price, S. L. *Phil. Mag.* **1994**.
23. Dang, L. X., Caldwell, J., Kollman, P. A. *preprint* **1990**.
24. Barker, J. A., Rettner, C. T. *J. Chem. Phys.* **1992**, *97*, 5844 - 5850.
25. Zhu, S., Philpott, M. R. *J. Chem. Phys.* **1994**, *100*, 6961-6968.
26. Pacchioni, G., Bagus, P. S., Nelin, C. J., Philpott, M. R. *Internat. J. Quantum Chem.* **1990**, *38*, 675-689.
27. Heinzinger, K. In *Computer Modeling of Fluids Polymers and Solids,*; C. R. A. Catlow, S. C. Parker, M. P. Allen, Ed.; Kluwer, Holland: 1990; v293, NATO ASI Series C. 357-404.
28. Reddy, M. R., Berkowitz, M. *Chem. Phys. Letters* **1989**, *155*, 173-176.
29. Raghavan, K., Foster, K., Motakabbir, K., Berkowitz, M. *J. Chem. Phys.* **1991**, *94*, 2110-2117.
30. Valette, G. *J. Electroanal. Chem.* **1981**, *122*, 285 - 297.
31. Sellers, H., Sudhakar, P. V. *J. Chem. Phys.* **1992**, *97*, 6644 - 6648.
32. Nazmutdinov, R. R., Probst, M. M., Heinzinger, K. *J. Electroanal. Chem.* **1994**, *369*, 227 - 231.
33. Spohr, E. *Chem. Phys.* **1990**, *141*, 87 - 94.
34. Rose, D. A., Benjamin, I. *J. Chem. Phys.* **1991**, *95*, 6856-6865.
35. Perera, L., Berkowitz, M. L. *J. Phys. Chem.* **1993**, *97*, 13803 - 13806.
36. Car, R., Parrinello, M. *Phys. Rev. Lett.* **1985**, *55*, 2471.
37. Löwen, H., Hansen, J., Madden, P. A. *J. Chem. Phys.* **1993**, *98*, 3275-3289.
38. Straus, J. B., Voth, G. A. *J. Phys. Chem.* **1993**, *97*, 7388-7391.
39. Marcus, R. M. *J. Chem. Phys.* **1965**, *43*, 679.
40. Halley, J. W., Hautman, J. *Ber. Bunsenges. Phys. Chem.* **1987**, *91*, 491-496.
41. Bader, J. S., Chandler, D. *J. Phys. Chem.* **1992**, *96*, 6423 - 6427.
42. Bennett, C. H. In *Algorithms for Chemical Computations*; R. E. Christoffersen, Ed.; ACS Symposium Series 46: Washington DC, 1977. 63-97.
43. Andersen, H. C. *J. Chem. Phys.* **1980**, *72*, 2384-2393.
44. Smit, B. In *Computer Simulation in Chemical Physics*; M. P. Allen D. J. Tildesley, Ed.; Kluwer, Holland: 1993; v397, NATO ASI Series C. 173-209.
45. Nicholson, D., Parsonage, N. G. *Computer Simulation and the Statistical Mechanics of Adsorption*; Academic Press: New York, 1982.
46. Cracknell, R. F., Nicholson, D., Quirke, N. *Mol. Phys.* **1993**, *80*, 885 - 897.
47. Laasonen, K., Sprik, M., Parrinello, M., Car, R. *preprint* **1993**.
48. Tuckermann, M., Laasonen, K., Sprik, M., Parrinello, M. *preprint* **1994**.
49. Bergsma, J. P., Gertner, B. J., Wilson, K. R., Hynes, J. T. *J. Chem. Phys.* **1987**, *86*, 1356-1376.
50. Brenner, D. W. *Phys. Rev. B* **1990**, *42*, 9458-9471.
51. Glosli, J. N., Belak, J., Philpott, M. R. *Electrochem. Soc. Symposium Proc.* **1995**, *95-4*, 25 - 37.
52. DePristo, A. E., Kara, A. *Adv. Chem. Phys.* **1990**, *77*, 163 - 253.
53. Brenner, D. W., Garrison, B. J. *Phys. Rev. B* **1985**, *34*, 1304.
54. Philpott, M. R. *unpublished theory* **1996**.

55. Stillinger, F. H., David, C. W. *J. Chem. Phys.* **1978**, *69*, 1473-1484.
56. Halley, J. W., Rustad, J. R., Rahman, A. *J. Chem. Phys.* **1993**, *98*, 2435-2438.
57. Glosli, J. N., Philpott, M. R. In *Microscopic Models of Electrode-Electrolyte Interfaces. Symp. Proc. 93-5*; J. W. Halley L. Blum, Ed.; Electrochem. Soc.: Pennington, New Jersey, 1993. 90-103.
58. Philpott, M. R., Glosli, J. N. In *Theoretical and Computational Approaches to Interface Phenomena*; H. Sellers J. T. Golab, Ed.; Plenum: New York, 1994. 75 - 100.
59. Glosli, J. N., Philpott, M. R. *J. Chem. Phys.* **1992**, *96*, 6962-6969.
60. Glosli, J. N., Philpott, M. R. *J. Chem. Phys.* **1993**, *98*, 9995-10008.
61. Berendsen, H. J., Grigera, J. R., Straatsma, T. P. *J. Chem. Phys.* **1987**, *90*, 6267.
62. Impey, R. W., Madden, P. A., McDonald, I. R. *J. Phys. Chem.* **1983**, *87*, 5071 - 5083.
63. Lee, C. Y., McCammon, J. A., Rossky, P. J. *J. Chem. Phys.* **1984**, *80*, 4448-4455.
64. Goodisman, J. *Electrochemistry: Theoretical Foundations*; Wiley-Interscience: New York, 1987.
65. Gouy, G. *Ann. phys.* **1917**, *7*, 129.
66. Chapman, D. L. *Phil. Mag.* **1913**, *25*, 475.
67. Bockris, J. O., Reddy, A. K. *Modern Electrochemistry*; Plenum Press: New York, 1973; Vol. 1.
68. Conway, B. E., Bockris, J. O., Ammar, I. A. *Trans. Faraday Soc.* **1951**, *47*, 756 - 766.
69. Landau, L. D., Lifschitz, E. M. *Electrodynamics of Continuous Media*; Addison-Welsey: 1960; 99 - 101.
70. Samec, Z., Johnson, B. W., Doblhofer, K. *Surface Sci.* **1992**, *264*, 440 - 448.
71. Stuve, E. M., Kizhakevariam, N. *Surface Sci.* **1993**, *A 11*, 2217 - 2224.
72. Wilson, M. A., Pohorille, A., Pratt, L. R. *J. Chem. Phys.* **1988**, *88*, 3281 - 3285.
73. Wilson, M. A., Pohorille, A., Pratt, L. R. *Chem. Phys.* **1989**, *129*, 209 - 212.
74. Wilson, M. A., Pohorille, A., Pratt, L. R. *J. Chem. Phys.* **1989**, *90*, 5211 - 5213.
75. Philpott, M. R., Glosli, J. N., Zhu, S. *unpublished* **1996**.

Chapter 3

Computer Simulation of the Structure and Dynamics of Water Near Metal Surfaces

E. Spohr

Abteilung für Theoretische Chemie, Universität Ulm,
Albert-Einstein-Allee 11, D–89069 Ulm, Germany

Structural and dynamic properties of pure water in contact with uncharged realistic metal surfaces are obtained by molecular dynamics simulations. The influences of adsorption energy, surface corrugation, electronic polarizability and surface inhomogeneity are investigated. The adsorption energy of water on the metal surface is found to be the most important parameter.

Computer modeling of electrochemical interfaces has become a well-established branch of interfacial electrochemistry. In recent years the interest shifted from studies of pure water near smooth model walls to increasingly realistic models of the metal phase (e. g., [1-11]), the properties of electrolyte solutions near the interface (e. g., [12-24]), and to free energy studies within the framework of the Marcus theory of electron transfer (e. g., [20, 25-27]), partial charge transfer [28] and ion transfer reactions [29].

It has been recognized that the structure of water near the interface determines the adsorption behavior of ions on the metal surface in specific ways [24, 30, 31]. Therefore, realistic models of the metal phase are needed in order to describe the inhomogeneity and orientational anisotropy in the aqueous phase adequately. Contrary to the situation for bulk liquid where reliable interaction potentials, from empirical parametrizations or from ab initio calculations, are available, the quantum chemical description of interactions between molecular adsorbates and metal substrates poses substantial problems due to the complexity of the system. Systematic studies contribute to the understanding of the key factors that determine the structure and dynamics at the electrochemical interface. In the present work the influence of water adsorption energy (for many transition metal surfaces a known experimental quantity [32]), surface corrugation, surface inhomogeneity (exemplified by the liquid mercury/liquid water interface), and electronic polarizability is investigated.

Models

Geometries.   The simulation of isolated interfaces is not possible. In every simulation at least two interfaces are present. Various system setups are possible. Conceptually simple is an alternating arrangement of liquid and solid phases with the minimum number of two equivalent interfaces, the properties of which can be averaged (Figure 1a). When applying periodic boundary conditions in all three directions of space this setup corresponds to a stack of infinitely many alternating thin liquid and solid slabs, which are infinitely extended in two dimensions (usually the x and y dimension). A rigorous treatment of the long-range interactions present in the liquid phase by, e. g., the Ewald method would, however, yield an undesired coupling of the different liquid slabs in such a system. Therefore, for the computation of the Coulomb interactions periodic boundary conditions are only to be taken into account in two dimensions, which is computationally more demanding than in three dimensions. If the solid phase is rigid and thick enough so that the water-metal interactions are fully converged such a system is equivalent to one with four interfaces, two vacuum/solid and two solid/liquid interfaces (fig 1b). In many cases it is convenient to simulate the liquid/solid interface together with a liquid/gas interface. In order to avoid the 'loss' of molecules from the liquid state into an infinitely large gas phase, the latter is confined on one side by a smooth wall. Such a system also consists of 4 interfaces, namely vacuum/solid, solid/liquid, liquid/gas and gas/wall (Figure 1c). All arrangements have been used in the literature.

Potential Functions.   In the studies discussed here the rigid TIP4P [33] and the flexible BJH [34] water models are used to model the water-water interactions. Both potential functions have been tested and used in many studies of the bulk phase of water and aqueous solutions.

Water-metal interactions consist of a non-electrostatic part and an electrostatic part that determines by and large the response of the metal to the charge

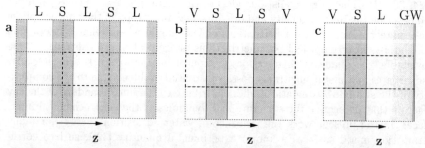

Figure 1: Geometries for simulations of liquid/solid interfaces. S, L, G, V, and W denote a solid phase, liquid phase, gas phase, vacuum, and confining wall, respectively.

distribution in the liquid phase on the basis of image charge interactions. The non-electrostatic part of the water-metal interactions describes the adsorption energy, adsorption site and geometry, and the surface translational modes on the basis of a set of pairwise additive functions between the atoms of the metal lattice and the atoms of the water molecule. Models for a vibrating Pt(100) surface, a rigid Hg(111) surface, a liquid mercury surface and an external potential with the periodicity of the Ni(100) surface have been used.

In the case of the Pt(100) surface the interaction potential is derived from semiempirical quantum chemical calculations of the interactions of a water molecule with a 5-atom platinum cluster [35]. The lattice of metal atoms is flexible and the atoms can perform oscillatory motions described by a single force constant taken from lattice dynamics studies of the pure platinum metal. The water-platinum interaction potential does not only depend on the distance between two particles but also on the projection of this distance onto the surface plane, thus leading to the desired property of water adsorption with the oxygen atoms on top of a surface atom. For more details see the original references [1, 2]. This model has later been simplified and adapted to the Pt(111) surface by Berkowitz and coworkers [3, 4] who used a simple corrugation function instead of atom-atom pair potentials.

The mercury-water interaction potential is derived in a similar manner from Hartree-Fock calculations of the interaction of a water molecule with 9-atom mercury clusters [36]. Potential functions fitted to the ab initio data were used to calculate the interactions of water molecules with a lattice of mercury atoms [7]. During the simulation this lattice was either kept rigid (Figure 6 and Figure 7) or the liquid water/liquid mercury interface was simulated (Figure 3).

A third model (termed 'model III' below) describes the metal surface as an external potential function [37], similar to the approach taken by Berkowitz in the case of the platinum surface [3, 4]. The external potential consists of a Morse function plus a corrugation term for oxygen-surface and a repulsive term for hydrogen-surface interactions:

$$V_{\text{water}-\text{surface}} = V_O(x_O, y_O, z_O) + V_H(z_{H1}) + V_H(z_{H2}) \qquad (1)$$

with

$$V_O(x, y, z) = D_O \left[\exp(-2\beta_O(z - z_1)) - 2 \cdot \exp(-\beta_O(z - z_1)) \right]$$

$$+ \; \alpha \cdot D_O \exp(-2\beta_O(z - z_1)) \cdot \left[\cos\left(\frac{10\pi x}{L_x}\right) + \cos\left(\frac{10\pi y}{L_y}\right) \right] \qquad (2)$$

and

$$V_H(x, y, z) = \gamma \cdot D_O \exp(-2\beta_H(z - z_2)). \qquad (3)$$

All parameters can be varied systematically but with the exception of D_O all parameters have been kept fixed in the different runs reported here. $\alpha = 0.1$ (except in Figure 5), $\gamma = 0.2$, $z_1 = 0$ Å, $z_2 = -4$ Å, $\beta_O = \beta_H = 1$ Å^{-1}, and the box lengths $L_x = L_y = 18$ Å. The corrugation (described by the parameter α) is felt only in the repulsive part of the Morse potential function and has a periodicity of 3.6 Å in both directions parallel to the surface, roughly corresponding to the periodicity on a Ni(100) surface. With the hydrogen-surface interactions being weakly repulsive (as characterized by positive γ), binding occurs predominantly through the oxygen atoms. This is in keeping with the other models of transition metal surfaces and *ab initio* calculations of water-metal interactions [35, 36, 38-41]. As α and γ are small, D_O is very nearly equal to the average adsorption energy of the water molecule. This external-field model of the interface makes it possible to easily vary key surface parameters in order to bring the adsorption behaviour of water into agreement with experimental evidence (e. g., [32]) for a particular metal surface. A similar route has been taken in [9].

In addition to the nonelectrostatic water-metal interactions, the response of the metal to the charge distribution in the liquid phase is taken into account by introducing image interactions. The static image plane is located at $z = 0$ for the platinum and mercury simulations and at $z = -3$ Å for the model III case. Electrostatic interactions have been calculated by means of a tabulated form of the two-dimensional Ewald summation. All simulations were performed at a temperature of 298±5 K. For the flexible BJH water model the Verlet algorithm with a time step length of 0.5 fs has been used, for the rigid TIP4P model the SHAKE algorithm with a time step of 2.5 fs.

Structure and Dynamics of Interfacial Water

Density Profiles. The interfacial water structure can be well characterized by the density profiles of oxygen and hydrogen atoms. The density profile is defined as the probability density to find an atom at a given distance from the surface, relative to the probability density of finding it at any place in the homogeneous bulk liquid. It thus describes the correlations between the phase boundary and the atoms. In interfacial problems, it plays a similar role as the pair correlation function in bulk liquids which describes the correlations between atoms. Like the pair correlation functions in the bulk, the density profile can be calculated by analytical theories based on a hierarchy of integral equations [42, 43]. It can also, at least in principle, be determined experimentally by X-ray reflectivity measurements [44, 45]. In the following we investigate the dependence of the density profile on the adsorption energy, the differences between liquid/liquid and liquid/solid water/mercury interfaces, the differences when using unpolarizable and polarizable water models, and the effect of corrugation.

Dependence on Adsorption Energy. The models described in the previous sections have been obtained either by adjusting an empirical potential function to available experimental data like adsorption energies and/or adsorbate frequencies, or by fitting of some set of analytical potential functions to semiempirical or ab initio quantum chemical calculations. Because the metal atom clusters in these calculations are small, and because of the semiempirical nature [35] or the neglect of electron correlation in the Hartree-Fock calculations [36] the interaction potentials can be regarded, at best, semiquantitatively correct. From UHV experiments, the adsorption energies of water on various transition metal surfaces are known experimentally [32]. It is, therefore, illustrative to investigate the changes in water structure with increasing adsorption energy. Model III described by eqs. (1-3) has been used, since the adsorption energy can be scaled easily in this model by scaling the parameter D_O.

Figure 2 shows the oxygen density profiles for four values of D_O equal to 12, 24, 36, and 48 kJ/mol. In the inset, the height of the first (diamonds) and second peak (crosses) of the density profile are shown for the pure water simulations corresponding to these values of D_O and several intermediate ones. A monotonous increase of peak heights is observed with increasing adsorption energy. Both curves level off at high adsorption energies. At low adsorption energies, peak heights approach 1 for $D_O \to 0$. This is in keeping with the observation that only small density variations are observed in simulations of water in contact with weakly attractive surfaces (see e.g. [46, 47]). The experimentally accessible adsorption energy of water on a transition metal surface can thus be regarded as an indicator for the degree of local structuring in the vicinity of the interface.

The density profiles obtained from "realistic" atomic models (platinum [2]

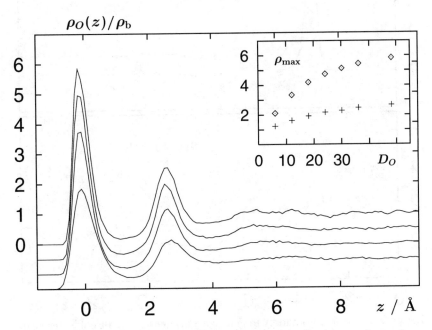

Figure 2: Normalized oxygen density profile perpendicular to the surface from simulations of pure water with adsorption energies of 12, 24, 36, and 48 kJ/mol (from bottom to top). The lower curves are shifted downwards by 0.5, 1.0, and 1.5 units. The inset shows the height of the first (diamonds) and second peak (crosses) as a function of adsorption energy.

and rigid mercury surfaces [7]) are qualitatively the same. Height and width
are correlated with the depth and force constant of the interaction potential. A
similar correlation holds for smooth interfaces (see [48] and references therein).

Liquid Water/Liquid Mercury Interface. In experimental studies the
liquid mercury electrode has been prototypical for a long time. Modeling the liquid mercury electrode is complicated by its liquid nature, which requires knowledge of mercury-mercury interactions. The liquid mercury/liquid water interface
was investigated by Heinzinger and coworkers [8, 22] using a pseudopotential
ansatz.

Figure 3 compares the oxygen and hydrogen density profiles for the interface
between pure water and rigid mercury (solid lines; taken from [7]) and water and

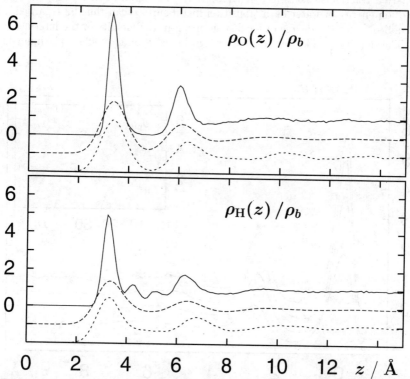

Figure 3: Oxygen (top) and hydrogen (bottom) density profiles of water near
a rigid mercury crystal (full line) and near liquid mercury (short dashes). The
long-dashed curves are the convolution of the density profiles near the rigid
crystal and the first maximum in the liqud mercury density profile according
to eq. 4 and 5. the dashed curves are shifted downward by one or two units
for better legibility.

liquid mercury (short dashes; taken from [8]) The features of the water density profiles at the liquid/liquid interface are washed out considerably relative to those at the liquid/solid interface. However, in the first layer this effect is almost entirely due to the roughness of the mercury surface: not all mercury atoms at the interface are in the same plane (at $z = 0$) but cover a range of approximately 1.3 Å (see Figure 1 in [8]). The width is larger than the width of the oxygen and hydrogen peaks near the solid surface. Consequently, in order to compare the liquid/liquid with the liquid/solid interface, the density distribution near the rigid surface must be convoluted with a width function $w(z)$ due to the mercury motion according to

$$\rho'(z) = \int_{-\infty}^{\infty} \rho(z')w(z - z')\mathrm{d}z'. \tag{4}$$

The width function is the normalized shape of the first mercury peak in the density profile, approximated as a Gaussian distribution of width $\sigma = 1.3$ Å

$$w(z) = \frac{\sqrt{2}}{\sqrt{\pi} \cdot \sigma} \exp[-2(z/\sigma)^2]. \tag{5}$$

The convoluted density profiles are plotted in Figure 3 as the long-dashed curves. The convolution accounts for almost all the structural differences in the range $z < 5$ Å. The heights of the first peaks of the oxygen and hydrogen density profiles after convolution are identical to those of the respective functions at the liquid/liquid interface. The intermediate maxima in the hydrogen density profiles around 4.3 and 5.1 Å vanish. Only in the second layer there are slight structural differences; the second peak is further away in the simulation of the liquid/liquid interface than in the corrected function.

Influence of Molecular Polarizability. Apart from the non-additive short-range intramolecular chemical interactions describing stretches, bends, and torsional motions, the most important many-body effect in polar liquid are induction interactions described by the dipole polarizability. The effective potential approach that is usually employed in liquid state simulations (see, e. g., [49]) works reasonably well as long as the microscopic electric fields are, on average, isotropic and independent of the local environment of the molecule. Recently it has been recognized that polarizable models are essential for the description of, e. g., ion solvation dynamics since the strong electric fields in the hydration shell of ions vary considerably with distance leading to a rather wide distribution of average molecular dipole moments. Since strong anisotropic electric fields can occur also in the vicinity of a metal electrode, the effect of the polarizability on the interfacial water structure cannot be neglected a priori. We have performed [50] comparative studies of the nonpolarizable TIP4P water model [33] and its polarizable extension [51] near the uncharged solid mercury surface and near the uncharged interface described by model III.

In Figure 4 the oxygen density profiles of the polarizable and the unpolarizable TIP4P model are compared for a film of water adsorbed on a model III surface ($D_O = 30$ kJ mol^{-1}). Near the metallic interface on the left the density profiles are almost identical. The liquid / gas interface on the right appears to

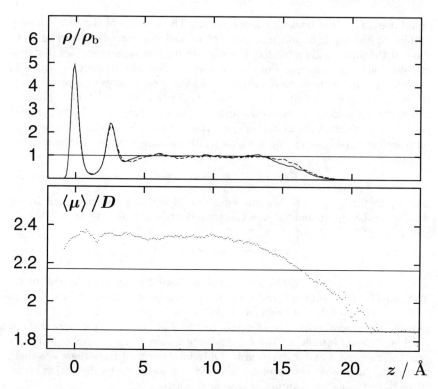

Figure 4: Top: Oxygen density profile of polarizable (solid line) and unpolarizable TIP4P water (dashed) in a water film near a model III surface.
Bottom: Distance dependence of the average molecular dipole moment of
the polarizable TIP4P water in the same system. The lower line indicates
the gas phase dipole moment of water, the upper one the dipole moment of
unpolarizable TIP4P water.

become slightly wider with the polarizable model. The bottom part of the figure
shows the distance dependence of the average dipole moment of the polarizable
water molecules. The dipole moment near the metal surface is almost unchanged
relative to the bulk phase. The range of the decrease from the average liquid
state dipole moment (≈ 2.35 D with this model) to the gas phase value of 1.85
D is of the order of 10 Å.

The Influence of Surface Corrugation. Figure 5 shows the dependence
of the oxygen density profiles in contact with a model III surface on the extent
of surface corrugation. The top part shows the results for adsorption energy
parameter $D_0 = 12$ kJ mol^{-1}, and in the lower part the results for $D_0 = 30$
kJ mol^{-1} are given. The corrugation parameter α is chosen as 0 (solid line),

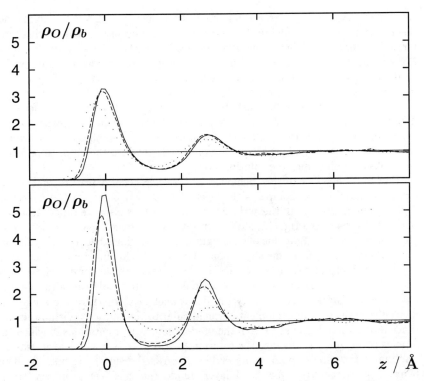

Figure 5: Top: Oxygen density profile of TIP4P water for a water film near a model (100) surface at adsorption energy $d_0 = 12$ kJmol^{-1} for different corrugation parameters $\alpha = 0$ (solid line), 0.1 (dashed) and 0.2 (dotted) (for definition see section). Bottom: The same curves, except for $d_0 = 30$ kjmol^{-1}.

0.1 (dashed), and 0.2 (dotted). $\alpha = 0$ corresponds to a completely flat surface. Typical corrugation energies on metal surfaces are estimated to be of the order of 20 % of the adsorption energy [32], corresponding to $\alpha = 0.1$ (in eq. 2). At small adsorption energies, the density profile is rather insensitive up to 20 % corrugation. At high corrugation a shift of the maximum to smaller distances is observed. The shift to smaller distances is a feature of the potential model used. The corrugation acts only on the repulsive branch of the water-surface potential, thus leading to a shift in the potential minimum. At the higher adsorption energy the density profiles are more sensitive to the value of α, mostly because a given value of α translates into a larger absolute energy difference between different sites. For $\alpha = 0.2$ the first maximum splits because the differences in equilibrium distances for different sites on the surface become too large.

Interfacial Polarization.   The orientational structure of water has been investigated in most studies of aqueous systems in inhomogeneous environments.

Almost everywhere a preference for orientations in which the water dipole moment is more or less parallel to the interface has been observed. The driving force for the avoidance of orientations that can lead to surface electrostatic polarization are, according to Lee *et al.* [46], the balance of (i) the packing forces which tend to produce a dense layer in contact with the surface and (ii) the tendency of molecules to maintain a maximal number of hydrogen bonds.

In ref. [2] the orientational distribution of water near the Pt(100) surface was investigated in great detail. A picture very similar to that near unpolar surfaces emerged. In spite of the preference for water adsorption through the oxygen atom relatively few configurations were observed, in which the dipole moment of the molecule points into the solution.

Figure 6 shows that this behavior is a rather general feature of the water/metal interface. It compares the orientational distribution of the molecular dipole moment relative to the surface normal in various distance ranges from the Pt(100) (left) and the Hg(111) surface (right). The reduced oxygen density profile is plotted on the right side of each figure. The base lines between distribution functions cut through the density profile. The distribution function in each panel is for those molecules that are located in the distance range beween these lines. By and large, the orientational distributions are quite similar in both systems. Over the first peak in the density profiles (panels *a* to *d*) there are almost no molecules with the dipole moment perpendicular to the surface. Within this distance range, there is a change from the predominance of orientations in which the dipoles point more or less into the solution (*a* and *b*) to one where a substantial fraction of the dipoles point more or less towards the surface (*c* and *d*). This behavior is characteristic for the "bilayer" model that has been proposed for the interpretation of the structure of water monolayers adsorbed on metal surfaces under ultrahigh vacuum conditions [32]. However, thermal motions wash out the ideal bilayer features considerably.

The orientational anisotropy ranges as far into the liquid phase as the density inhomogeneities do (roughly up to panel *m*), with increasingly less pronounced features. Slightly beyond the second maximum in the density profile the orientational distribution is isotropic, as it has to be the case for a bulk-like liquid.

Residence Times.   The dynamic behavior of water is frequently characterized by the self diffusion coefficient (sdc) D, which can be calculated from the particle mean square displacements via the Einstein relation or from the velocity autocorrelation functions (acf) via the Kubo relation. Near an interface this quantity D is not the self diffusion coefficient, since there are no free boundary conditions for the surface layer. Sonnenschein and Heinzinger [52] calculate a property called residence autocorrelation function

$$R(r, t) = \frac{1}{N_r} \sum_{i=1}^{N} [\theta_i(r, 0) \cdot \theta_i(r, t)], \tag{6}$$

where $\theta_i(r, t)$ is a function which is equal to 1 if molecule i is in region r at time t and equal to 0 otherwise. N_r is the average number of molecules in the region r. The residence acf is then fitted to the solution of the diffusion equation for the proper boundary conditions under the assumption that the density is homogeneous over the layer. With this method, Sonnenschein and Heinzinger find isotropic diffusion in the surface layer of the water slab confined between (12-6) Lennard-Jones walls. The sdc increases near the Lennard-Jones wall because the hydrogen bond network is weakened.

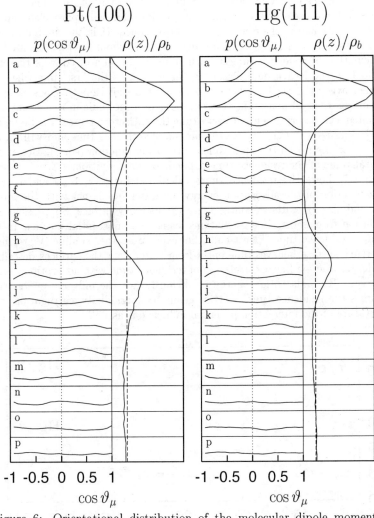

Figure 6: Orientational distribution of the molecular dipole moment on pt(100) (left) and hg(111) (right). $\cos \vartheta_\mu$ is the angle between the water dipole vector and the surface normal that points into the water phase. Panels a to p on the left are sampled from the distance intervals which are indicated by the cuts through the density profile $\rho(z)/\rho_b$ on the right.

The assumption that the adsorbed layer is homogeneous is certainly not fulfilled in the strongly inhomogeneous region near a metallic electrode. Therefore, only the residence times are calculated for water molecules in the adsorbed layer. In order to fit a single exponential function to the calculated residence acf reliably, the correlation time has to be chosen quite large, up to 200 picoseconds. After 200 picoseconds, however, there is a substantial probability that a molecule has left the surface for an extended time interval and then returns. The acf consequently will not decay to zero in the long time limit but will approach a limiting value, which is equal to the fraction of adsorbed molecules in the simulation cell. The contribution of molecules re-entering the layer should not be counted for the calculation of residence times. Each term in the sum of eq. 6 is therefore multiplied by the Heaviside function $H_{t_1}(t_0 - t)$. t_0 is the first time when the molecule leaves the layer for a period longer than a wait time t_1. Chosing, e. g., $t_1 = 0$ means that all non-diffusive oscillations of a molecule in its solvent cage are counted as a desorption event every time the trajectory crosses the boundary line. Chosing $t_1 > 0$ will increase the measured residence time and will, in the limit $t_1 \to \infty$, approach the unmodified correlation function (eq. 6). Hence, the numerical value of the residence time depends on its definition.

Figure 7 shows the residence times calculated for $t_1 = 0$ ps and $t_1 = 4$ ps as a function of the adsorption energy for water in contact with the model surface and also for water in contact with the mercury surface. Suppressing the contribution from short time oscillations leads to a substantial increase in the calculated residence time. The trends, however, are rather similar. Water near the mercury surface behaves differently from water near the surfaces described by model III where a linear correlation of residence time with adsorption energy is observed. The adsorption energy is obviously not the only parameter that determines the residence time. Lattice geometry, periodicity, corrugation, and the curvature of the interaction potential influence it as well.

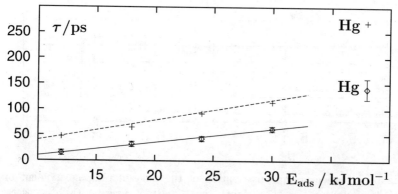

Figure 7: Residence times of water in the adsorbed layer near corrugated model surfaces. The corresponding values for adsorbed water near the mercury surface are also given, as indicated. Data are plotted as a function of water adsorption energy. Diamonds (with error bars) are for the choice $t_1 = 0$ ps, plus signs are for $t_1 = 4$ ps (for definition see text). The lines are merely drawn to guide the eye.

Summary and Conclusions

A systematic study of physical effects that influence the water structure at the water/metal interface has been made. Water structure, as characterized by the atom density profiles, depends most strongly on the adsorption energy and on the curvature of the water-metal interaction potential. Structural differences between liquid/liquid and liquid/solid interfaces, investigated in the water/mercury two-phase system, are small if the the surface inhomogeneity is taken into account. The properties of a polarizable water model near the interface are almost identical to those of unpolarizable models, at least for uncharged metals. The water structure also does not depend much on the surface corrugation.

Using values characteristic of water adsorbed on transition metal surfaces the liquid phase is inhomogeneous in a region of about 10 Å thickness. Two or three pronounced atomic layers can be observed. Orientational anisotropy is observed on the same length scale. The layering and the anisotropy are indicative of metal 'hydration', similar to that of large ions. For neutral metal surfaces, the hydration takes place in such a way as to maximize hydrogen bonding. The dynamics of water depends mostly on the adsorption energy but also on other properties such as curvature of the interaction potential energy. Realistic residence times of water molecules on a transition metal surface are in the range of hundred picoseconds to nanoseconds on uncharged surfaces.

Acknowledgments

Helpful discussions with A. Kohlmeyer and financial support by the Fonds der chemischen Industrie is gratefully acknowledged.

Literature Cited

1. Spohr, E.; Heinzinger, K., *Ber. Bunsenges. Phys. Chem.* **1988**, *92*, 1358.
2. Spohr, E., *J. Phys. Chem.* **1989**, *93*, 6171.
3. Foster, K.; Raghavan, K.; Berkowitz, M., *Chem. Phys. Lett.* **1989**, *162*, 32.
4. Raghavan, K.; Foster, K.; Berkowitz, M., *Chem. Phys. Lett.* **1991**, *177*, 426.
5. Raghavan, K.; Foster, K.; Motakabbir, K.; Berkowitz, M., *J. Chem. Phys.* **1991**, *94*, 2110.
6. Heinzinger, K., *Pure & Appl. Chem.* **1991**, *63*, 1733.
7. Böcker, J.; Nazmutdinov, R. R.; Spohr, E.; Heinzinger, K., *Surf. Sci.* **1995**, *335*, 372.
8. Böcker, J.; Spohr, E.; Heinzinger, K., *Z. Naturforsch.* **1995**, *50a*, 611.
9. Zhu, S.-B.; Philpott, M. R., *J. Chem. Phys.* **1994**, *100*, 6961.
10. Siepmann, J. I.; Sprik, M., *Surf. Sci. Lett.* **1992**, *279*, L185.
11. Siepmann, J. I.; Sprik, M., *J. Chem. Phys.* **1995**, *102*, 511.
12. Benjamin, I., *J. Chem. Phys.* **1991**, *95*, 3698.
13. Rose, D. A.; Benjamin, I., *J. Chem. Phys.* **1991**, *95*, 6956.
14. Rose, D. A.; Benjamin, I., *J. Chem. Phys.* **1993**, *98*, 2283.
15. Seitz-Beywl, J.; Poxleitner, M.; Heinzinger, K., *Z. Naturforsch.* **1991**, *46a*, 876.
16. Glosli, J. N.; Philpott, M. R., *J. Chem. Phys.* **1992**, *96*, 6962.

17. Glosli, J. N.; Philpott, M. R., in *Microscopic Models of Electrode-Electrolyte Interfaces*, edited by Halley, J. W.; Blum, L. (Electrochemical Society Inc., Pennington, 1993), No. 93-5, pp. 90–103.
18. Glosli, J. N.; Philpott, M. R., *J. Chem. Phys.* **1993**, *98*, 9995.
19. Spohr, E., *Chem. Phys. Lett.* **1993**, *207*, 214.
20. Rose, D. A.; Benjamin, I., *J. Chem. Phys.* **1994**, *100*, 3545.
21. Philpott, M. R.; Glosli, J. N., *J. Electrochem. Soc.* **1995**, *142*, L25.
22. Böcker, J.; Gurskii, Z.; Heinzinger, K., submitted to J. Phys. Chem.
23. Tóth, G.; Heinzinger, K., *Chem. Phys. Lett.* **1995**, *245*, 48.
24. Spohr, E., *Acta Chem. Scand.* **1995**, *49*, 189.
25. Benjamin, I., *J. Phys. Chem.* **1991**, *95*, 6675.
26. Straus, J. B.; Voth, G. A., *J. Phys. Chem.* **1993**, *97*, 7388.
27. Smith, B. B.; Halley, J. W., *J. Chem. Phys.* **1994**, *101*, 10915.
28. Nazmutdinov, R. R.; Spohr, E., *J. Phys. Chem.* **1994**, *98*, 5956.
29. Pecina, O.; Schmickler, W.; Spohr, E., *J. Electroanal. Chem.* **1995**, *394*, 29.
30. Heinzinger, K., in *Structure of Electrified Interfaces, Frontiers of Electrochemistry*, edited by Lipkowski, J.; Ross, P. N. (VCH, New York, 1993), Chap. 7. Molecular Dynamics of Water at Interfaces, p. 239.
31. Heinzinger, K., *Mol. Sim.* **1996**, *16*, 19.
32. Thiel, P. A.; Madey, T. E., *Surf. Sci. Reports* **1987**, *7*, 211.
33. Jorgensen, W. L.; Chandrasekhar, J.; Madura, J. D.; Impey, R. W.; Klein, M. L., *J. Chem. Phys.* **1983**, *79*, 926.
34. Bopp, P.; Jancsó, G.; Heinzinger, K., *Chem. Phys. Lett.* **1983**, *98*, 129.
35. Holloway, S.; Bennemann, K. H., *Surf. Sci.* **1980**, *101*, 327.
36. Nazmutdinov, R. R.; Probst, M.; Heinzinger, K., *J. Electroanal. Chem.* **1994**, *369*, 227.
37. Spohr, E., *J. Mol. Liquids* **1995**, *64*, 91.
38. Ribarsky, M. W.; Luedtke, W. D.; Landman, U., *Phys. Rev. B* **1985**, *32*, 1430.
39. Rosi, M.; Bauschlicher Jr, C. W., *J. Chem. Phys.* **1989**, *90*, 7264.
40. Yang, H.; Whitten, J. L., *Surf. Sci.* **1989**, *223*, 131.
41. Sellers, H.; Sudhakar, P. V., *J. Chem. Phys.* **1992**, *97*, 6644.
42. Vossen, M.; Forstmann, F., *J. Chem. Phys.* **1994**, *101*, 2379.
43. Booth, M. J.; Duh, D. M.; Haymet, A. D. J., *J. Chem. Phys.* **1994**, *101*, 7925.
44. Toney, M. F.; Howard, J. N.; Richter, J.; Borges, G. L.; Gordon, J. G.; Melroy, O. R.; Wiesler, D. G.; Yee, D.; Sorensen, L. B., *Nature* **1994**, *368*, 444.
45. Magnussen, O. M.; Ocko, B. M.; Adzic, R. R.; Wang, J. X., *Phys. Rev. B* **1995**, *51*, 5510.
46. Lee, C. Y.; McCammon, J. A.; Rossky, P. J., *J. Chem. Phys.* **1984**, *80*, 4448.
47. Spohr, E., submitted to Chem. Phys. Lett.
48. Spohr, E.; Heinzinger, K., *Electrochim. Acta* **1988**, *33*, 1211.
49. Allen, M. P.; Tildesley, D. J., *Computer Simulations of Liquids* (Oxford University Press, New York, 1987).
50. Kohlmeyer, A.; Witschel, W.; Spohr, E., submitted to Chem. Phys. Lett.
51. Rick, S. W.; Stuart, S. J.; Berne, B. J., *J. Chem. Phys.* **1994**, *101*, 6141.
52. Sonnenschein, R.; Heinzinger, K., *Chem. Phys. Lett.* **1983**, *102*, 550.

Chapter 4

Fundamental Thermodynamic Aspects of the Underpotential Deposition of Hydrogen, Semiconductors, and Metals

Alireza Zolfaghari and Gregory Jerkiewicz[1]

**Département de chimie, Université de Sherbrooke,
Sherbrooke, Québec J1K 2R1, Canada**

The paper presents theoretical methodology that allows determination of thermodynamic state functions of the under-potential deposition of hydrogen, UPD H, and semiconductor or metallic species, UPD M. The experimental approach involves temperature dependence of the UPD by application of cyclic-voltammetry or chronocoulometry. The theoretical approach is based on a general electrochemical adsorption isotherm and numerical calculations which lead to determination of the Gibbs free energy of adsorption, ΔG°_{ads}, as a function of T and θ. Temperature dependence of ΔG°_{ads} (for θ = const) leads to appraisal of the entropy of adsorption, ΔS°_{ads}, whereas coverage dependence of ΔG°_{ads} (for T = const) allows assessment of the nature of the lateral interactions between the adsorbed species; knowledge of ΔG°_{ads} and ΔS°_{ads} leads to determination of ΔH°_{ads}. The paper presents new approach which permits elucidation of the bond energy between the substrate, S, and H_{UPD} or S and M_{UPD}, $E_{S-H_{UPD}}$ and $E_{S-M_{UPD}}$, respectively. Comprehension of $E_{S-H_{UPD}}$ is essential in assessment of the strength of the $S-H_{UPD}$ bond and the adsorption site of H_{UPD}. Knowledge of $E_{S-M_{UPD}}$ is of importance in: (i) evaluation of the strength of the cohesive forces acting between S and M_{UPD} that are responsible for the adhesion of the adsorbate to the substrate; and (ii) comparison of the $S-M_{UPD}$ bond with that observed for the 3-D bulk deposit of M. The UPD H on Rh and Pt electrodes from aqueous H_2SO_4 solution is discussed as an example of application of this methodology.

The under-potential deposition, UPD, of hydrogen, H, and semiconductors and metals, abbreviated by M, of the *p* and *d* blocks of the periodic table on transition-metal has been a subject of intense studies in electrochemical surface science (*1-15*).

---

[1]Corresponding author

The UPD refers to the phenomenon of deposition of H or M on a foreign metallic substrate, S, at potentials positive to the equilibrium potential of the hydrogen evolution reaction, HER, E°_{HER}, or to the equilibrium potential of the bulk deposition of M, $E^\circ_{M^{z+}/M}$. The UPD of H is known (*1-13,16-19*) to take place on Rh, Pt, Ir and Pd at potentials positive roughly between 0.05 and 0.40 V versus the reversible hydrogen electrode, RHE. The UPD of M takes place on the above mentioned noble-metal substrates as well as on other transition metals such as Au, Ag and Cu on which the UPD H is not observed (*20-28*). The UPD M appears to be a more general phenomenon which precedes 3D bulk-type deposition at potentials positive with respect to $E^\circ_{M^{z+}/M}$ and it always takes place at a metal substrate more noble with respect to the species undergoing the under-potential deposition. Thus the UPD adlayer acts as a precursor for the formation and growth of the 3D bulk-type phase on the foreign metal substrate. Various UPD systems are reviewed in refs. *20, 21, 24* and *29* with description of different substrates and metal ions deposited from aqueous or non-aqueous solutions. The UPD M is usually limited to a monolayer and the process resembles chemisorption of a submonolayer or a monolayer, ML, of a metal or a semiconductor from the gas phase. The origin of the UPD can be explained in terms of the existence of stronger attractive, chemisorptive forces between the foreign metal substrate, S, and the depositing species, here H or M, than those between like atoms within the 3D deposit of M. In other words, the $S-M_{UPD}$ bond energy, $E_{S-M_{UPD}}$, is greater than that of the $M-M$ bond in the 3D lattice of M, E_{M-M}.

Kolb et al. (*23*) observed that the potential of stripping of the bulk deposit is shifted towards lower potentials with respect to the potential of desorption of the UPD layer; this difference was defined as the underpotential shift, ΔE_P. The underpotential shift was correlated to the difference in work functions, Φ, between the substrate, S, and the UPD species, M, and expressed by the following equation:

$$\Delta E_P = \alpha\,\Delta\Phi \tag{1}$$

where $\alpha = 0.5$ V eV^{-1} and $\Delta\Phi = \Phi_S - \Phi_M$. The consequence of equation 1 is that the work function of the substrate, Φ_S, should be greater the work function of the UPD species, Φ_M, thus $\Phi_S \rangle \Phi_M$ if the under-potential deposition is to occur.

It is well recognized in electrochemical surface science that anions coadsorbed on the electrode surface with the UPD species influence their cyclic-voltammetry, CV, and chronocoulometry characteristics (*30-32*). Structural changes associated with M and anion coadsorption are investigated by electrochemical, radiochemical, and UHV techniques as well as by scanning tunneling microscopy, STM, and atomic force microscopy, AFM (*14,15,33-40*). These changes can be related to such thermodynamic state functions as the Gibbs free energy, entropy and enthalpy of adsorption (*18,19,30,32*).

Whereas the origin of the UPD H and UPD M as well as the 3D bulk deposition of M is quite well understood, there is a lack of knowledge of the $S-M_{UPD}$ bond energy, $E_{S-M_{UPD}}$, as well as thermodynamic state functions for the UPD. This paper addresses this issue and demonstrates a theoretical methodology which allows determination of the state functions and $E_{S-M_{UPD}}$ on the basis of experimental data. The experimental methodology involves application of cyclic-voltammetry over a wide temperature range, or chronocoulometry whereas the theoretical treatment is based on a general electrochemical isotherm and adsorption thermodynamics.

Under-Potential Deposition of Hydrogen

The UPD H from an acidic aqueous solution may be represented by the following single-electrode processes:

working electrode $\quad H^+ + e + S \;=\; S - H_{UPD}$ (2)

reference electrode $\quad 1/2\,H_2 \;=\; H^+ + e$ (3)

The activity of the solvated proton in the bulk of the electrolyte of the working-electrode and reference-electrode compartments is the same and it equals a_{H^+}. The pressure of the H_2 in the reference-electrode compartment is $P_{H_2}^r$. If the temperature is the same in both compartments, then the process is described by the following general electrochemical adsorption isotherm:

$$\frac{\theta_{H_{UPD}}}{1-\theta_{H_{UPD}}} = a_{H^+}\,\exp\!\left(-\frac{FE_{SHE}}{RT}\right)\exp\!\left(-\frac{\Delta G^\circ_{ads}(H_{UPD})}{RT}\right)$$ (4)

where $\theta_{H_{UPD}}$ is the surface coverage of H_{UPD}, E_{SHE} is the potential measured versus the standard hydrogen electrode, SHE, $\Delta G^\circ_{ads}(H_{UPD})$ is the standard Gibbs free energy of adsorption, T is the temperature and R and F are physico-chemical constants. The above formula is a specific form of the following general relation for the conditions mentioned above:

$$\frac{\theta_{H_{UPD}}}{1-\theta_{H_{UPD}}} = \frac{a_{H^+}^w}{a_{H^+}^r}\sqrt{P_{H_2}^r}\,\exp\!\left(-\frac{FE_{eq}}{RT}\right)\exp\!\left(-\frac{\Delta G^\circ_{ads}(H_{UPD})}{RT}\right)$$ (5)

where $a_{H^+}^w$ and $a_{H^+}^r$ are the activities of H^+ in the working-electrode and reference-electrode compartments, and E_{eq} is the equilibrium potential related to $P_{H_2}^r$ through the Nernst formula. When the temperatures of both compartments are the same and the activities of H^+ are equal, then equations 4 and 5 reduce to the following formula:

$$\frac{\theta_{H_{UPD}}}{1-\theta_{H_{UPD}}} = \sqrt{P_{H_2}^r}\,\exp\!\left(-\frac{FE_{RHE}}{RT}\right)\exp\!\left(-\frac{\Delta G^\circ_{ads}(H_{UPD})}{RT}\right)$$ (6)

where E_{RHE} is the potential measured versus the reversible hydrogen electrode, RHE, immersed in the same electrolyte.

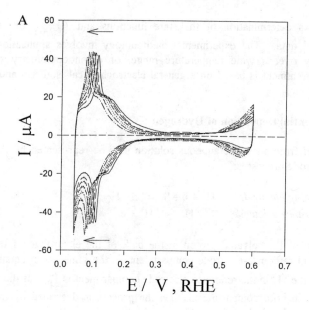

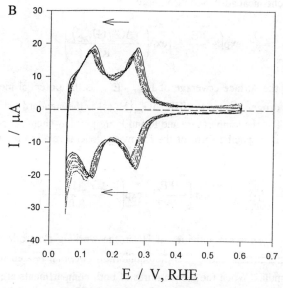

Figure 1. Series of cyclic-voltammetry, CV, profiles for the under-potential deposition of H, UPD H, from 0.50 M aqueous solution of H_2SO_4 for a temperature range between 273 and 343 K, with an interval of 10 K, and recorded at the sweep rate s = 20 mV s^{-1}. **A.** For Rh; the electrode surface area $A_r = 0.70 \pm 0.01$ cm^2 (Adopted from ref. *19*). **B.** For Pt; the electrode surface area $A_r = 0.72 \pm 0.01$ cm^2. The arrows indicate the shift of the adsorption and desorption peaks upon the temperature increase.

It should be added that equation 4 is neither the Langmuir nor the Frumkin isotherm and $\Delta G_{ads}^{\circ}(H_{UPD})$ refers to the Gibbs free energy of adsorption at given $\theta_{H_{UPD}}$ and T *(18,19)*. Assessment of the relation between $\Delta G_{ads}^{\circ}(H_{UPD})$ and $(\theta_{H_{UPD}}, T)$ may clarify whether the process follows either of the two electrochemical adsorption isotherms. However, even in the case of H chemisorption under gas-phase conditions, in absence of the electrified double layer, the relations between $\Delta G_{ads}^{\circ}(H_{UPD})$ and $\theta_{H_{UPD}}$ are more complicated and rarely follow the two most fundamental cases *(41)*. Thus it is reasonable to conclude that H chemisorption does not follow the common isotherms due to presence of complex first-nearest and second-nearest neighbor lateral interactions between the adsorbed species *(41-43)*.

It should be stressed that both equations 4 and 6 may be applied for determination of $\Delta G_{ads}^{\circ}(H_{UPD})$, but equation 4 requires precise knowledge of the activity coefficient of the proton whereas equation 6 demands knowledge of the hydrogen gas pressure in the reference electrode compartment which is more accessible. In other words, it is easier to apply equation 6 to determine $\Delta G_{ads}^{\circ}(H_{UPD})$ than equation 4. It is evident that experimental appraisal of E at which H_{UPD} reaches a given $\theta_{H_{UPD}}$ at a given T allows numerical determination of $\Delta G_{ads}^{\circ}(H_{UPD})$. Such calculations may be performed for a series of coverages of H_{UPD} and for various temperatures, and they results in evaluation of $\Delta G_{ads}^{\circ}(H_{UPD})$ as a function of $\theta_{H_{UPD}}$ and T, $\Delta G_{ads}^{\circ}(H_{UPD})$ versus $(\theta_{H_{UPD}}, T)$.

Cyclic-voltammetry, CV, and chronocoulometry *(44)* are experimental techniques that can be applied to evaluate the dependence of the H_{UPD} surface coverage on temperature variation and to determine $\Delta G_{ads}^{\circ}(H_{UPD})$. CV is in many respects the electrochemical equivalent of temperature programmed desorption, TPD, although CV results in determination of different thermodynamic parameters than TPD. CV allows one to study the surface coverage of the adsorbed species during potential-stimulated adsorption and desorption at various temperatures. Theoretical treatment of CV experimental results leads to elucidation of important thermodynamic state functions such as ΔG_{ads}°, ΔH_{ads}° and ΔS_{ads}°. Thus it is sensible to refer to it as *potential-stimulated adsorption-desorption*, PSAD.

Figure 1 shows two series of CV adsorption-desorption profiles for UPD H on Rh and Pt from 0.50 M aqueous H_2SO_4 solution. In the case of Rh there is only one adsorption-desorption peak where as in the case of Pt there are two peaks. Upon the temperature increase, the peaks shift towards less-positive potentials and there is a slight redistribution of the charge between the two peaks for Pt *(18,19)*. These data allow calculation of $\Delta G_{ads}^{\circ}(H_{UPD})$ based on equation 5 and the results are shown in Figure 2 as 3D plots of $\Delta G_{ads}^{\circ}(H_{UPD})$ versus $(\theta_{H_{UPD}}, T)$. $\Delta G_{ads}^{\circ}(H_{UPD})$ assumes the most negative values in at the lowest T and the lowest $\theta_{H_{UPD}}$. In the case of Rh, $\Delta G_{ads}^{\circ}(H_{UPD})$ has values between -18 and -8 kJ mol^{-1} whereas in the case of Pt, it varies between -25 and -11 kJ mol^{-1}. For a given constant T, $\Delta G_{ads}^{\circ}(H_{UPD})$ increases towards less-negative values with $\theta_{H_{UPD}}$ augmentation indicating that the lateral interactions between H_{UPD} adatoms are repulsive. The $\Delta G_{ads}^{\circ}(H_{UPD})$ versus $\theta_{H_{UPD}}$

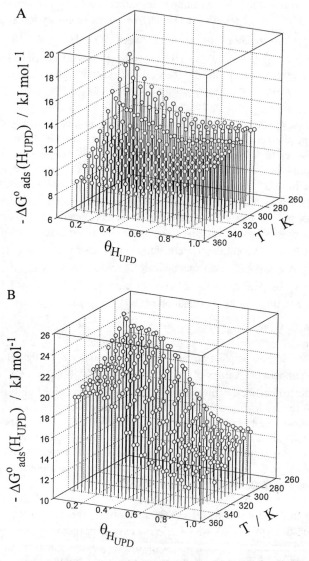

Figure 2. 3D plots showing the Gibbs free energy of the under-potential deposition of H, $\Delta G^{\circ}_{ads}(H_{UPD})$, versus $\theta_{H_{UPD}}$ and T, $\Delta G^{\circ}_{ads}(H_{UPD}) = f(\theta_{H_{UPD}}, T)$, for adsorption from 0.50 M aqueous solution of H_2SO_4. **A.** Rh (Adopted from ref. *19*). **B.** Pt. Augmentation of $\Delta G^{\circ}_{ads}(H_{UPD})$ with increase of $\theta_{H_{UPD}}$ for T = const points to the repulsive nature of lateral interactions between H_{UPD} adatoms. The $\Delta G^{\circ}_{ads}(H_{UPD})$ versus T relations for $\theta_{H_{UPD}}$ = const allow elucidation of the entropy of adsorption, $\Delta S^{\circ}_{ads}(H_{UPD})$.

relations are non-linear indicating that the adsorption process is complex and may not be simply described by the Langmuir or the Frumkin isotherm (*18,19*). For a given constant $\theta_{H_{UPD}}$, the relation between $\Delta G^\circ_{ads}(H_{UPD})$ and T is linear and it describes the entropy of adsorption, $\Delta S^\circ_{ads}(H_{UPD})$, through the following relation:

$$\Delta S^\circ_{ads}(H_{UPD}) = -\left(\frac{\partial \Delta G^\circ_{ads}(H_{UPD})}{\partial T}\right)_{\theta_{H_{UPD}}=const} \tag{7}$$

Figure 3 shows $\Delta S^\circ_{ads}(H_{UPD})$ for UPD H on Rh and Pt as a function of $\theta_{H_{UPD}}$. In the case of Rh $\Delta S^\circ_{ads}(H_{UPD})$ has values between -125 and -30 J mol^{-1} K^{-1} whereas in the case of Pt it varies between -75 and -40 J mol^{-1} K^{-1}. The $\Delta S^\circ_{ads}(H_{UPD})$ versus $\theta_{H_{UPD}}$ plot for Pt forms two waves which are associated with two peaks in the CV adsorption-desorption profiles. The enthalpy of adsorption is readily determined based on the experimental values of $\Delta G^\circ_{ads}(H_{UPD})$ and $\Delta S^\circ_{ads}(H_{UPD})$ and equation 8:

$$\Delta H^\circ_{ads}(H_{UPD}) = \Delta G^\circ_{ads}(H_{UPD}) + T \Delta S^\circ_{ads}(H_{UPD}) \tag{8}$$

Figure 4 shows values of $\Delta H^\circ_{ads}(H_{UPD})$ for UPD H on Rh and Pt as a function of $\theta_{H_{UPD}}$ determined on the basis of the experiment data presented in Figures 2 and 3. The results demonstrate that in the case of Rh, $\Delta H^\circ_{ads}(H_{UPD})$ varies between -52 and -20 kJ mol^{-1} whereas in the case of Pt, it falls between -45 and -28 kJ mol^{-1}. The $\Delta H^\circ_{ads}(H_{UPD})$ versus $\theta_{H_{UPD}}$ relation for Pt reveals two waves which again are associated with the two peaks in the CV adsorption-desorption profiles.

It is essential to elaborate on the $\Delta S^\circ_{ads}(H_{UPD})$ versus $\theta_{H_{UPD}}$ and $\Delta H^\circ_{ads}(H_{UPD})$ versus $\theta_{H_{UPD}}$ plots. An analysis of the data shown in Figures 3 and 4 reveals that the enthalpy and entropy variations are mirror images. Such thermodynamic dependences are well known in catalysis (*45-46*) and show that variation of the entropy of adsorption is always counterbalanced by alteration of the enthalpy of adsorption. This phenomenon is recognized as *a compensation effect* (*45*) and the data presented in the present paper indicate that it is also observable for electrochemical systems.

An alternative approach that may be applied to determine $\Delta H^\circ_{ads}(H_{UPD})$ involves combination of equation 4 with the Gibbs-Helmholtz relation which leads to the following formula (*18,19*):

$$\frac{\partial(E/T)}{\partial(1/T)} = -\left[\frac{\Delta H^\circ_{ads}(H_{UPD})}{F}\right]_{\theta_{H_{UPD}}=const} \tag{9}$$

Thus by experimental determination of pairs of values of E and T at which the H_{UPD} coverage is constant, $\theta_{H_{UPD}}=const$, and by plotting E/T versus 1/T one obtains linear relations and from their slope one may evaluate $\Delta H^\circ_{ads}(H_{UPD})$. The authors applied

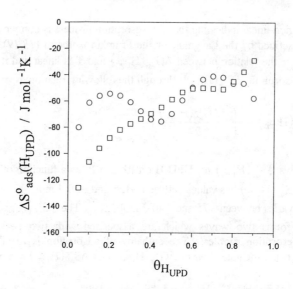

Figure 3. Dependence of $\Delta S^o_{ads}(H_{UPD})$ on $\theta_{H_{UPD}}$ for the under-potential deposition of H, UPD H, from 0.50 M aqueous H_2SO_4 solution. □ refers to Rh; $\Delta S^o_{ads}(H_{UPD})$ has values between −125 and −30 J mol^{-1} K^{-1}. O refers to Pt; $\Delta S^o_{ads}(H_{UPD})$ has values between −75 and −40 J mol^{-1} K^{-1}.

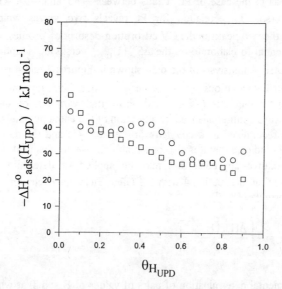

Figure 4. Dependence of $\Delta H^o_{ads}(H_{UPD})$ on $\theta_{H_{UPD}}$ for the under-potential deposition of H, UPD H, from 0.50 M aqueous H_2SO_4 solution. □ refers to Rh; $\Delta H^o_{ads}(H_{UPD})$ has values between −52 and −20 kJ mol^{-1}. O refers to Pt; $\Delta H^o_{ads}(H_{UPD})$ has values between −45 and −28 kJ mol^{-1}.

this methodology and found that such determined values $\Delta H^{\circ}_{ads}(H_{UPD})$ agreed to within 1.5 kJ mol^{-1} with those shown in Figure 4.

The energy of the $S - H_{UPD}$ bond (here $Rh - H_{UPD}$ and $Pt - H_{UPD}$), $E_{S-H_{UPD}}$, may be determined on the basis of Born-Haber thermodynamic cycles for the respective single-electrode processes (*18,19*). Summation of the two single-electrode reactions shown in equations 2 and 3 leads to the following overall process:

$$1/2\, H_2 + S \xrightarrow{\;\Delta H^{\circ}_{ads}(H_{UPD})\;} S - H_{UPD} \tag{10}$$

Addition of the Born-Haber cycles (*18,19*) leads to the the following relation for the $S - H_{UPD}$ bond energy, $E_{S-H_{UPD}}$:

$$E_{S-H_{UPD}} = \frac{1}{2} D_{H_2} - \Delta H^{\circ}_{ads}(H_{UPD}) \tag{11}$$

where D_{H_2} is the dissociation enerhy of the H_2 molecule. Figure 5 demonstrates relations between $E_{Rh-H_{UPD}}$ or $E_{Pt-H_{UPD}}$ and $\theta_{H_{UPD}}$ based on the values of $\Delta H^{\circ}_{ads}(H_{UPD})$ shown in Figure 4. The $Rh - H_{UPD}$ bond energy is between 240 and 270 kJ mol^{-1} and that of the $Pt - H_{UPD}$ bond is between 250 and 265 kJ mol^{-1}; both bond energies vary slightly with the H_{UPD} surface coverage. The results presented in Figure 5 show that $Rh - H_{UPD}$ and $Pt - H_{UPD}$ bond energies fall close to that for the respective bond energies between the same substrate and the chemisorbed H, H_{chem}, $E_{Rh-H_{chem}}$ and $E_{Pt-H_{chem}}$, respectively, which is 255 kJ mol^{-1} for Rh and $243 - 255$ kJ mol^{-1} for Pt (*41*). Proximity of these values indicates that H_{UPD}, is an energetic equivalent of H_{chem}. Moreover, if the bond energies for H_{UPD} and H_{chem} are so close, then it is reasonable to assume that H_{UPD} similarly to H_{chem} is strongly embedded in the surface lattice of the metal substrate.

It is apparent on the basis of equation 11 that the $E_{S-H_{UPD}}$ versus $\theta_{H_{UPD}}$ plots follow the variations of $\Delta H^{\circ}_{ads}(H_{UPD})$ since $1/2\, D_{H_2}$ represents a constant.

An aspect that the authors would like to emphasize is that the present treatment, being the first approach, does not take into account the specific adsorption of anions and their contribution to the overall adsorption charge. Indeed, there is a certain anion contribution and it may affect the values of $\Delta G^{\circ}_{ads}(H_{UPD})$ but the present approximation, which is meant to verify the validity of the theoretical treatment, does not take it into account. However, experimental research on Pt and Rh single crystals is under way and the results will be presented in subsequent papers which will take into account the anion effect.

Under-Potential Deposition of Semiconductors and Metals

The under-potential deposition of semiconductors and metals, UPD M, from an aqueous solution containing a monovalent cation, M^+, on a substrate more noble than

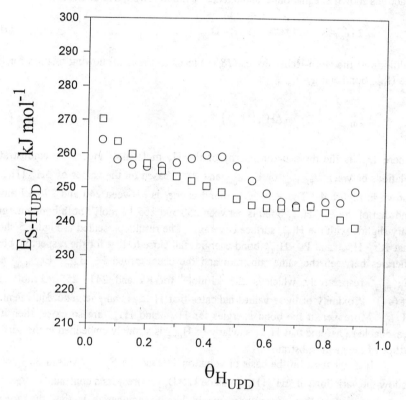

Figure 5. Dependence of $E_{S-H_{UPD}}$ on $\theta_{H_{UPD}}$ for the under-potential deposition of H, UPD H, from 0.50 M aqueous H_2SO_4 solution. $\square$ refers to the $Rh-H_{UPD}$ bond energy, $E_{Rh-H_{UPD}}$, which have values between 240 and 270 kJ mol^{-1}. $\bigcirc$ refers to the $Pt-H_{UPD}$ bond energy, $E_{Pt-H_{UPD}}$, which assumes values between 250 and 265 kJ mol^{-1}. The variation of $E_{S-H_{UPD}}$ versus $\theta_{H_{UPD}}$ follows the $\Delta H_{ads}^{\circ}(H_{UPD})$ versus $\theta_{H_{UPD}}$ relations.

the species undergoing the UPD may be represented by the following single-electrode processes:

$$working\ electrode \quad S + M^+ + e = S - M_{UPD} \tag{12}$$
$$reference\ electrode \quad M = M^+ + e \tag{13}$$

Summation of the above relations leads to the following overall equation which represents the formation of the UPD layer of M on S:

$$S + M = S - M_{UPD} \tag{14}$$

The treatment presented here is based on the supposition that M^+ undergoes a complete discharge to M. If the activities of M^+, a_{M^+}, in the working-electrode and reference-electrode compartments are the same and if they are maintained at the same temperature, T, then the relation between the surface coverage of M, $\theta_{M_{UPD}}$, and the applied potential, E, is described by the general electrochemical adsorption isotherm:

$$\frac{\theta_{M_{UPD}}}{1 - \theta_{M_{UPD}}} = a_{M^+} \exp\left(-\frac{FE}{RT}\right) \exp\left(-\frac{\Delta G^\circ_{ads}(M_{UPD})}{RT}\right) \tag{15}$$

where E is the potential measured versus the standard potential of the M^+/M reference electrode, $\Delta G^\circ_{ads}(M_{UPD})$ is the standard Gibbs free energy of adsorption of the process, thus the formation of UPD layer of M on S as shown in equation 14, and R and F are explained above. Equation 15 is a specific form of the following relation:

$$\frac{\theta_{M_{UPD}}}{1 - \theta_{M_{UPD}}} = \frac{a^w_{M^+}}{a^r_{M^+}} \exp\left(-\frac{FE_{eq}}{RT}\right) \exp\left(-\frac{\Delta G^\circ_{ads}(M_{UPD})}{RT}\right) \tag{16}$$

where $a^w_{M^+}$ and $a^r_{M^+}$ are the activities of M^+ in the working-electrode and reference-electrode compartments, and E_{eq} is the equilibrium electrode potential. When T is the same in both compartments and when $a^w_{M^+} = a^r_{M^+}$, one obtains equation 15.

There are many similarities between the theoretical treatment of the UPD H and the UPD M. For instance, it is evident on the basis of the above formula that experimental evaluation of the potential at which $\theta_{M_{UPD}}$ has a given value for a series of temperatures allows numerical determination of $\Delta G^\circ_{ads}(M_{UPD})$. Such calculations may be performed for all values of $\theta_{M_{UPD}}$ and T, and one may plot $\Delta G^\circ_{ads}(M_{UPD})$ as a function of $\theta_{M_{UPD}}$ and T, thus $\Delta G^\circ_{ads}(M_{UPD})$ versus $(\theta_{M_{UPD}}, T)$. Evaluation of the $\Delta G^\circ_{ads}(M_{UPD})$ versus $\theta_{M_{UPD}}$ relations (for T = const) allows one to assess the energy of lateral interactions between the M_{UPD} adatoms. The entropy of adsorption, $\Delta S^\circ_{ads}(M_{UPD})$, may evaluated on the basis of equations 7 and subsequently, the enthalpy of adsorption, $\Delta H^\circ_{ads}(M_{UPD})$, is readily determined based on equation 8.

An alternative approach that may be applied to determine $\Delta H^\circ_{ads}(M_{UPD})$ is similar to the one presented above for the enthalpy of adsorption of H_{UPD} (see equation 9) and it is based on the Gibbs-Helmholtz formula and the general adsorption isotherm presented in equation 15. Thus by experimental determination of pairs of values of E and T at which the M_{UPD} coverage is constant, $\theta_{M_{UPD}} = const$, and by plotting E/T versus 1/T one obtains linear relations and from their slope one may evaluate $\Delta H^\circ_{ads}(M_{UPD})$. It should be added that there is a limited amount of data on thermodynamics of the UPD of semiconductors and metals, and at present that authors are unaware of any temperature dependence measurements of the UPD M. The chronocoulometry methodology applied by Lipkowski et al. (*30,32,44*) results in evaluation of the surface coverage of the adsorbed metallic and anionic species and their ΔG°_{ads}. The approach proposed in this paper and based on temperature-dependence measurements followed by theoretical treatment represents an extension to the approach developed and applied by the laboratory of Lipkowski and it will allow one to assess consistency of the surface thermodynamic data.

Knowledge of the $S-M_{UPD}$ bond energy is essential in evaluation of the cohesive forces acting between the metal substrate and the under-potential deposited species that are responsible for the adhesion of the deposited monolayer to the substrate. They are of different nature than the forces acting between alike atoms in the 3D lattice of S or M. Thus, the magnitude of $E_{S-M_{UPD}}$ is a good measure of the adhesion the deposit to the substrate. The energy of the $S-M_{UPD}$ bond, $E_{S-M_{UPD}}$, may be determined on the basis of the following Born-Haber thermodynamic cycles for the respective single-electrode processes shown in equations 12 and 13:

$$M^+_{(gas)} \; + \; e_{(gas)} \; + \; S \; \xleftarrow{\;I_M\;} \; M_{(gas)} \; + \; S$$

$$\downarrow \Delta H^\circ_{solv} \qquad \uparrow \Phi + EF \qquad\qquad\qquad \uparrow E_{S-M_{UPD}} \qquad\qquad (17)$$

$$M^+_{(aq)} \; + \; e_{S(E\neq0)} \; + \; S \; \xrightarrow{\Delta H^\circ_{WE}} \; S-M_{UPD}$$

$$M_{(gas)} \; \xrightarrow{\;I_M\;} \; M^+_{(gas)} \; + \; e_{(gas)}$$

$$\uparrow \Delta H^\circ_{atom(M)} - \Delta H^\circ_{surf(M)} \, \sigma_m \quad \downarrow \Delta H^\circ_{solv} \qquad \uparrow \Phi \qquad\qquad (18)$$

$$M \; \xrightarrow{\Delta H^\circ_{RE}} \; M^+_{(aq)} \; + \; e_{M(E=0)}$$

where $\Delta H^\circ_{surf(M)}$ the standard surface enthalpy defined as the energy required to create 1 cm^2 of the surface (the unit being J cm^{-2}) and σ_m is the molar surface of M, thus the number of cm^2 occupied by 1 mole of M surface atoms (the unit being

$cm^2 \, mol^{-1}$); the unit of the product $\Delta H^{\circ}_{surf(M)} \, \sigma_M$ is $J \, mol^{-1}$. The surface enthalpy is related to the surface tension, γ, through the following relation $(45,46)$:

$$\Delta H^{\circ}_{surf(M)} = \gamma - T \frac{\partial \gamma}{\partial T} \tag{19}$$

Thus even if $\Delta H^{\circ}_{surf(M)}$ is not well known, one may evaluate it based on the temperature dependence of the surface tension.

By adding the above Born-Haber cycles (equations 17 and 18) and bearing in mind that there is a Volta potential difference between the metal and the solution so that M^+ and e extracted from two solid electrodes are at different electrostatic potentials and that this difference compensates exactly the work function variation shown in equation 17 for the working electrode, EF, one obtains the following relation for the $S-M_{UPD}$ bond energy, $E_{S-M_{UPD}}$:

$$E_{S-M_{UPD}} = \Delta H^{\circ}_{atom(M)} - \Delta H^{\circ}_{surf(M)} \, \sigma_m - \Delta H^{\circ}_{ads}\left(M_{UPD}\right) \tag{20}$$

There are several significant implications of equation 20. The above formula indicates that: (i) $E_{S-M_{UPD}}$ follows changes of $\Delta H^{\circ}_{ads}\left(M_{UPD}\right)$; (ii) the surface enthalpy, $\Delta H^{\circ}_{surf(M)}$, depends on the surface structure thus it may be concluded that the value of $E_{S-M_{UPD}}$ varies with the structure of the metal substrate; and (iii) the molar surface of M, σ_m, is surface-geometry dependent at it affects the magnitude of $E_{S-M_{UPD}}$.

Recent studies on the UPD M show that the adsorption-desorption CV profiles often reveal a hysteresis whose origin has been assigned to lateral interactions between the UPD species and the coadsorbed anions, and to surface reconstruction or surface compression processes $(30,32-40)$. It ought to be emphasized that the cathodic component of the CV profiles allows determination of $\Delta G^{\circ}_{ads}\left(M_{UPD}\right)$ whereas the anodic one refers to $\Delta G^{\circ}_{des}\left(M_{UPD}\right)$. If the CV profiles are symmetrical, thus if there is no hysteresis, then $\Delta G^{\circ}_{ads}\left(M_{UPD}\right) = -\Delta G^{\circ}_{des}\left(M_{UPD}\right)$ and the adsorption and desorption processes are energetically equivalent, as it is the case of the UPD H. On the other hand, if a hysteresis effect is observable in the CV adsorption-desorption profiles, then integration of the $\Delta G^{\circ}_{ads}\left(M_{UPD}\right)$ versus $\theta_{M_{UPD}}$ relation provides the Gibbs free energy of *adsorption* of one monolayer (or submonolayer, depending on the system) of the UPD species. Integration of the $\Delta G^{\circ}_{des}\left(M_{UPD}\right)$ versus $\theta_{M_{UPD}}$ relation gives the Gibbs free energy of *desorption* of one monolayer (or submonolayer) of the UPD species. Considering that a hysteresis effect implies that $\Delta G^{\circ}_{ads}\left(M_{UPD}\right) \neq -\Delta G^{\circ}_{des}\left(M_{UPD}\right)$, then their difference, $\delta \Delta G^{\circ}\left(M_{UPD}\right)$, defined by equation 21, is a measure of the Gibbs free energy associated with the lateral interactions between M_{UPD} and the anions during the adsorption and desorption processes, as well as the surface reconstruction and surface compression processes.

$$\delta \Delta G^{\circ}\left(M_{UPD}\right) = \Delta G^{\circ}_{ads}\left(M_{UPD}\right) + \Delta G^{\circ}_{des}\left(M_{UPD}\right) \tag{21}$$

Conclusions

The authors present thermodynamic methodology which can be applied to the phenomenon of the under-potential deposition on metal electrodes. It allows determination of the Gibbs free energy, entropy and enthalpy of adsorption, ΔG^o_{ads}, ΔS^o_{ads} and ΔH^o_{ads}. New theoretical approach is presented which leads to determination of the bond energy between the metal substrate and the under-potential deposited species, here H or M, $E_{S-H_{UPD}}$ and $E_{S-M_{UPD}}$, respectively. Knowledge of $E_{S-M_{UPD}}$ is essential in evaluation of the binding forces acting between the substrate and the deposit, thus the forces that are responsible for adhesion of the deposit to the substrate. The authors also discuss the hysteresis effect often observed in the CV adsorption-desorption profiles for UPD M and indicate that the difference between $\Delta G^o_{ads}(M_{UPD})$ and $\Delta G^o_{des}(M_{UPD})$ is a measure of the Gibbs free energy associated with the lateral interactions between M_{UPD} adatoms and the coadsorbed anions as well as the surface reconstruction and surface compression processes.

Acknowledgments

Acknowledgment is made to the NSERC of Canada and the FCAR du Québec for support of this research project. A. Zolfaghari acknowledges a graduate fellowship from MCHE of Iran. The authors are indebted to Prof. A. Lasia of l'Université de Sherbrooke for discussions and comments on thermodynamics of the UPD.

References

1. Will, F.G. *J. Electrochem. Soc.* **1965**, *112*, 451-455.
2. Boeld, W.; Breiter, M. W. *Z. Elektrochem.* **1960**, *64*, 897-902.
3. Breiter, M. W.; Kennel, B. *Z. Elektrochem.* **1960**, *64*, 1180-1187.
4. Frumkin, A. N. In *Advances in Electrochemistry and Electrochemical Engineering*; Delahey, P., Ed.; Interscience Publishers: New York, 1963; Vol. 3, pp 287-391.
5. Woods, R. In *Electroanalytical Chemistry*; Bard, A., Ed.; Marcel Dekker: New York, 1977; Vol. 9, pp 27-162.
6. Conway, B. E. *Theory and Principles of Electrode Processes*; Ronald Press: London, 1965.
7. Enyo, M. In *Modern Aspects of Electrochemistry*; Conway, B. E.; Bockris, J. O'M.; Eds.; Plenum Press: New York, 1975; Vol. 11, pp 251-314.
8. Enyo, M. In *Comprehensive Treatise of Electrochemistry*; Conway, B. E.; Bockris, J. O'M.; Eds.; Plenum Press: New York, 1983; Vol. 7, pp 241-300.
9. Conway, B. E.; Angerstein-Kozlowska, H.; Dhar, H. P. *Electrochim. Acta* **1974**, *19*, 455-460.
10. Conway, B. E.; Angerstein-Kozlowska, H. *Acc. Chem. Res.* **1981**, *14*, 49-56.
11. Conway, B. E.; Angerstein-Kozlowska, H.; Ho, F. C. *J. Vac. Sci. Technol.* **1977**, *14*, 351-364.

12. Conway, B. E.; Angerstein-Kozlowska, H.; Sharp, W. B. A. *J. Chem. Soc., Faraday Trans. I* **1978**, *74*, 1373-89; see also Conway, B. E.; Currie, J. C. *J. Chem. Soc., Faraday Trans. I* **1978**, *74*, 1390-1402.
13. Conway, B. E. *Sci. Prog. Oxf.* **1987**, *71*, 479-510.
14. Gregory, B. W.; Stickney, J. L. *J. Electroanal. Chem.* **1990**, *300*, 543-561.
15. Suggs, D. W.; Stickney, J. L. *Surface Sci.* **1993**, *290*, 362-374; 375-387.
16. Conway, B. E.; Jerkiewicz, G. *J. Electroanal. Chem.* **1993**, *357*, 47-66; see also Conway, B. E.; Jerkiewicz, G. *Zeit. Phys. Chem. Bd.* **1994**, *183*, 281-286.
17. Jerkiewicz, G.; Borodzinski, J. J.; Chrzanowski, W.; Conway, B. E. *J. Electroanal. Soc.* **1995**, *142*, 3755-3763.
18. Jerkiewicz, G.; Zolfaghari, A. *J. Electrochem. Soc.* **1996**, *143*, 1240-1248.
19. Jerkiewicz, G.; Zolfaghari, A. *J. Phys. Chem.* **1996**, *100*, 8454-8461.
20. Kolb, D. M. In *Advances in Electrochemistry and Electrochemical Engineering*; Gerischer, H.; Tobias, C. W.; Eds.; Wiley Interscience: New York, 1978, Vol. 11, pp 125-271.
21. Despic, A. R. In *Comprehensive Treatise of Electrochemistry*; Conway, B. E.; Bockris, J. O'M.; Yeager, E.; Khan, S. U. M.; White, R. E.; Eds.; Plenum Press: New York, 1983, Vol. 7, pp 451-528.
22. Conway, B. E.; Chacha, J. S. *J. Electroanal. Chem.* **1990**, *287*, 13-41.
23. Kolb, D. M; Przasnycki, M.; Gerischer, H. *J. Electroanal. Chem.* **1974**, *54*, 25-38.
24. Adzic, R. R. In *Advances in Electrochemistry and Electrochemical Engineering*; Gerischer, H.; Tobias, C. W.; Eds.; Wiley Interscience: New York, 1978, Vol. 13, pp 159-260.
25. Van der Eerden, J. P.; Staikov, G.; Kashchiev, D.; Lorenz, W. J. *Surface Sci.* **1979**, *82*, 364-382.
26. Jüttner, K.; Lorenz, W. J. *Z. Phys. Chem.* **1980**, *122*, 163-185
27. Hanson, M. E.; Yeager, E. In *Electrochemical Surface Science*; Soriaga, M. P.; Ed.; ACS Symposium Series: Washington, 1988, Vol. 378, pp 141-153.
28. Gileadi, E. *Electrode Kinetics*; VCH: New York, 1993.
29. Budevski, E. B.; Staikov, G. T.; Lorenz, W. J. *Electrochemical Phase Formation and Growth*; VCH: New York, 1996.
30. Lipkowski, J. Stolberg, L., Yang, D.-F., Pettinger, B., Mirwald, S., Henglein, F.; Kolb, D. M. *Electrochim. Acta* **1994**, *39*, 1045-1056.
31. Mrozek, P.; Sung, Y.-E.; Han, M.; Gamboa-Aldeco, M.; Wieckowski, A.; Chen, C.-H.; Gewirth, A. A. *Electrochim. Acta* **1995**, *40*, 17-28.
32. Shi. Z.; Lipkowski, J. *J. Phys. Chem.* **1995**, *99*, 4170-4175; see also Savich, W.; Sun, S.-G.; Lipkowski, J.; Wieckowski, J. *J. Electroanal. Chem.* **1995**, *388*, 233-237.
33. Zelenay, P.; Rice-Jackson, L. M.; Wieckowski, A.; Gawlowski, J. *Surface Sci.* **1991**, *256*, 253-263; see also Varga, K.; Zelenay, P.; Wieckowski, A. *J. Electroanal. Chem.* **1992**, *330*, 453-467.
34. Zelenay, P; Wieckowski, A. In *Electrochemical Interfaces*; Abruña, H.; Ed.; VCH: New York, 1991, pp 479-527.
35. Zelenay, P.; Gamboa-Aldeco, M.; Horanyi, G.; Wieckowski, A. *J. Electroanal. Chem.* **1993**, *357*, 307-326.

36. Gamboa-Aldeco, M.; Herrero, E.; Zelenay, P. S.; Wieckowski, A. *J. Electroanal. Chem.* **1993**, *348*, 451-457.
37. Krauskopf, E. K.; Wieckowski, A. In *Adsorption of Molecules at Metal Electrodes*; Lipkowski, J.; Ross, P. N.; Eds.; VCH: New York, 1992, pp 119-169.
38. Sawaguchi, T.; Yamada, T.; Okinaka, Y.; Itaya, K. *J. Phys. Chem.* **1995**, *99*, 14149-14155.
39. Wan, L.-J.; Yau, S.-L.; Itaya, K. *J. Phys. Chem.*, **1995**, *99*, 9507-9513.
40. Manne, S.; Hansma, P. K.; Massie, J.; Elings, V. B.; Gewirth, A. A. *Science*, **1991**, *251*, 183-186; see also Chen, C.; Gewirth, A. A. *J. Am. Chem. Soc.*, **1992**, *114*, 5439-5440.
41. Christman, K. *Surface Sci. Rep.* **1988**, *9*, 1-163.
42. Langmuir, I. *J. Am. Chem. Soc.* **1918**, *40*, 1361-1403.
43. Fowler, R. H.; Guggenheim, F. A. *Statistical Thermodynamics*; Cambridge University Press: London, 1939.
44. Lipkowski, J.; Stolberg, L. In *Adsorption of Molecules at Metal Electrodes*; Lipkowski, J.; Ross, P. N., Eds.; VCH: New York, 1992, pp 171-238.
45. Adamson, A. W. *Physical Chemistry of Surfaces*; John Wiley and Sons: New York, 1990.
46. Somorjai, G. A. *Introduction to Surface Chemistry and Catalysis*; John Wiley and Sons: New York, 1994.

Chapter 5

Micrometer-Scale Imaging of Native Oxide on Silicon Wafers by Using Scanning Auger Electron Spectroscopy

Mikio Furuya

**Kanagawa Industrial Technology Research Institute,
705–1 Shimoimaizumi, Ebina-shi, Kanagawa 243–04, Japan**

Intriguing patterns of the ultra thin native oxide film were observed by SEM, AFM and SAM methods. The ultra thin native oxide film were formed on the surface of the silicon wafer during photo-lithographic process. The pattern might be formed by the influence of the residual photo sensitive resin. The investigations on the electron beam-induced damage had been performed to find out the conditions to reduce the reduction of the oxide film by using thermally grown silicon oxide film as a reference sample. Micrometer-scale images of the ultra thin native oxide film were observed by using SAM methods. Thickness estimation for the native oxide film on the silicon wafers was achieved by the intensity ratio of LVV spectra of the elemental silicon and the silicon oxide. Calculated thickness for the ultra thin native oxide film was $0.8 \sim 1.2$ nm.

1. Introduction

In recent development of the semiconductor industries, thermal oxide film thickness of less than 5 nm has been used in semiconductor devices such as metal-oxide-semiconductor (MOS) structures. Thickness of less than 5 nm is almost near the thickness of a native oxide film on the surface of silicon wafer. Therefore the characterization of ultra thin native oxide film is important in the semiconductor process technology. The secondary electron microscopy (SEM), the scanning Auger electron microscopy (SAM), the atomic force microscopy (AFM) and the X-ray photoelectron spectroscopy (XPS) might be the useful characterization method for the surface of the silicon wafers.

A number of works have been performed about the thickness estimation of native oxide film on a silicon wafers by using XPS, and obtained good results to characterize the raw surface of the silicon wafers (1-2). However, XPS is an analytical method which is applicable to a relatively large area, due to the primary X-ray probe cannot focus on a small spot. Therefore a micrometer-scale structure of the native oxide has not been investigated by using XPS.

AFM method is available to observe a nanometer scale roughness on the surface of the silicon wafer with native oxide. As a matter of course, topographical images of AFM could not imply the thickness information of the over layer film, and AFM methods might be unavailable for the area over than 10^{-2} cm due to the instrumentation limits.

SAM is a representative method of the surface analysis aimed at below the micron area. The micro structure of the native oxide film might be able to observe by means of SAM method. However, the severe electron beam-induced damage might be arisen on a silicon oxide film by the high electron current density of primary electron beam (3-7). The silicon oxide is reduced to the elemental silicon due to Joule's heat in an ultra high vacuum analyzer chamber of SAM equipment. The conditions to reduce the reduction of the silicon oxide film due to the electron beam-induced damage have been preferred by the investigations on the compulsive reduction of the silicon oxide film.

In this report, the author presents the investigative study of the preferred analytical conditions to observe a micrometer-scale structure of the native oxide film on a silicon wafer and the thickness estimation method by using LVV Auger spectra intensity ratio for the elemental silicon and the silicon oxide.

2. Experiments

SEM observation and SAM measurement had been performed by using scanning Auger electron spectroscopy (PHI-SAM4300). The cylindrical mirror analyzer (CMA) and the primary electron beam of this equipment were settled in coincident axes. In order to obtain clear contrast images, the normal of the sample surface was tilted 30 degrees in respect to the primary electron beam. The observation were performed under the primary electron beam conditions of accelerating voltage $1 \sim 7$ kV and beam current $1 \sim 1000$ nA.

In order to find out the region of reduced electron beam-induced damage, the investigations on the compulsive reduction of the silicon oxide film were performed by using thermally grown silicon oxide film as a reference sample. The electron beam-induced damage were evaluated as a function of LVV spectra intensity ratio for the elemental silicon and the silicon oxide.

The relative sensitivity factor (RSF) of LVV spectra of the silicon were obtained for the calculation of the thickness estimation of the ultra thin native oxide film. Although the characteristics nature of the thermal silicon oxide film on the silicon wafers might be different from that of the ultra thin native oxide film, RSF of LVV spectra for the silicon were obtained by the measurement of Auger signal intensity using thermally grown silicon oxide film of 100 nm thickness as a reference sample

The variations of Auger signal intensity for Si-LVV, O-KLL and C-KLL were measured across the surface of the native oxide film. In order to obtain the clear element map images of SAM methods, the normal of the sample surface was also tilted as in the SEM observations, and integral Auger spectra were used in lieu of the conventional derivative Auger spectra.

On the measurement of the intensity of Auger electron spectra for the thickness calculation of the native oxide film, in order to set the incident angle of the primaryelectron beam and the collection angle of Auger electron constant, the normal of the sample surface was settled coincident with the axes of CMA and the primary electron beam.

AFM measurement was performed by using scanning tunneling microscope (Digital Instruments, Inc. Nanoscope-II).

The ultra thin native oxide film on the silicon wafers was formed in an aqueous acid solution during the photo-lithographic process in the semiconductor fabrication system. The mirror polished silicon wafer was spread with the photo sensitive resin, exposed the resin to the ultraviolet light, developed the resin in the organic solvent, strip off the residual photo sensitive resin from the surface of the silicon wafers in a hot acid solution and rinsed in the ultra-pure water.

3. Results and Discussion

Figure 1 shows SEM images of the ultra thin native oxide film on the silicon wafers. The excellent contrast pattern was observed by the primary electron beam condition of low energy and high current, rather than high energy and low current. The observed conditions of Figure 1 were accelerating voltage 3 kV, electron current 20 nA and the tilt angle of the sample 30 degrees. The intriguing pattern on the surface of the silicon wafers was observed using low energy SEM method. The excellent contrast pattern was not obtained on the condition of tilt angle 0 degrees. The light and darkness pattern might be considered as the lateral variations of the electron charge quantity across the surface of the silicon wafers with native oxide. The light region of this pattern corresponds to the relatively thick oxide region and the darkness to the thin oxide.

Figure 2 shows the topographical images on the surface of the native oxide film observed using AFM methods, a considerable undulation was obviously recognized compared with that of mirror polished silicon wafers. However AFM methods might be unavailable for the lateral range over than 10^{-2} cm, due to the limits of the actuator of the sample stage by using the piezoelectric effect. In addition, AFM method could not measure the total thickness of the over layer film but only measure the surface roughness.

Figure 3 shows the LVV spectrum of the native oxide film on the silicon wafer. The horizontal axis represents the kinetic energy of Auger electron, and vertical the differentiated signal intensity. LVV spectra of the elemental silicon and the silicon oxide show the different kinetic energy because of the difference of the chemical state in valence band electron, higher kinetic energy peak correspond to LVV spectrum of elemental silicon and lower silicon oxide. The peak intensity of kinetic energy 92 eV *Is* represents the intensity of LVV spectrum of the elemental silicon ,passing through the native oxide film from the silicon substrate. The peak intensity of kinetic energy 76 eV *Iso* represents the LVV spectrum intensity of the silicon native oxide film. The ratio *Iso/Is* is related to the thickness of native oxide film on the silicon wafers.

In order to find out the region of reduced damage, the investigation on the ratio of *Iso/Is* as a function of the irradiation time of the primary electron beam was carried out. The measurements were performed using thermally grown silicon

Figure 1. The intriguing pattern of the ultra thin native oxide film on the silicon wafers were observed using low energy SEM methods. The light and darkness pattern might be speculated the deference of the electron charge.

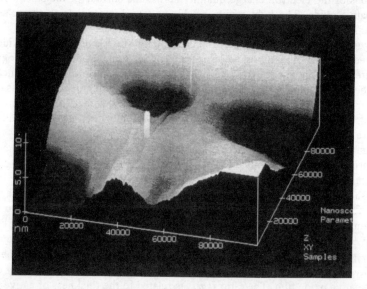

Figure 2. The topographical images of the surface of the native oxide film was observed using AFM methods. A considerable undulation was recognized compared with the mirror polished silicon wafers.

oxide film thickness of 100 nm as a reference sample. Figure 4 shows the ratio of *Iso/Is* as a function of the irradiation time of the electron beam. Accelerating voltage of the primary electron beam was kept constant 3 kV. The diameter of the primary electron beam was defined by 84-16% transition width using a conventional knife edge method. As the electron current was changed from 109 to 2.7 nA, the diameter of electron beam was varied from 2.0×10^{-4} to 1.34×10^{-4} cm in this experiments and electron dosage was varied from $1.25 \times 10^{+21}$ to $2.7 \times 10^{+21}$ e/cm^2. As the diameter of the primary electron beam is changed with the electron current, the product of electron current and irradiation time were kept constant at 6.1×10^{-6} coulomb for each measurement in Figure 4. Although the product of the current and the irradiation time is different from the electron dosage, but the value is a meaningful index in practical use. The regions of reduced electron beam-induced damage (a) and (c) were clearly observed on both sides of the damage region (b) in Figure 4.   The reduced damage region of (a) is a region of short irradiation time and high current, and the region of (c) long irradiation time, low current and reduce the beam diameter.

Also the investigation on the ratio of *Iso/Is* as a function of the accelerating voltage of the primary electron beam was carried out. The accelerating voltage of the primary electron beam was changed from 2.0 to 7.0 kV. The conditions of the electron probe current 50 nA, the diameter of electron beam 1.8×10^{-4} cm, the electron dosage $1.43 \times 10^{+21}$ e/cm^2 and electron current density 2.0 A/cm2 were kept constant. The intensity ratio of *Iso/Is* was almost kept constant value of 4.0 in this accelerating voltage range. The intensity ratio of *Iso/Is* was no participation with the accelerating voltage in this experiment.

It could be concluded from the above-mentioned experiments on the compulsive reduction of silicon oxide that the conditions of the long irradiation time and the low current is the preferable analytical conditions to reduce the reduction of the oxide film by the electron beam-induced damage.

Figure 5 shows the element map images of, (a) LVV spectrum of silicon oxide, (b) LVV spectrum of elemental silicon, (c) KLL spectrum of oxygen and (d) KLL spectrum of carbon respectively. Analytical conditions to obtain respective element map images were, accelerating voltage 3 kV, primary electron beam current 1.0 nA, diameter of the electron beam 1.0×10^{-4} cm, current density 1.27×10^{-1} A/cm2, acquisition time 0.11 sec/pixel of the SAM images and electron dosage $8.7 \times 10^{+16}$ e/cm2. The measured value of reduction index *Is/Iso* was less than 1 % on this finally adopted analytical conditions. The sample was tilted 30 degrees, and the integral Auger spectrum was used in lieu of the derivative Auger spectrum to obtain respective element map images of SAM. Figure 5 (a) and (c) show the identical light and darkness pattern, (a) and (b) the reverse pattern and (d) no participation with (a), (b) or (c). Although the mechanisms of the pattern formation was not investigated, the pattern might be formed due to the unequal affection of the residual photo sensitive resin at the interface with the silicon wafers in the strip off process by using the hot acid solution. These patterns represent not only the distribution of the elements on the surface of native oxide film, but also the lateral variation of the thickness of the native oxide film across the surface of the silicon wafers.

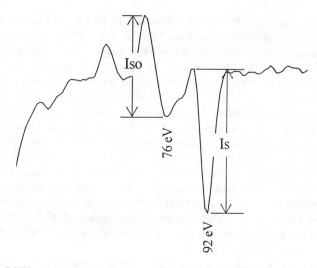

Figure 3. LVV spectrum of the native oxide film on silicon wafers. The peak of 92 eV is LVV spectrum of the elemental silicon from the silicon substrate and 76 eV the silicon oxide.

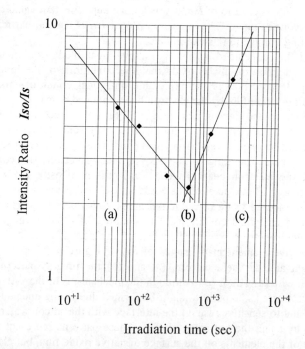

Figure 4. The investigations of compulsive reduction were performed by the thermally grown silicon oxide. The products of electron current and irradiation time were kept constant to 6.1×10^{-6} coulomb in respective measurements. The reduce damage regions (a) and (c) were found on both sides of the damage region (b).

Figure 6 shows a cross sectional schematic diagram of a surface of the silicon wafers with the native oxide film thickness of d. Auger electron intensity of LVV spectrum of silicon oxide Iso is given by the integral of Equation (1) from 0 to d, and elemental silicon Is is given by the integral of Equation (2) from d to ∞. Let us divide Equation (1) by Equation (2). The factors ks and kso are the sensitivity factor of the LVV spectrum of elemental silicon and silicon oxide respectively. The sensitivity factors of ks and kso are the function of the ionization cross section, Auger transition probability and the density of element. In general, those values could not obtained independently, but the sensitivity factor ks and kso are obtained easily by the measurement of RSF for the bulk materials.

The inelastic mean free path (IMFP) of Auger electron could be calculated by the theoretical formula (8). The calculated IMFP values for the LVV spectrum of the elemental silicon in the matrix of elemental silicon is 0.51 nm, of the silicon oxide in the matrix of silicon oxide is 0.56 nm, and of the elemental silicon in the matrix of silicon oxide is 0.59 nm, in addition, of the silicon oxide in the matrix of elemental silicon is 0.48 nm. This theoretical formula of IMFP for the Auger electrons gives considerably correct value in the relatively high kinetic energy region. However, the correctness of the calculated IMFP are inadequate in this low kinetic energy region. Because of the kinetic energy of the LVV spectrum of the elemental silicon is close to that of silicon oxide, IMFP of the elemental silicon is close to that of silicon oxide. Then the assumption, $\lambda so = \lambda s = \lambda = 0.5$ nm, could be made for a convenience on the calculation of native oxide film thickness d in the Equations (1-2).

The sensitivity factor of the bulk silicon oxide $kso=Iso(\infty)$ is the integral of Equation (1) from $x=0$ to $x=d=\infty$. The sensitivity factor of the bulk elemental silicon $ks=Is(0)$ is the integral of Equation (2) from $x=d=0$ to $x=\infty$. The ratio of relative sensitivity factor $ks/kso = Is(0)/Iso(\infty) = 5.7$ was obtained by the measurement using thermally grown silicon oxide and silicon wafer as a reference sample. Though the used angle θ was the collection angle of the CMA 42.3 degrees in this calculation. The actual collection angle of CMA used in this study was 42 ± 6 degrees, the error of the calculation from this condition may be less than ± 10 %.

After the rearrangement of the formula, Equation (3) gives the relation of the film thickness d with the intensity ratio Iso/Is for LVV spectrum of the native oxide on the silicon wafer. As shown in Table I, the native oxide film thickness of d for Figure 1 or Figure 5 were calculated by equation (3) from the measured intensity ratio Iso/Is. The mean value of the thickness was 0.8 nm on the relatively thin region and was 1.2 nm on the relatively thick region.

$$I_{so} = \int_{0}^{d} kso \exp\left(-\frac{x}{\lambda so \cos\theta}\right) dx$$

Equation (1)

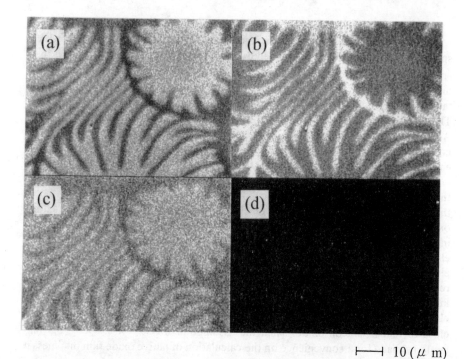

┣━━┫ 10 (μ m)

Figure 5. SAM images of the native oxide film on the silicon wafer was observed by using, (a) LVV spectrum of silicon oxide, (b) LVV spectrum of silicon, (c) O-KLL and (d) C-KLL respectively.

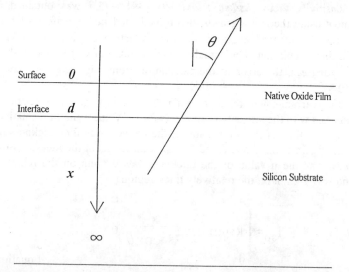

Figure 6. Cross sectional schematic diagram for the calculation of the native oxide film thickness of d. The normal of the sample surface was set coincident with the axes of the CMA and the primary electron beam.

$$Is = \int_d^\infty k s \exp\left(-\frac{x}{\lambda s \cos\theta}\right) dx$$

Equation (2)

$$d = 0.74 \; \lambda \; \ln(5.7(Iso/Is)+1)$$

Equation (3)

Table I . Thickness of the native oxide film calculated by equation (3)

Relatively thick region				Relatively thin region			
Iso(kc/s)	*Is*(kc/s)	*Iso/Is*	*d* (nm)	*Iso*(kc/s)	*Is*(kc/s)	*Iso/Is*	*d* (nm)
183	31	5.8	1.31	184	119	1.55	0.85
188	34	5.5	1.29	188	123	1.52	0.84
191	46	4.2	1.19	166	119	1.39	0.81
175	48	3.6	1.14	181	110	1.15	0.75

Conclusion

(1)A micro structure of the ultra thin native oxide film on a silicon wafer was observed by using SEM, AFM and SAM methods.

(2)The patterns of ultra thin native oxide film was formed in aqueous acid solution during photo-lithographic process in a semiconductor fabrication system.   Although the mechanisms of the pattern formation was not investigated, it might be due to the unequal affection of the residual photo sensitive resin on the strip off process using a hot acid solution.

(3)The study on the electron beam-induced damage of the silicon oxide film was carried out using thermally grown silicon oxide film as a reference sample. Although the reduction of the oxide film is sensitive to the electron beam density, the long acquisition time reduce the electron beam-induced damage because of the heat dissipation in the thin oxide films.

(4)The conditions to obtain the element map of the native oxide film were, accelerating voltage 3 kV, primary beam current 1.0 nA, diameter of the electron beam 1.0×10^{-4} cm, current density 1.27×10^{-1} A/cm^2, irradiation time 0.11 sec/pixel and electron dose $8.7 \times 10^{+16}$ e/cm2.

(5)Thickness estimation of the native oxide film was achieved by using the measured intensity ratio *Iso/Is* for LVV spectrum of the elemental silicon and the silicon oxide.   The calculated thickness of the native oxide film was 0.8 nm on the relatively thin region, and was 1.2 nm on the relatively thick region.

References

(1) M. F. Hochella, Jr.; Surf. Sci., 197 (1988) L260-L268

(2) A. Ishizaka, S. Iwata and Y. Kamigaki; Surf. Sci., 84 (1979) 355

(3) S. Ichimura and R. Shimizu; J. Appl. Phys., 50 (1979) 6020

(4) C. G. Pantano and T. E. Madey; Appl. Surf. Sci., 7 (1981) 115-141

(5) Simon Thomas; J. Appl. Phys., 45 (1974) 161

(6) A.V.Oostrom; Surf. Sci., 89 (1979) 615

(7) Y. E. Strausser and J. S. Johannessen; NBS Special Publication, 400-23 (1976)

(8) S. Tanuma, C. J. Powell and D. R. Penn; Surf. Interf. Anal., 21 (1994) 165

Chapter 6

Growth Kinetics of Phosphate Films on Metal Oxide Surfaces

T. S. Murrell[1,3], M. G. Nooney[2], L. R. Hossner[1], and D. W. Goodman[2]

Departments of [1]Soil and Crop Sciences and [2]Chemistry,
Texas A&M University, College Station, TX 77843

The kinetics of phosphate uptake by hematite, titania, and alumina were examined by exposing freshly prepared thin films to phosphate solutions at incremental times and subsequently analyzing the surfaces by Auger electron spectroscopy (AES), X-ray photoelectron spectroscopy (XPS), and temperature programmed desorption (TPD). Thin film hematite exposed to a sodium phosphate solution demonstrated initially rapid phosphate uptake during the first 10 minutes of solution exposure, followed by a 10 min. induction period after which phosphate accumulated as a species different than that adsorbed initially. For titania and alumina exposed to a calcium phosphate solution, the initially rapid reaction was completed after 1 and 3 hours, respectively. Subsequent rapid accumulations occurred after 3-4 hours for titania and 25-30 hours for alumina. The data for these two oxides were comparable to *in-situ* studies using the same solutions. We postulate that phosphate chemisorbs initially as an inner-sphere complex, but with continued exposure, it forms large polymeric PO_4 structures.

The interest in the uptake of phosphate by metal oxides such as iron, alumina, and titania arises from current problems found in the study of soils, corrosion, and biomimetic materials. Phosphate reactions with iron and aluminum oxides and hydroxides have been extensively studied by soil chemists because these soil components are the most abundant of the naturally occurring metal oxides (*1, 2*) and are the inorganic soil constituents primarily responsible for phosphate reactions in

---

[3]Current address: Potash and Phosphate Institute, P.O. Box 1275, Burnsville, MN 55337–9998

soils (3, 4). Phosphate reactions with titania have been investigated to promote bone growth on titanium implants (5).

Phosphate adsorption onto variable charge surfaces is commonly known to influence surface electrostatic potential. For variable charged surfaces, such as metal oxides, the surface charge is determined by adsorbed H^+, making surface charge dependent upon pH. The adsorption of negatively charged solution phosphate species increases the negativity of the net surface charge, effectively shifting the point of zero charge to lower pH values. To understand the mechanisms of such interactions, the nature of the adsorbed phosphate must be determined. It has long been suspected that phosphate surface speciation is dependent upon sorption kinetics, but conclusive evidence is still lacking. Consequently, this chapter emphasizes the use of spectroscopic techniques to investigate the products of liquid-solid reactions of metal oxides with solution-phase phosphate. If speciation can be determined, the mechanisms of the effects of adsorbed phosphate upon surface electrostatic potential can be identified.

Phosphate Reactions with Iron Oxides. Kinetic studies have divided phosphate reactions into two categories (6): an initially rapid first layer chemisorption, and reactions characterized by slower uptake kinetics thought to arise from solid state diffusion, diffusion through surface pores, migration from within aggregated particles to surface sites, or precipitation of insoluble phosphates at the surface (7). Most of the studies in the literature have focused upon the initial chemisorption reaction, particularly of phosphate with goethite (α-FeOOH). The models developed from studies of this iron oxide have served as a paradigm for phosphate reactions with other oxides.

One of the pivotal early investigations into phosphate reactions with iron oxides postulated that ionic solution phosphate species reacted with OH groups at the surface by protonating them to $-OH_2$ and then exchanging with them to form bonds to structural iron (8). Later investigations using *ex-situ* transmission infrared spectroscopy showed that the hydroxyl groups involved in phosphate ligand exchange reactions were those singly coordinated to iron (9). Analyses of the bulk-terminated structure of the most predominant surface of goethite, the (100) face, have shown that each of these hydroxyl groups occupies 0.305 nm^2 (10). Studies of phosphate reactions with goethite have calculated that phosphate adsorbs with a packing area of one phosphate tetrahedron per 0.66 nm^2 (11). From this, researchers have postulated that phosphate adsorbs by replacing hydroxyls and forming a binuclear bridging configuration with two iron atoms (11). Assuming a binuclear bridging complex, a mean phosphate adsorption of 2.5 μmol P m^{-2} is calculated, a value approached by many studies (12).

Calculations of hydroxyl packing densities for various faces have also been performed for hematite (12). The faces considered theoretically suitable for binuclear bridging complexes of phosphate were the (110), (100), and (223) planes. Calculations of phosphate uptake, assuming a binuclear complex, predicted that phosphate uptake for the (110), (100), and (223) planes would be 4.2, 4.8, and 3.6 μmol P m^{-2}, respectively. Reported values from adsorption studies range from 0.31 - 3.3 μmol P m^{-2} (13, 14). A reason for the difference between the measured and calculated phosphate concentrations is that hematite, unlike goethite, can vary widely in crystal morphology, and the planes amenable to phosphate adsorption may not be the most dominant faces (*14*).

Support for the formation of binuclear bridging configurations has come from studies using transmission infrared spectroscopy to investigate the first layer chemisorption of phosphate on goethite (*15, 16*). However, the bands associated with such bonds are similar to those reported for orthophosphates in solution as well as phosphate minerals (*10*). More recent studies using *in-situ* attenuated total reflection Fourier transform infrared techniques have shown that the bonding configuration of phosphate on goethite is dependent upon solution pH and surface coverage (*17*). Phosphate has been found to exist in singly protonated as well as deprotonated bidentate configurations. In addition, monodentate bonds may occur concurrently with these configurations.

Surface science techniques have not been fully exploited by investigators interested in phosphate adsorption at iron oxides surfaces. In one study, the surface composition of phosphated synthetic goethite was examined by X-ray photoelectron spectroscopy (XPS). From estimates of sampling depth and atomic composition, phosphate adsorbed in a binuclear complex was considered plausible (*18*). A later study, examining phosphate reactions with naturally occurring goethite, investigated both the initially rapid first layer chemisorption reaction and the reactions characterized by slower uptake kinetics (*19*). Findings indicated that phosphate in the first layer was bound as $H_2PO_4^-$ and comprised ~1.3 atomic % of the surface. Phosphate present at the surface after long reaction times existed in crystallites ranging in size from 0.1 - 4 μm on the goethite surface. Thus, after long reaction times, phosphate precipitation at the surface was considered likely, with the composition of the bulk phase phosphate dependent upon solution composition.

In the area of corrosion science, orthophosphates are well known corrosion inhibitors for iron. A study using AES depth profiling revealed that when Fe was exposed to a phosphate solution devoid of calcium, phosphate oxidized the iron substrate, forming a thin (7-8 monolayers thick) iron oxide layer (*20*). However, when iron was exposed to a calcium phosphate solution, a thicker (14 monolayers) oxide layer formed. The phosphate and calcium signals in this layer tracked each other, indicating a composition of coprecipitated calcium and phosphate.

Recently, atomic force microscopy (AFM) and scanning tunneling microscopy (SEM) have been incorporated to investigate the nature of hematite surfaces. AFM analyses of the cleaved (001) surface revealed an unreconstructed termination of the bulk structure (*21*). Later investigations with STM indicated that surface oxygens on this plane may undergo relaxation in response to reduced iron coordination (*22*).

Phosphate Reactions with Aluminum and Titanium Oxides. Phosphate reactions with aluminum oxides are considered to be analogous to those on iron oxides. In one study using gibbsite (α-$Al_2O_3 \cdot 3H_2O$), the bonding of phosphate was interpreted to be dependent on solution pH, forming a monodentate configuration at lower solution pH and bonding in a binuclear bridging structure at higher pH (*23*). Recently, ^{31}P magic angle spinning nuclear magnetic resonance (MAS NMR) has been used to investigate phosphate reactions with synthetic amorphous aluminum hydroxides of two different particle sizes (24). The investigators found that for larger particle sizes, phosphate was limited to surface reactions, and did not become incorporated into the bulk structure. However, for smaller particle sizes, phosphate did react to form aluminum phosphate crystallites.

Investigators in the area of heterogeneous catalysis have examined alumina and titania into which phosphoric acid had been incorporated. The phosphoric acid structure was investigated using Fourier transform infrared spectroscopy after activating the catalysts by evacuation in the IR cell at 773 K (25). From reactions with bases, it was found that POH groups at the surfaces of both oxides act as Brønsted acids. Results from CO surface reactions showed that phosphate incorporated in titania may reduce Ti^{4+} to Ti^{3+}. Reactions with CO_2, producing coordinated molecular species, confirmed the presence of Lewis-acid sites on surfaces of both alumina and titania oxides with phosphoric acid incorporation. Investigators posited that upon activation of the catalysts in vacuo by heating, the incorporated phosphate may have lost a proton which combined with OH at the oxide surface to form H_2O. This, they argued, would have created an ionic bond between the phosphate and the oxide surface.

Few surface science studies have been conducted to examine phosphate uptake from solution by aluminum and titanium oxide surfaces. One investigation, examining the reaction of gibbsite (α-$Al_2O_3\cdot3H_2O$) with $Ca(H_2PO_4)_2$, showed that surface concentrations of Ca and P had a ratio of Ca/P = 1/20 (26). Another study by the same group investigated this same system by XPS angular distributions (27). The authors concluded that phosphate accumulated at the surface of gibbsite and did not seem to be incorporated into the bulk structure.

From this discussion, it is evident that phosphate interactions with oxides are complex. In-situ measurements are the ideal methods for analysis of phosphate films, since the effects of the liquid environment on film structure and composition can be studied. However the relative paucity of available techniques have limited the progress toward an understanding of the chemistry of phosphate uptake on metal oxides. Controversy will continue until more investigations are conducted that directly analyze the nature of the phosphate-oxide surface complexes. The research presented here is an attempt to conduct surface sensitive measurements of phosphate adsorption reactions under well controlled conditions. We assume that the low vapor pressure of phosphate films prevents loss of phosphate from the oxide surface upon exposure to vacuum.

Materials and Methods

All of the experiments described below were performed in a combined ultrahigh vacuum chamber and solution reaction apparatus, which has been described previously (28, 29). This chamber had a base operating pressure of ~7×10^{-10} Torr and was equipped with AES and TPD. Another chamber was used for XPS measurements, and has also been described before (30). Modifications were made to the original electrochemical cell for the adsorption and desorption experiments. A solution delivery column was constructed of Teflon to allow investigations of a wide range of acidic and basic solutions. A peristaltic pump maintained a constant solution flow rate of 5 ml min.$^{-1}$, providing fresh solution to the sample throughout the experiments. A further modification was the installation of a nitrogen jet which was used to purge the cell during experiments and to remove excess solution remaining on the surface of the sample before withdrawal from the adsorption cell.

Metal oxides were first synthesized and characterized in ultrahigh vacuum (UHV). The freshly prepared oxide surfaces were then translated to the liquid reaction cell for exposure to phosphate-containing solutions. After the reactions had occurred, the sample was brought back into an antechamber, pumped to high vacuum, and then translated back into the UHV chamber for analysis. In this way, freshly prepared oxide surfaces were exposed to phosphate solutions with minimal atmospheric contamination.

Kinetics of Phosphate Adsorption on Hematite. Thin films of α-Fe$_2$O$_3$ were synthesized by a modification of the in-situ oxidation technique described by Corneille, et al. (*30*). Iron was deposited on Re(0001) at ~300K in a background of ~5×10^{-5} Torr O$_2$ under static flow maintained by a turbomolecular pump. The Fe$_2$O$_3$ stoichiometry was confirmed by comparison with reference Auger spectra for Fe in the 40-60 eV kinetic energy range (*31*) and by comparison with XPS reference spectra of Fe(2p) and O(1s) binding energies (*30*).

The kinetics of phosphate adsorption were investigated by exposing freshly prepared thin film Fe$_2$O$_3$ to phosphate-containing solutions for various times at ~19 °C. The solution contained 3 μg P ml^{-1}, added as H$_3$PO$_4$, an ionic strength of 0.01 M using NaCl as the background electrolyte, and a pH adjusted to 6.7 using NaOH. Exposure times included .13, 5, 10, 20, 40, and 160 min.

Kinetics of Phosphate Adsorption on Titania and Alumina. Thin film TiO$_2$ was deposited on Ta foil by evaporation of Ti in an oxygen background of 5×10^{-7} Torr oxygen. Al$_2$O$_3$ was deposited on Ta foil by heating aluminum wire in 2×10^{-6} Torr oxygen.

Phosphate adsorption kinetics for both of these oxide surfaces were studied by immersing the oxide sample in a ~19 °C solution containing 3.5 mM CaCl$_2$, 119 μg P ml^{-1} as KH$_2$PO$_4$, a pH adjusted to 6.5 with NaOH, and an ionic strength of 0.01 M, using NaCl as the background electrolyte.

Results and Discussion

Phosphate Adsorption on Hematite. The presence of P as PO$_4$ at the surface was confirmed in AES by principal minima occurring near 94 and 110 eV kinetic energy (*32*). In addition, the minima occurring at ~43 and 52 eV demonstrate that the Fe$_2$O$_3$ stoichiometry of the thin film oxide was maintained after solution exposure. To determine the accumulation of phosphate at the surface, AES P/Fe intensity ratios were calculated from measurements of the P(LMM) transition at 110 eV and the Fe(LMM) transition at 651 eV. A plot of this ratio with reaction time (Figure 1) shows that phosphate accumulates rapidly at the surface of thin film Fe$_2$O$_3$ during the first 10 min.

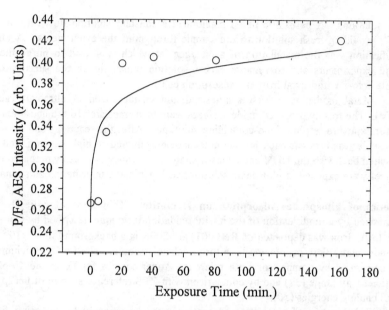

Figure 1. P/Fe AES intensity ratios for incremental exposure times of thin film α-Fe$_2$O$_3$ to a sodium phosphate solution (3 μg P ml^{-1}, pH 6.7, and 0.01 M NaCl).

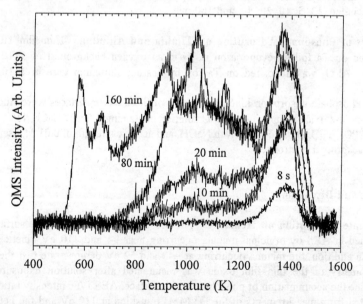

Figure 2. TPD spectra representing the desorption of PO$_3$ groups from the α-Fe$_2$O$_3$ surface as a function of exposure to sodium phosphate solution, as in Fig. 1. Reproduced with permission from reference 39.

of exposure to a phosphatic solution. However, after this initially rapid reaction, phosphate surface coverage shows no increase, within experimental error.

In addition to AES, TPD was also used to determine the amount of phosphate accumulating at the surface (*33*). The high sensitivity of the quadrupole mass spectrometer (QMS) makes TPD capable of detecting sub-monolayer quantities of thermally desorbed species. The masses monitored in these experiments were PO, PO_2, PO_3, and PO_4, corresponding to mass/charge ratios of 47, 63, 79, and 95 respectively. Heating a phosphate-covered iron oxide surface resulted in TPD spectra consisting of a sharp high temperature state (~1380 K) and broad low temperature states occurring at temperatures from ~500 - 1380 K (Figure 2). Redhead analysis, assuming first order desorption kinetics and a prefactor of 10^{13} s^{-1} (*34*), was used to calculate the activation energy of desorption (*E*) of these peaks. The high temperature feature was characterized by $E \approx 88$ kcal mol^{-1}, while the lower temperature desorption peaks corresponded to *E* ranging from 36 - 73 kcal mol^{-1}. These spectra show that at low solution exposure times, phosphate desorbs primarily as the high temperature state. However, as phosphate reactions proceed, more phosphate begins to desorb at lower temperatures. The appearance of the lower temperature desorption features with increasing solution exposure times was representative of the TPD spectra for mass/charge ratios corresponding to PO_2, PO_3, and PO_4. However, for PO, the spectra consisted primarily of the high temperature desorption state (Figure 3). The relative amount of phosphate desorbing after each reaction time was determined by integrating the TPD spectra for each mass monitored (Figure 4). This analysis, like the AES data, revealed an initially rapid increase in phosphate accumulation at the iron oxide surface during the first 10 min. of solution exposure. During the following 10 min. of solution exposure, reduced phosphate accumulation at the surface was observed for 10 min. in the spectra for PO, PO_2, and PO_3. However, The spectra for PO_4 showed a continued increase during this time. After this 10 min. induction period, phosphate accumulation again became measurable for all PO_x species. Most of the phosphate desorbing at the high temperature is associated with PO, while most of the phosphate desorbing at lower temperatures is associated with PO_2, PO_3, and PO_4 species. The PO_4 signal in particular had the largest intensity at the longer solution exposure times. A comparison between the desorption behavior of phosphate and that of iron is shown in Figure 5. This figure shows that at temperatures close to the high temperature phosphate desorption feature, some iron desorbs from the oxide surface as a result of an interfacial reaction between the oxide thin film and the Re substrate. After heating to ~1500 K, Auger analysis indicates that no iron oxide existed on the surface.

TPD spectra of mass 47 (PO) were integrated and converted to surface populations of phosphate molecules (Figure 6). To perform this calculation, nanomolar quantities of phosphate were deposited by micropipette on freshly prepared Fe_2O_3 thin films. Integration of the QMS signal corresponding to the thermal desorption of the known quantity of deposited phosphate provided the standard for converting integrated mass spectrometer signals to surface densities of phosphate molecules. The curve resulting from this calculation shows an initially rapid increase in phosphate adsorption during the first 10 min. of exposure. After this period, no increase in intensity of the 1380 K feature was observed for ~10 min. The coverage at this plateau has been calculated to be ~3×10^{14} molecules cm^{-2}, and represents a surface coverage of ~1/3 of

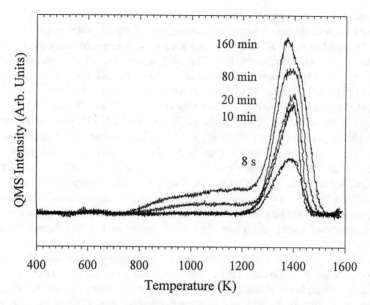

Figure 3. TPD spectra representing the desorption of PO groups from the α-Fe$_2$O$_3$ surface as a function of exposure to sodium phosphate solution, as in Fig. 1.

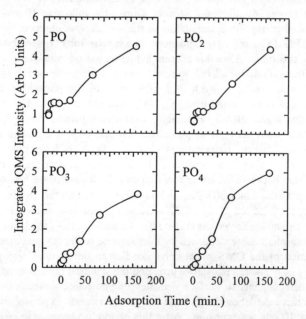

Figure 4. Integrated TPD spectra of phosphate films on thin α-Fe$_2$O$_3$ films as a function of exposure time to sodium phosphate solution, as in Fig. 1.

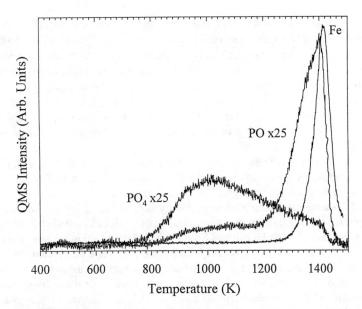

Figure 5. TPD spectra of the α-Fe$_2$O$_3$ surface (Fe) and the adsorbed phosphorus (PO and PO$_4$) for an exposure of 40 min. to sodium phosphate solution, as in Fig. 1. Reproduced with permission from reference 39.
Copyright © 1996 American Institute of Physics.

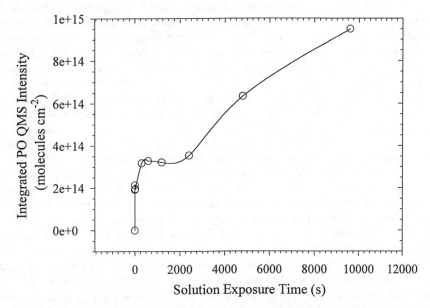

Figure 6. Surface phosphate coverage, derived from integrated PO TPD peak areas, as a function of α-Fe$_2$O$_3$ exposure time to sodium phosphate solution, as in Fig. 1. Reproduced with permission from reference 39.
Copyright © 1996 American Institute of Physics.

the available iron oxide surface sites, assuming $\sim 1 \times 10^{15}$ such sites exist (*35*). Assuming that the surface area of the thin film iron oxide is approximately that of the metal substrate (~ 1 cm^2), the accumulation of phosphate may be calculated as one phosphate molecule per 0.33 nm^2, or ~ 5 µmol P m^{-2}. This coverage is, within experimental error, close to the 3.6 - 4.8 µmol P m^{-2} coverage predicted for a binuclear complex on hematite (*12*) and may indicate that phosphate has completely exchanged with the singly coordinated hydroxyl groups. Continued solution exposure resulted in rapid phosphate accumulation, with a major fraction of the desorption species occurring in the ~ 500 - 1380 K range.

The AES and TPD data from the kinetic studies suggest that for exposures less than 10 minutes, the high temperature feature accounts for the entire first layer species. The phosphate which adsorbs initially is bound very tightly to the iron oxide surface. During TPD analyses, as more thermal energy is provided to the system, this tightly bound phosphate thermolyzes into small mass units, such as PO, before it desorbs, and it may in fact desorb with Fe evolved from the interfacial reaction between the thin film oxide and the Re substrate. After the initial, rapid phosphate adsorption, an induction period ensues, lasting 10 min., in which phosphate accumulation is observed predominantly in PO$_4$ spectra. It is possible that this particular mass/charge ratio is the most sensitive to the phosphate species dominant at longer exposure times. At the end of this induction period, phosphate again accumulates at the surface. TPD analyses have shown that this phosphate thermally desorbs at lower temperatures and as larger fragments than that adsorbed initially. This indicates that this phosphate is not bound as strongly and does not fully thermolyze before desorbing. Accumulation of this phosphate species is detected by TPD analyses, but not by AES analyses.

When analyzing plots of AES intensity ratios, it is important to distinguish between two types of breaks: those due to the exponential dependence of substrate attenuation by an overlayer, and those arising from changes in the rate of overlayer adsorption. Roll over associated with the first type of break can be ruled out in the spectra presented here, since this signal is approximately 80 - 90% of the clean surface signal intensity. Thus, the Auger breaks are probably due to changes in the rate of adsorption. The causes for this may be changes in the growth mode of phosphates at the surface, such as the initiation of phosphate incorporation into the oxide layer or three dimensional growth. Consequently, the discrepancy between AES and TPD in detecting phosphate accumulating at these longer solution exposure times may indicate that this phosphate is following one of these different growth modes.

To test for possible differences in the chemical states of phosphate during the kinetic experiments, freshly prepared thin film Fe$_2$O$_3$ was exposed to the same phosphatic solution mentioned previously and analyzed using XPS. However, to aid in the detection of different chemical states of adsorbed phosphate, the range of exposure times was extended to 1.5, 20, 360, 420, 735, and 1570 min. (*33*). For the shorter solution exposure times in this series, a single P(2p) peak occurred in the spectrum. The P(2p) binding energy of this peak increased from ~ 133.2 at 1.5 min. to ~ 133.6 at 1570 min. However, at the 735 and 1570 min. exposure times, two additional peaks emerged at ~ 138.7 and ~ 143.3 eV. This shift in binding energy indicates that phosphorus at the surface of the iron oxide becomes more oxidized. This suggests that the environment surrounding the surface phosphorus becomes more electronegative with continued exposure to solution. An explanation for this phenomenon comes from a study which examined the shifts in P(2p) binding energies with various degrees of phosphate condensation (*36*). Condensed phosphates are anionic entities characterized by corner-sharing PO_4 tetrahedra (*37*). The condensation phenomenon results from the removal of H and OH from individual phosphate tetrahedra to produce water and phosphoric anions with various geometries. The condensed phosphates range in composition from two tetrahedra sharing a single oxygen (diphosphates) to chains of linked tetrahedra with molecular weights up to several million (long chain polyphosphates). Within this range, phosphate tetrahedra can be arranged in strands (polyphosphates), rings (cyclophosphates), and three dimensional branched networks (ultraphosphates). XPS analyses of a monophosphate (Na_3PO_4), a diphosphate ($Na_4P_2O_7$), a long chain polyphosphate with eight formula units ($NaPO_3$), and an ultraphosphate with four formula units (P_4O_{10}) show that as the degree of condensation increases, the binding energy of the P(2p) photoelectron also increases (*36*). This may result from changes in the occupation of the d_π - p_π orbitals with structural distortions caused by condensation (*36*). These results suggest that in this work, longer exposure times to phosphatic solutions result in the formation of more condensed phosphates at the surface.

Phosphate Adsorption on Titania. Titania surface quality and thickness were monitored by AES (*33*). All of the titania surfaces used in these experiments were approximately 15-20 monolayers thick and were free of contaminants such as carbon. The intensities of the kinetic energy peaks at 110 eV for phosphorus and 387 eV for titanium were used to calculate the AES intensity ratios in Figure 7. These ratios were plotted as a function of exposure time to phosphate solution. A clear break can be seen at approximately one hour. Breaks in the Ca/Ti and K/Ti AES intensity ratios were also observed at 1 hr. An examination of the Auger data reveals that the Auger breaks occur before significant attenuation of the substrate oxide Auger peaks. Thus, it is likely that the Auger break reflects a change in the growth mode of the surface phosphate rather than an artifact of the escape depth of the Auger electrons.

Analysis of TPD spectra revealed that the intensity of a feature at 950 K depended upon the amount of solution dried on the surface (*33*). Therefore, the relative fraction of the cracking patterns for this peak were assumed to reflect the composition of the reagent solution. Thus K:P = 1 and Ca:P = 1.35 for this desorption feature. This allowed relative sensitivity factors to be determined for calcium, potassium, and phosphorus in the TPD spectra. When analyzing other desorption features, these sensitivity factors were used to adjust the ratios of areas from integrated peaks to ratios of surface concentrations.

The appearance of a desorption feature arising from deposited salt was thought to result from using a solution with a phosphate concentration nearly 40 times that of the solution used in the iron oxide study. Consequently, in all of these experiments, the samples were rinsed for 5 - 10 seconds in doubly-deionized ultrapure water after solution exposure. This minimized the amount of salt deposited at the surface after removal from the reactant solution. Such deposits, if allowed to remain, could have introduced error in calculating adsorption isotherms. Unlike the desorption feature at 950 K, desorption features at 1200 K and 1400 K did not appear to be reduced by

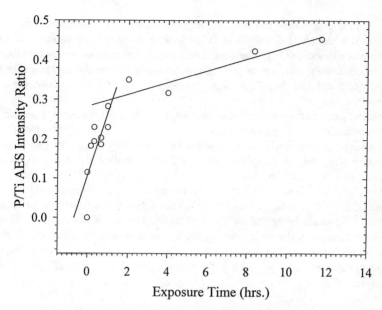

Figure 7. P/Ti AES peak intensity ratios as a function of exposure time of the titania surface to a calcium phosphate solution (3.5 mM CaCl$_2$, 119 µg P ml^{-1} as KH$_2$PO$_4$, pH 6.5, 0.01 M NaCl).

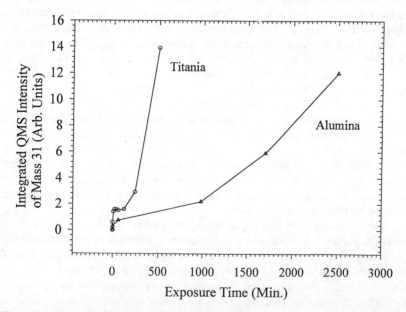

Figure 8. Adsorption isotherms of phosphorus on titania and alumina constructed from the integrated TPD spectra resulting from exposure to calcium phosphate solution, as in Fig. 7.

rinsing, within error of the experiment. They were therefore considered representative of surface reaction products rather than accumulated salts.

By using the sensitivity ratios discussed previously, the Ca:P ratio was calculated to be 0.6 for the titania surface exposed for 500 min. to calcium phosphate solution. For exposures as long as 15 hr., the Ca:P ratio decreased to 0.12. Thus, the composition of the adsorbed species increased in phosphorus concentration relative to calcium and potassium with continued exposure to the calcium phosphate solution. Common mineral forms of calcium phosphate have stoichiometries of Ca:P between 0.5 and 2 (*38*). The formation of phosphate tetrahedral structures devoid of cations could account for the measured Ca:P ratios. Consequently, it is possible that a condensed phase of PO_4 units formed with prolonged solution exposures (*37*).

An adsorption isotherm was constructed from integration of TPD peak desorption features for mass 31 (P) associated with increasing exposure time to calcium phosphate solution (Figure 8). The rapid uptake of phosphorus that begins at approximately 200 min. is a result of the large desorption feature at 1200 K (*33*). The rapid rise at early exposure times is primarily a result of the peak at 950 K and may be considered a result of residual salts accumulated at the surface, despite the rinsing procedure.

Phosphate Adsorption on Alumina. AES data was used to verify the growth of the Al_2O_3 surface and to calibrate surface coverage (*33*). Fully oxidized aluminum can be identified by a single peak at 60 eV, which has a 10 eV peak shift to lower energy than the aluminum metal transition. From the attenuation of the substrate Ta peaks, the coverage of the alumina films used for all the experiments was estimated to be ~15 monolayers. The Ca/Al, K/Al and P/Al AES intensity ratios taken for a series of exposure times to calcium phosphate solution all show breaks at 3-4 hr.

Figure 9 shows several TPD spectra of mass 31 (P) at different exposure times to calcium phosphate solution (*33*). Phosphorus desorption from alumina shows broad features from 450-700 K, the residual salt peak at 980 K, and a high temperature feature, which occurs between 1400 and 1560 K. Variation in the temperature associated with the high temperature desorption feature may be a result of variations in heating rate or thermocouple placement. Unlike the titania spectra, the large feature at 1200 K is absent. In addition, at least one new feature can be observed at approximately 450-700 K. Figure 9b shows the corresponding calcium and potassium desorption features. The ratio of calcium and potassium ions to phosphorus is comparable, but for longer exposures, this ratio decreases dramatically. For a 30 hr. exposure, the calcium and potassium to phosphorus ratios drop to less than 1% of the solution concentration. Integration of several phosphorus desorption spectra provide the data for an adsorption isotherm shown in Figure 8. Onset of rapid uptake of phosphate is observed between 20-25 hr.

Conclusions

For thin film hematite exposed to a sodium phosphate solution, initially rapid uptake occurs during the first 10 minutes of solution exposure in which phosphate completes first layer adsorption. A short induction period ensues, lasting 10 min., which

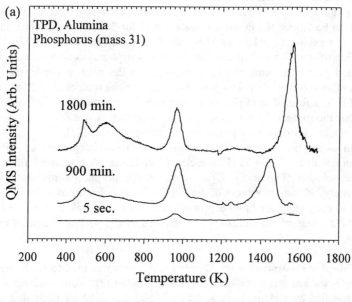

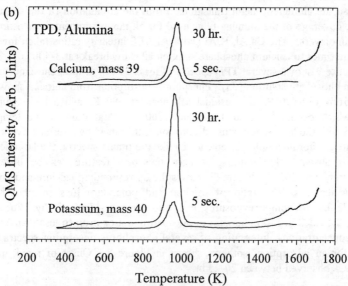

Figure 9. TPD spectra showing the desorption of phosphorus (a) and a comparison of calcium and potassium desorption for a 5 second exposure and a 30 hour exposure (b) for Al_2O_3 exposed to calcium phosphate solution, as in Fig. 7.

corresponds to a saturation of surface sites that comprise a TPD desorption feature at 1380 K. After this induction period, phosphate species form which are distinct from those involved in the initial reactions.

The rate of uptake of calcium and potassium phosphate on the alumina and titania surfaces can be directly compared from their respective adsorption isotherms. The onset of rapid uptake is between 200-250 min. for the titania surface and between 1500 and 2000 min. for the alumina surface. This is comparable to *in-situ* studies where the onset of rapid growth is 160 min. for titania powders and 1800 min. for alumina powders (Campbell, A. A., Battelle, Pacific Northwest Laboratories, personal communication, 1996). This correlation between *in-situ* and *ex-situ* measurements further validates our technique for making kinetic measurements on thin metal oxide films that have been exposed to solution and returned to vacuum.

For all the oxides studied, the rapid growth at longer solution exposure times shown in the TPD adsorption isotherms does not follow the break in the Auger peak ratios. This discrepancy reflects a change in the growth mode of surface phosphate. The Auger roll over corresponds to the rapid completion of the first layer of adsorbed phosphate, which can be defined as the coverage where subsequently adsorbed phosphate does not occupy metal oxide surface sites. Consequently, phosphate can either accumulate three dimensionally on the first layer, or diffuse into the bulk oxide layer as subsurface phosphate. XPS data for hematite and calculated surface concentration ratios of adsorbed species on titania and alumina indicate that phosphate condensation may be the growth mode at longer solution exposures.

Acknowledgments

Support for this work was provided by Laboratory Directed Research and Development (LDRD) funding from Pacific Northwest National Laboratories. One of us (TSM) gratefully acknowledges the support of the International Minerals Corporation and the Potash and Phosphate Institute.

References

1 Schwertmann, U.; Taylor, R. M. In *Minerals in Soil Environments 2nd Edition*; Dixon, J. B.; Weed, S. B., Eds.; Soil Science Society of America Book Series, Number 1; SSSA: Madison, WI, 1989, pp 379-438.

2 Colombo, C.; Barrón, V.; Torrent, J. *Geochim. et Cosmochim. Acta* **1994**, *58*, 1261.

3 Soper, R. J.; Racz, G. J. In *The Role of Phosphorus in Agriculture*; Khasawneh, F. E.; Sample, E. C.; Kamprath, E. J., Eds.; ASA, CSSA, SSSA: Madison, WI, 1980, pp 263-310.

4 Borggaard, O. K. *J. Soil Sci.* **1983**, *34*, 333.

5 Bunker, B. C.; Rieke, P. C.; Tarasevich, B. J.; Campbell, A. A.; Fryxell, G. E.; Graff, G. L.; Song, L.; Liu, J.; Virden, J. W.; McVay, G. L. *Nature* **1994**, *264*, 48.

6 Barrow, N. J. *Adv. Agron.* **1985**, *38*, 183.

7 Torrent, J. *Aust. J. Soil Res.* **1991**, *29*, 69.

8   Hingston, F. J.; Atkinson, R. J.; Posner, A. M.; Quirk, J. P *Nature*, **1967**, *215*, 1459.
9   Russell, J. D.; Parfitt, R. L.; Fraser, A. R.; Farmer, V. C. *Nature* **1974**, *248*, 220.
10   Goldberg, S.; Sposito, G. *Commun. in Soil Sci. Plant Anal.* **1985**, *16*, 801.
11   Atkinson, R. J.; Posner, A. M.; Quirk, J. P. *J. Inorg. Nucl. Chem.* **1972**, *34*, 2201.
12   Barrón, V.; Herruzo, M.; Torrent J. *Soil Sci. Soc. Am. J.*, **1988**, *52*, 647.
13   Borggaard, O. K. *Clays Clay Miner.*, **1983**, *31*, 230.
14   Colombo, C.; Barrón, V.; Torrent, J. *Geochim. Cosmochim. Acta*, **1994**, *58*, 1261.
15   Atkinson, R. J.; Parfitt, R. L.; Smart, R. St. C. *J. Chem. Soc. Faraday I.* **1974**, *70*, 1472.
16   Parfitt, R. L.; Atkinson, R. J.; Smart, R. St. C. *Soil Sci. Soc. Am. Proc.* **1975**, *39*, 837.
17   Tejedor-Tejedor, M. I.; Anderson, M. A. *Langmuir*, **1990**, *6*, 602.
18   Martin, R. R.; Smart, R. St. C. *Soil Sci. Soc. Am. J.* **1987**, *51*, 54.
19   Martin, R. R.; Smart, R. St. C., Tazaki, K. *Soil Sci. Soc. Am. J.*, **1988**, *52*, 1492.
20   Kamrath, M.; Mrozek, P.; Wieckowski, A. *Langmuir*, **1993**, *9*, 1016.
21   Johnsson, P. A.; Eggleston, C. M.; Hochella, M. F. Jr. *Am. Mineral.*, **1991**, *76*, 1442.
22   Eggleston, C. M.; Hochella, M. F. Jr. *Am. Mineral.*, **1992**, *77*, 911.
23   Rajan, S. S. S. *Nature*, **1975**, *253*, 434.
24   Lookman, R.; Grobet, P.; Merckx, R.; Vlassak, K. *European J. Soil Sci.*, **1994**, *45*, 37.
25   Busca, G.; Ramis, G.; Lorenzelli, V.; Rossi, P. F. *Langmuir*, **1989**, *5*, 911.
26   Alvarez, R.; Fadley, C. S.; Silva, J. A.; Uehara, G. *Soil Sci. Soc. Am. J.*, **1976**, *40*, 615.
27   Alvarez, R.; Fadley, C. S; Silva, J. A. *Soil Sci. Soc. Am. J.*, **1980**, *44*, 422.
28   Leung, L. -W.; Gregg, T. W.; Goodman, D. W. *Rev. Sci. Instrum.* **1991**, *62*, 1857.
29   Murrell, T. S.; Corneille, J. S.; Nooney, M. G.; Vesecky, S. M.; Hossner, L. R.; Goodman, D. W., *Rev. Sci. Instrum.* (submitted).
30   Corneille, J. S.; He, J. -W.; Goodman, D. W. *Surf. Sci.* **1995**, *338*, 211.
31   Smentowski, V. S.; Yates, J. T. *Surf. Sci.* **1990**, *232*, 113.
32   Bernett, M. K.; Murday, J. S.; Turner, N. H. *J. Elect. Spectrosc. Rel. Phenom.* **1977**, *12*, 375.
33   Murrell, T. S.; Nooney, M. G.; Hossner, L. R.; Goodman, D. W. *Langmuir* (submitted).
34   Redhead, P. A. *Vacuum*, **1962**, *12*, 203.
35   Woodruff, D. P.; Delchar, T. A. *Modern Techniques of Surface Science*; Cambridge University Press: New York, NY, 1986.
36   Chassé, T.; Franke, R.; Urban, C.; Franzheld, R.; Streubel, P.; Meisel., A. *J. Elect. Spectrosc. Rel. Phenom.* **1993**, *62*, 287.
37   Durif, A. *Crystal Chemistry of Condensed Phosphates*; Plenum: New York, NY, 1995.
38   Lindsay, W. L., *Chemical Equilibria in Soils*; Wiley: New York, NY, 1979.
39   Nooney, M. G.; Murrell, T. S., Corneille, J. S.; Rusert, E. I.; Hossner, L. R.; Goodman, D. W. *J. Vac. Sci. Technol. A.* **1996**, *A14*, 1357.

Chapter 7

The Effects of Bromide Adsorption on the Underpotential Deposition of Copper at the Pt(111)—Solution Interface

Nenad M. Marković, C. A. Lucas, Hubert A. Gasteiger[1], and Philip N. Ross, Jr.

Materials Sciences Division, Lawrence Berkeley National Laboratory, University of California, 1 Cyclotron Road, Berkeley, CA 94720

The Pt(111)-Cu-Br system was examined by utilizing *ex-situ* UHV and *in-situ* surface x-ray scattering (SXS) spectroscopic techniques in combination with the rotating ring-disk measurements. In copper free solution at 0.2 V, we observe a hexagonal, close-packed bromide monolayer that is incommensurate, but aligned with the underlying platinum lattice. The nearest neighbor distance is a continuous function of the electrode potential; bromide adsorption begins at ca. -0.15 V and reaches a coverage of $\approx 0.42 \pm 5\%$ ML at 0.5 V. Cu UPD on Pt(111) in the presence of bromide is a two-step process, with the total amount of Cu deposited at underpotentials $\approx 0.95 \pm 5\%$ ML. The RRDE analysis shows that Br coverage changes relatively little (< 0.05 ML) over the entire Cu UPD region. We propose a model wherein the first stage of deposition occurs by displacement of close packed Br ad-species by Cu adatoms through a "turn-over" process in which Cu is sandwiched between the Pt surface and the Br overlayer, leading to the formation of an ordered Pt(111)-Cu-Br bilayer intermediate phase that closely resembles the (111) planes of the Cu(I)Br crystal. The coverage of both Cu and Br in this intermediate phase is ≈ 0.5 ML. The second stage is the filling-in of the Cu plane of the bilayer to form a pseudomorphic (1×1) Cu monolayer and a disordered Br adlayer with a coverage of $\approx 0.4 \pm 5\%$ ML.

Underpotential deposition (UPD) of foreign metal adatoms on platinum single crystals in different acid electrolytes occupies a special position in interfacial electrochemistry. Of various systems examined the underpotential deposition of Cu on a Pt(111) surface has been of particular interest since the interpretation of the nature of the Pt(111)-Cu-Anion structure have been the subject of considerable controversy. Overviews with some different perspectives can be find in the References. [1-13].

[1]Current address: Institute of Surface Chemistry and Catalysis, Ulm University, D–89069 Ulm, Germany

Examination of the effects of halide ions on Cu UPD onto Pt(hkl) surfaces is an important part of current research in our laboratory. We review here some results on the adsorption of Br^- on Pt(111), including some new data for the effects of Br^- on UPD of Cu on the Pt(111) surface. Recently, combining *ex-situ* and *in-situ* spectroscopic techniques with traditional electrochemical methods has enabled us to resolve both the nature of the structure and the kinetics of monolayer formation of Cu onto the Pt(111) surface in the presence of ClO_4^-, HSO_4^-, and Cl^- anions. By utilizing the rotating ring-disk method $(RRD_{Pt(111)} E)$ we have shown that a slow, kinetically controlled Cu deposition in perchloric acid (and to some extent in sulfuric acid) is significantly promoted in the presence of Cl^-. The mechanism of enhancement is correlated with the tendency of weakly hydrated chloride ions to perturb the hydration shell of strongly solvated Cu^{2+}, thereby enhancing the electron transfer reaction. In addition, $RRD_{Pt(111)}E$ measurements clearly demonstrated that the UPD of Cu on Pt(111) is a two-step process, and the total amount of Cu deposited at underpotential is $\approx 0.9 \pm 5\%$ ML [*11, 13*]. A careful analysis of x-ray scattering data enabled us to deduce that a (4×4) phase, also observed by low energy electron diffraction (LEED) after emersion of Pt(111) electrode from solution containing Cu^{2+} and Cl^- ions, contained both Cu and Cl in the unit cell [*12*]. Although it was clear that the structure formed by Cu on Pt(111) is strongly influenced by Cl^-, the lack of information on the coverage, structure and electrosorption valency of adsorbed Cl^- (in solution free of copper) a imposes significant limitation on our understanding of the mechanism of Cu UPD. It is obvious, therefore, that in order to explore the Pt(111)-Cu-halide system in more detail one should replace Cl^- with another halide anion which might give new insight into the complex surface chemistry of Cu UPD on the Pt(111)-halide surface. An obvious choice was to replace Cl^- with Br^-, since the coverage of bromide on Pt(111) can be determined independently using the RRDE technique.

In this paper, bromide adsorption on the Pt(111) and the Cu UPD on the Pt(111) surface in the presence of Br^- anions is examined by employing similar tactics to our recent study of the Cu UPD on Pt(111) in the presence of Cl^-. In this case, the RRDE technique will in addition be used to establish the bromide adsorption isotherm on Pt(111) during the deposition of Cu. We show that the *in-situ* SXS and *ex-situ* UHV measurements provide sufficient information to generate a detailed model of the Cu UPD process on Pt(111) in solutions containing Cl^- and Br^-.

Experimental

RRDE measurements. The pretreatment and assembling of the Pt(111) crystal (0.283 cm^2) in a $RRD_{Pt(111)}E$ configuration was fully described in our previous paper [*14*]; *i.e.*, following flame-annealing, the single crystal was mounted in the disk position of an insertable rotating ring-disk electrode assembly (Pine Instruments). In the present study the validity of the crystal preparation and its clean transfer into a interchangeable disk (Pt)-ring electrode arbor was verified by the voltammetry recorded on the disk electrode. A typical voltammogram of the Pt(111) surface in 0.1M $HClO_4$ recorded at 1600 rpm (insert (a) in Figure 1) was identical with the curve obtained at 0 rpm indicating that the electrode remained clean even at high rotation rate over extended periods of time. The cyclic voltammogram of Pt(111) in 0.1 M $HClO_4$ with $8 \cdot 10^{-5}$ M Br$^-$ under a rotation rate of 900 rpm (Figure 1) is also in excellent agreement with the data in the literature [15, 16] and clearly demonstrates the high quality of the Pt(111) surface in the $RRD_{Pt(111)}E$ configuration even at a low sweep rate of 10 mV/s.

Details of *ring-shielding* experiments are provided in References [13] and [17]. Briefly, in the Cu UPD study the ring electrode was potentiostated at $E_r = -0.275$ V, *i.e.*, at the potential where Cu^{2+} is reduced to Cu^0 under diffusion control. In RRDE experiments examining the bromide adsorption isotherm the ring electrode was potentiostated at 1.08 V, such that the oxidation of Br$^-$ to Br_2 occurs under pure diffusion control. The *ring-shielding* property of the RRDE was employed to assess the *mass flux* of both the Cu^{2+} and the Br$^-$ from and to the Pt(111) disk electrode, as shown previously for the copper deposition on the (111) surface in the presence of Cl$^-$ [13] and for the Br$^-$ adsorption in the absence of Cu in solution [17]. If either Cu or Br are adsorbed onto the Pt(111) disk, the surface coverage ($\theta_{Br,Cu}$) can be assessed from ring currents in either potentiodynamic or potentiostatic experiments. In this report, however, the adsorption isotherm will be assessed based only on the potentiodynamic experiments;

$$\Delta\theta_{Cu,Br} = \frac{1}{Q} \frac{\frac{1}{v}\int(i_r - i_r^\infty)dE}{AnN} \tag{1}$$

where i_r^∞ and i_r are unshielded and shielded ring currents, N is collection efficiency (N=0.22±5% [11]), v refers to the sweep rate, n is the number of electrons (n = 2 for Cu^{2+} reduction to Cu^0 and n = 1 for Br$^-$ oxidation to Br_2 on the ring electrode) and Q is the charge corresponding to monolayer formation of adatoms on Pt(111) based on the surface atomic density of Pt(111)-(1 × 1) ($1.53 \cdot 10^{15}$ atoms/cm^2), assuming one completely discharged adatom per platinum atom. (*e.g.*, 208 µC/cm^2 for Br_{ad} and 480 µC/cm^2 for Cu_{upd}). Equation 1 constitutes the basis for the assessment of the copper and bromide adsorption isotherm, for more details see References [13] and [14].

Ex-situ UHV measurements. The UHV-electrochemistry system has been described in detail elsewhere [3]. Surface cleanliness was monitored by Auger electron spectroscopy (AES) and the surface structure determined using low energy electron diffraction (LEED). After UHV preparation and characterization, the Pt(111) single crystal was transferred into an electrochemical environment using technique designed to minimized the possibility of contamination. Following

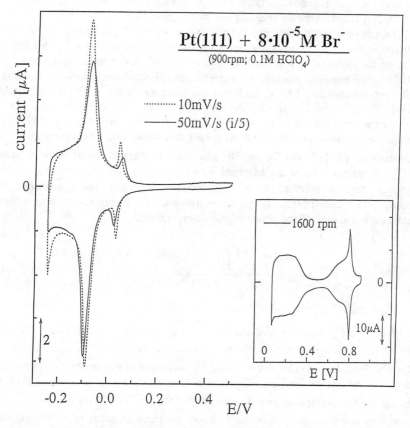

Figure 1. Cyclic voltammograms of Pt(111) mounted in the disk position of the RRDE in 0.1 M HClO₄ (insert) and 0.1 M HClO₄ with $8 \cdot 10^{-5}$ M Br⁻.

electrochemical characterization, the Pt(111) crystal was emersed from the electrolyte and returned to the UHV environment for post-electrochemical analysis.

In-situ x-ray scattering measurements. X-ray measurements were conducted on beam-line 7-2 at the Stanford Synchrotron Radiation Laboratory (SSRL) and on beam-line X-20A at the National Synchrotron Light Source (NSLS) using monochromatic X-ray beams of wavelengths 1.240 Å and 1.409 Å, respectively. In each case, the detector resolution was defined by 2 mm (vertical) and 8 mm (horizontal) slit at distance of ≈ 0.5 m from the sample position. The (111) face is indexed to a hexagonal lattice described in Reference [18]. Using this scheme, the surface normal lies along the $(00L)_{hex}$ direction, and the $(H00)_{hex}$ and $(0K0)_{hex}$ vectors lie in the plane of the surface and subtend 60°. In this coordinate system $(0,0,3)_{hex} = (1,1,1,)_{cubic}$ and $(0,1,2)_{hex} = (0,0,2)_{cubic}$.

Solutions and Reagents. Solutions of 0.1 M HClO$_4$ (EM Science, *Suprapur*) were made up with pyrolytically triply distilled water. 0.1 M KBr (Baker, *Ultrex*) and Cu^{2+} (from CuO, Aldrich, *Puratronic*) were injected into supporting electrolytes to the desired concentrations. All potentials are referenced to a normal calomel electrode (NCE), the potential of which is +0.268 V versus a standard hydrogen electrode [19]. A closed bridge was placed between the NCE reference electrode and the cell to prevent Cl$^-$ contamination.

Results and Discussion.

Adsorption isotherm of Br$^-$ on Pt(111); RRDE measurements. Figure 2 shows a ring-shielding experiment on Pt(111) in 0.1 M HClO$_4$ with $8 \cdot 10^{-5}$ M Br$^-$ at 50 mV/s (E_r=1.08 V). Starting from the negative potential limit, the ring current below approximately -0.15 V equals i_r^{∞} (Figure 2b) indicating the absence of bromide adsorption on the Pt(111) disk. This could similarly be inferred from the correspondence of the disk voltammograms with and without bromide in solution (Figure 2a) and is in agreement with the threshold potential for bromide adsorption found in *ex-situ* experiments [15]. As the Pt(111)-disk potential is swept across the first voltammetric peak (Figure 2a), the associated adsorption of bromide is demonstrated by the concomitant decrease of the ring current below its unshielded value (Figure 2b). The same observation is made for the subsequent, more positive peak, which clearly relates this process to further bromide adsorption on the Pt(111)-disk. Following these two characteristic voltammetric peaks, the Pt(111)-disk current diminishes to a double-layer-like structure above ≈ 0.2 V. At the same time, the ring current remains below i_r^{∞} until the positive potential limit is reached, manifesting the continuous adsorption of bromide on the Pt(111)-disk.

The qualitative correspondence between ring and disk currents may be evaluated quantitatively in terms of θ_{Br} according to Equation 1. To avoid the mass transport resistances the bromide adsorption isotherm will be extracted from the negative-going sweep. The resulting Coulombic charge of the Pt(111) disk in bromide

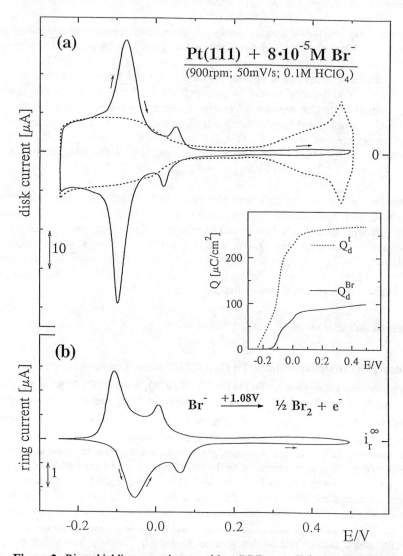

Figure 2. Ring-shielding experiment with a RRD$_{Pt(111)}$E in 0.1 M HClO$_4$ without and with bromide. (a) (——) Cyclic voltammogram on the Pt(111)-disk; (- - - -) base voltammogram without bromide in solution. **(b)** Corresponding ring shielding currents. <u>Insert</u> (- - - -) the disk charging curve in the presence of Br⁻; (——) the charge contribution to the disk due to bromide adsorption/desorption.

solution, Q_d^t (≈ 265 $\mu C/cm^2$), as well as the charge contribution which is due only to bromide adsorption, Q_d^{Br} (≈ 105 $\mu C/cm^2$), *versus* disk potential is shown in an insert of Figure 2. Even though the major adsorption wave coincides with the more negative voltammetric peak (*ca.* -0.1 V), ≈ 0.05 ML of bromide are being adsorbed during the second peak (≈ 0.05 V), clearly discernible by the inflection point of the isotherm at ≈ 0.2 V ($\theta_{Br} = 0.38$ ML). We assign this voltammetric feature to a change in θ_{Br} producing a bromide coverage necessary for the transition from a disordered to an ordered phase. It is important to note that the difference between Q_d^t and Q_d^{Br} amounts to ≈ 160 $\mu C/cm^2$ (insert of Figure 2). Therefore, a saturated coverage of H_{ads} on Pt(111) corresponds to the value associated with "weakly" adsorbed hydrogen on Pt(111) in the absence of bromide, a quite substantial proof that the anomalous feature is not adsorbed hydrogen as assumed originally [20-22] and in agreement with CO displacement experiments [16]. Furthermore, since the anomalous feature occurs at the same potential with respect to the reversible hydrogen electrode in both HClO₄ and KOH [23], it must be related to the adsorption of oxygen-containing species (OH_{ads}).

Surface structure of Br_{ad} on Pt(111); x-ray measurements. In-situ x-ray scattering offers an unique opportunity to obtained detailed information on the structure of the electrode/electrolyte interface [24-26]. In this section, we describe some of x-ray measurements for Br/Pt(111) system, more details could be found in References [27-29]. With the electrode potential held at ≈ 0.2V, we found four symmetry independent in-plane peaks at (0.683, 0), (0.683, 0.683), (1.366, 0) and (1.366, 1.366). The position of the observed peaks are shown at the top of Figure 3. The measurements were performed at l=0.15 which is at sufficiently high incident angle to allow penetration of x-ray beam through the liquid overlayer. The location of these peaks and the absence of any intensity at smaller scattering vectors indicates that the adsorbed bromide ions form a hexagonal monolayer structure that is aligned along, but incommensurate with, underlying Pt lattice. The integrated intensities of the four reflections decreased monotonically with the scattering vector and this is consistent with the structure schematically illustrated in Figure 3. The proximity of the peaks to (2/3n, 2/3m), n, m = 0,1,2 probably resulted in the assignment of a (3 × 3) structure in the ex-situ LEED study [15]. There is, however, no indication that the overlayer locks into the commensurate structure, but rather undergoes a continuous change in interatomic distance with potential. For convenience we will in the following refer to the incommensurate hexagonal structure as the "(3 × 3)" phase at all potentials where it forms. The domain size of the bromide adlayer, inferred from the width of rocking scans through different surface reflections, has been found to be in the range 100-150 Å. The domain size of the bromide adlayer is considerably smaller than the average Pt(111) terrace size, *ca.* 600 Å. The relatively poor ordering of the adlayer is possibly the result of the slow kinetics of surface ordering, as will be discussed in the last section.

From the positions of the adsorbate diffraction peaks the near-neighbor spacing of the bromide anions and hence the surface density can be calculated. For the lowest order diffraction peak (h, 0), the bromide near-neighbor spacing is given by

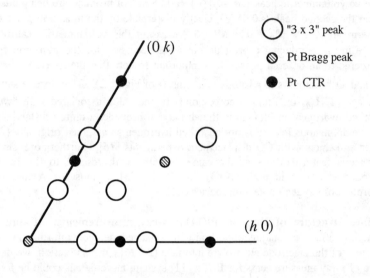

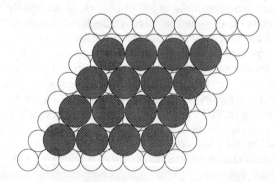

Figure 3. Top: A representation of the in-plane scattering in reciprocal space. The location of CTR's and Pt Bragg reflections are shown by filled circles. The open circles are at position where diffraction from the bromide adlayer was observed. Bottom: Proposed structure in which the Pt atoms are shown by open circles and the bromide overlayer atoms by filled circles.

$a_{Br} = a_{Pt}/h$. At 0.2 V this gives a Br = 4.07 Å, corresponding to surface density of $\approx 7.0 \times 10^{14}$ cm^{-2}, or $\theta = 0.43$ ML (1ML $\equiv$ 1 Br : 1 Pt). The near-neighbor distance can be compared to the van der Waals diameter, $a_{vdw} = 3.70$-4.00 Å, for the bromide ion [30]. At higher potentials (≈ 0.7 V) a contraction of the Br adlayer was observed ($\theta_{Br} = 0.47$ ML).

In order to ascertain the surface normal structure of the bromide adlayer and determine the coverage to confirm our structural model, we carried out specular x-ray ray reflectivity measurements. The interpretation of CTR measurements have also been described in several papers and readers are refereed to these papers for details [24, 25, 27]. Briefly summarizing, the CTR analyses gave the surface coverage of bromide, $\theta_{Br} = 0.43$ ML at 0.2 V and $\theta_{Br} = 0.48$ ML at 0.7 V. The platinum-bromide layer spacing (≈ 2.7 Å) suggest that bromide is covalently bonded to the platinum (the covalent radii for Pt and Br are 1.30 and 1.14 Å, respectively).

Cu-Br structure on Pt(111); ex-situ UHV measurements. Figure 4 shows the cyclic voltammogram of Pt(111) in 10 mM HClO$_4$ in the presence of 1 mM of Cu^{+2} and Br$^-$, recorded in a thin-layer electrochemical cell with an attached UHV transfer system. The sample was emersed at the indicated potential (E_e, between the two anodic peaks) during the positive-going sweep as to avoid the mass transfer effects of the deposition process. The LEED pattern after immersion is shown in detail in the insert (b) of Figure 4. The pattern has three set of spots, insert (c), the fundamental beams for (111)-(1 × 1), superlattice beams near the (3/4n, 3/4m), n, m = 0, 1, 2 positions and a set of weaker beams which are multiple scattering spots. This pattern was observed previously [4] and was denoted (4 × 4), which is the nearest commensurate unit cell. As in the case of LEED analysis for Cu-Cl system, due to the non-uniqueness of LEED patterns [31], and from the uncertainties involved in *ex-situ* measurements *per se*, it is difficult to resolve the true nature of the Cu UPD process in the presence of co-adsorbed bromide anions. Therefore in order to unambiguously resolve the observed (4 × 4) LEED pattern it is imperative to devise *in-situ* measurements which can provide information of the coverage of both copper and Br ad-species, and to provide atomically resolved structural information on codeposeted Cu-Br layers. Based on our new RRDE measurements as well as x-ray scattering experiments we will be able to resolve the nature of the (4 × 4) LEED structure for Cu UPD on Pt(111) in the presence of bromide anions, as will be discussed in the next two sections.

Adsorption isotherm of Cu^{2+} on Pt(111)-Br$^-$ surface; RRDE measurements. The voltammetry of Pt(111) containing Cu^{2+} and Br$^-$ ions is illustrated in Figure 5. While monitoring the Cu UPD on the disk electrode, the ring electrode was potentiostated at -0.275 V, a potential at which a solution phase of Cu^{2+} is deposited onto the ring electrode at the diffusion controlled rate. When there is no deposition of Cu on the disk, this rate is referred to as the unshielded ring current, i_r^{∞} [11, 13]. Note also that at this potential no Cu$^+$ is produced, *i.e.*, a complete conversion of Cu^{2+} to Cu^0 occurs [13]. Starting at 0.5 V and scanning the disk

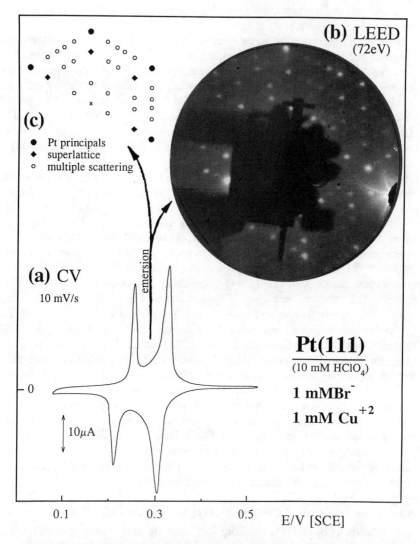

Figure 4. (a) Cyclic voltammetry of UHV-prepared Pt(111) in 10 mM HClO$_4$ in the presence of bromide and copper ions. (b) LEED after emersion at E$_e$ during the positive-going sweep. (c) Schematic of the LEED pattern shown in insert (b).

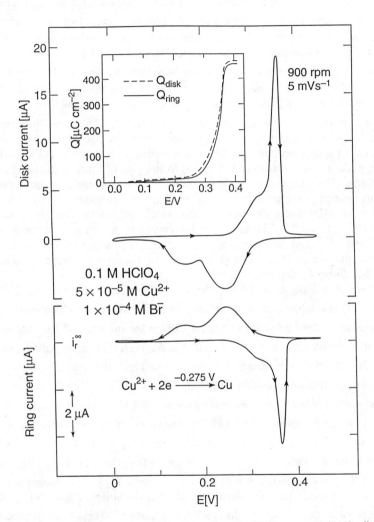

Figure 5. Top: Cyclic voltammogram for Cu UPD on a Pt(111) disk electrode. Bottom: Ring electrode currents recorded with the ring being potentiostated at -0.275 V. Insert: Charges on the disk, Q_d, and the ring, Q_r electrode.

potential negatively, the ring current above *ca.* 0.3 V equals i_r^∞ indicating the absence of Cu deposition on the Pt(111) disk electrode. As the disk potential is swept across the Cu UPD region both Cu UPD peaks recorded on the disk electrode, first at 0.25 V and the second at 0.15 V, are mirrored by the decrease in the ring current from its i_r^∞ value, Figure 5. Following these two characteristic Cu UPD peaks, the Pt(111)-disk current diminishes to a capacitive current below 0.1 V, manifesting the completion of the Cu monolayer on the Pt(111) disk electrode; consistent with the return of i_r to i_r^∞ in the same potential region. The fact that the ring shielding current, i_r, parallels with the Cu UPD on the disk electrode clearly indicates that both cathodic processes are primarily related with the Cu UPD, as was in the case for Cu UPD in the presence of Cl^- [*11, 13*]. The same observation is made for the subsequent positive sweep, where there are two distinctive processes seen from the ring electrode; the first as a shoulder at *ca.* 0.375 V and second as a sharp peak at 0.425 V, both clearly related to stripping of Cu from the Pt(111) disk electrode. The total amount of Cu deposited by UPD was evaluated quantitatively by integration of the ring current, yielding a maximum Cu coverage of 0.95 ± 5% ML (almost identical with the value inferred from disk measurements, the insert in Figure 5). This rather small difference in the charge assessed from the disk and the ring measurements (*ca.* 5 $\mu C/cm^2$) suggests that the anion co-adsorption contributes insignificantly to the Coulombic charge passing the interface during Cu deposition.

The above experiment indicates that the voltammetric feature for the Cu UPD on Pt(111) in solution containing Br^- is very similar with the results obtained under identical experimental conditions for Cu UPD in the presence of Cl^- anions [*11, 13*]. In particular, the splitting of Cu UPD peaks, characteristic for Cu/Cl^- system, was also recorded in the presence of Br^-. In addition, the total charge associated with the Cu UPD from solution containing Br^-, *ca.* 0.95 ± 5% ML, is close to the value of 0.9 ± 5% ML assessed in the presence of Cl^-. Although the Cu UPD on Pt(111) in the presence of Br^- and Cl^- anions shows a similar behavior, a close inspection of these two systems indicated that the roles of Cl^- and Br^- in Cu UPD is not identical. First of all, the onset of Cu UPD in the presence of Br^- anions occurs at a more negative potential (≈ 0.05 V). Secondly, Cu/Br^- system is a more irreversible process (*i.e.*, the separation between the stripping and UPD peaks is more pronounced for Cu/Br^- than for Cu/Cl^- system). These two differences are determined mainly by the strength of interaction of these two halide anions with clean Pt(111). Since the interaction of Br^- with the Pt(111) surface is stronger than interaction of Cl^- [*32*], one can reasonably conclude that the shape of Cu UPD peaks in the solution containing different halides is mainly determined by Pt(111)-halide interaction, *i.e.*, Pt(111)/Br^- being stronger than Pt(111)/Cl^-, as we discussed in our recent paper [*32*].

Adsorption isotherm of Br⁻ on Pt(111)-Cu$_{upd}$ surface; RRDE measurements. In the following we will use the ring shielding properties of the RRDE to assess the mass flux of Br⁻ from and to the Pt(111) disk electrode *during the Cu UPD formation*. In this experiment the ring electrode is potentiostated at 1.08 V, such that Br⁻ oxidation to Br$_2$ occurs under diffusion control. Figure 6 shows a ring shielding experiment on Pt(111) in 0.1 M HClO$_4$ with *ca.*, 1×10^{-4} M of both Br⁻ and Cu^{2+} ions. Starting from the positive potential limit, the ring current at ≈ 0.45 V equals i_r^∞ indicating the absence of either Br⁻ adsorption or desorption on the disk. In fact, at this potential the surface is almost fully covered by Br$_{ad}$, as indicated by the bromide adsorption isotherm obtained in a solution free of copper, insert of Figure 6. In the potential region between 0.45 and 0.3V, a small desorption of Br⁻ from the platinum (≈ 2 µC/cm^2) just preceding Cu deposition is demonstrated by the concomitant increase of the ring current above its unshielded value. As the disk potential is swept across the first UPD peak, the i_r^∞ remains almost constant, manifesting an absence of either Br⁻ adsorption or desorption. While the formation of the second Cu UPD peak is accompanied by an increase in the ring current from i_r^∞, completion of the Cu monolayer caused i_r^∞ to return back to its unshielded value. Therefore, close to the formation of full monolayer of copper, a small amount of Br ions are leaving the interface, confirming that the completion of the Cu monolayer is indeed associated with the desorption of Br⁻; the amount of the desorbed Br⁻, however, is relatively small (*ca.*, 7 µC/cm^2). Upon the sweep reversal, the i_r^∞ remains constant until the onset of the Cu stripping is observed on the disk electrode. As the disk potential is swept across the first Cu UPD stripping peak, the associated adsorption of Br⁻ (≈ 8 µC/cm^2) is mirrored by the decrease of the ring current below its unshielded value. A different observation is made for the subsequent, more positive stripping peak, where the stripping of Cu adatoms from the (111) sites is associated by desorption of Br⁻ from the surface (≈ 4 µC/cm^2). The resulting θ_{Br} *versus* the disk potential on the Pt(111) surface modified by Cu adatoms, and the corresponding adsorption isotherm for Br⁻ on Pt(111) in the solution free of Cu^{2+}, are shown in the insert of Figure 6. To avoid the mass transport resistances the bromide adsorption isotherm will be extracted from the negative-going sweep. Clearly, the initial stripping of copper is associated with simultaneous adsorption of additional Br⁻; note that the θ_{Br} in between the two Cu UPD peaks (point 2 in insert) is almost higher than corresponding coverage on clean Pt(111), *i.e.*, ≈ 0.45 ML. The complete stripping of copper from Pt(111) causes a desorption of Br⁻ (sweeping the electrode potential from point 2 to 1, ≈ 4 µC/cm^2), and therefore the total contribution of charge due to adsorption/desorption of Br⁻ during the Cu UPD is rather low, *ca.*, 4 µC/cm^2. Interestingly, after a complete stripping of copper from the (111) surface, the bromide surface coverage is exactly the same as in the case when the adsorption of

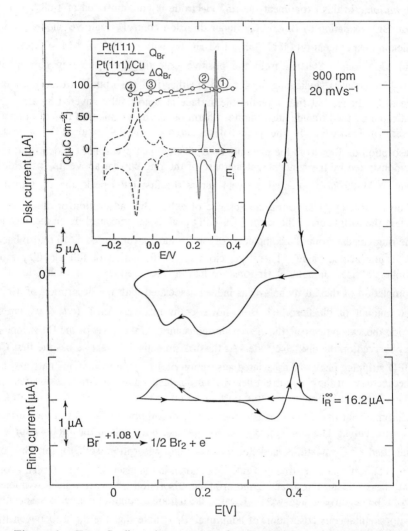

Figure 6. <u>Top</u>: Cyclic voltammogram for Cu UPD on a Pt(111) disk electrode. <u>Bottom</u>: Simultaneous ring shielding current. <u>Insert</u>: Integrated charges of bromide flux to the disk as a function of potential with and without copper present in a solution.

Br⁻ on Pt(111) was monitored in a solution free of copper (insert of Figure 6). It should also be noted that in potential range where an ordered (4 × 4) bilayer structure was observed by *ex-situ* LEED there is an increase in the surface coverage by Br *versus* the Pt surface in the Cu free solution. The true nature of this structure, however, can only be resolved by SXS measurements, as we discuss below.

Cu-Br structure on Pt(111); x-ray scattering measurements. In this section, we present some x-ray scattering measurements of the Cu UPD layer on Pt(111) in a solution containing bromide anions. Selected results are summarized in Figure 7. As in the case of Cu UPD on Pt(111) in the presence of Cl⁻ [*12*], in the solution containing Br⁻ four symmetry independent peaks were observed at (0, 0.741), (0.741, 0.741), (0, 1.482), (0.741, 1.482). These positions are shown in Figure 7 (top). The changes in the scattering intensity at (0.741, 0.741, 0.1) (not shown) suggested that the structure responsible for this peak is potential dependent and present only in the potential region between the two Cu UPD peaks. The location of these peaks indicates that the adsorbed species form a hexagonal superstructure which is incommensurate with the underlying platinum lattice; the nearest commensurate unit cell is (4 × 4), as was seen by *ex-situ* LEED (Figure 5). For convenience, we shall in the following discussion refer to this structure as the "(4 × 4)" phase, although it is not actually a commensurate structure. The inserts in Figure 7 show two rocking scans (φ scans) through the (1, 0, 0.1) and (0, 0.741, 0.1) peaks, respectively. Both peaks have Lorentzian lineshape with the with of the (0.741, 0.741) peak corresponding to a domain size of ≈200 Å. The relative intensities of the four in-plane peaks can be used to derive a structural model for the adsorbed layer (for more details see reference [*12*]). Additional information, however, can first be obtained by utilizing "anomalous scattering" methods, *i.e.*, measuring the peak intensities close to the Cu K adsorption edge (8979 eV). In general the intensities of the relatively strong (0.741, 0.741) and (0, 1.482) peaks were weaker close to the Cu edge. Without further analysis, this result indicates that the adsorbed superstructure does not consist of only Cu atoms (which would results in the reduction of the intensity of all in plane peaks near the Cu edge) but must also include Br. The simplest real space structure consistent with the data has a hexagonal unit cell with the lattice spacing 3.74 Å. The unit cell contains one copper and one bromide atom arranged in two hexagonal layers, one of Cu and one of Br, each with the coverage of ≈0.55 ML. The proposed model is in good agreement with the RRDE studies, which clearly showed that within the potential region where the "(4 × 4)" structure was observed, a close packed layer of bromide is coadsorbed with ≈0.5 ML of Cu.

Although the in-plane structure of the Br layer in the Pt(111)-Cu-Br phase is similar with that in the Cu free Pt(111)-Br structure, the incommensurate hexagonal Pt(111)-Cu-Br bilayer phase is more complex, reflecting the different chemistry of the underlying atoms. Returning to Pt(111)-Br system, the "(3 × 3)" structure noted above is stable between 0.2 to 0.7 V. Sweeping the potential to - 0.2 V always caused the in-plane peak to disappear, consistent with the RRDE coverage measurements which indicate desorption of bromide within this potential region.

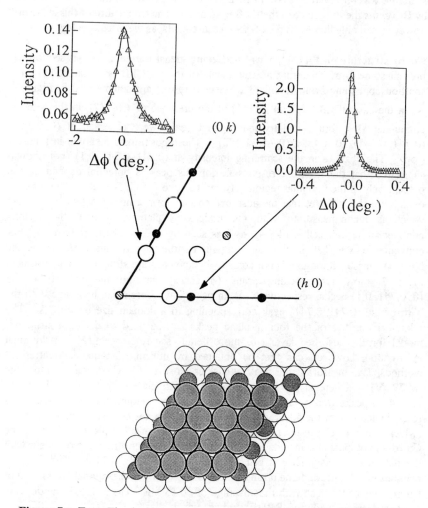

Figure 7. Top: The in-plane scattering intensities in reciprocal space at 0.275 V; the locations of CTR's are shown by open circles crosses and the superstructure peaks by circles. Bottom: Representation of the proposed structure in which the Pt atoms are shown by open circles and the Cu-Br overlayer atoms by filled circles. Insert: Shows two φ scans through the (0, 0.741, 0.1) (left) and (1, 0, 0.1) (right) peaks, respectively.

Sweeping the potential from -0.2 to 0.6 V did not, however, usually lead to the reappearance of in-plane peaks (in one of the experiments the structure could be recovered), implying that kinetics of ordering of bromide adatoms on Cu free Pt(111) is an "unpredictable" and rather slow process. In the solution containing Cu^{2+}, starting at 0.6 V and sweeping the potential across the Cu UPD peaks ($\approx 0.25 < E < 0.3$ V), we observed the formation of the "(4 × 4)" structure, and a further decrease in the potential ($\approx E < 0.2$ V) caused the "(4 × 4)" structure to disappear. Reversing the sweep in the positive direction, the "(4 × 4)" phase reappears at a potential corresponding to the first anodic peak, and remains to potentials just above the second anodic peak.

The results in Figure 6 indicate beyond any doubt that the surface coverage by Br undergoes only small (< 0.05 ML) changes upon the deposition of Cu even up to a nominal monolayer. We therefore propose that the mechanism of Cu UPD occurs by displacement of the surface bromide adatoms by Cu adatoms through a "turn-over" process in which Cu is sandwiched between the Pt surface and the bromide overlayer, *i.e.*, a Pt(111)-Cu-Br bilayer structure. Depending on the surface coverage by Cu adatoms, bilayer is either disordered ($\theta_{Cu} < 0.5$ ML) or an ordered incommensurate structure which is aligned with the underlying platinum lattice ($\theta_{Cu} \approx 0.5$ ML), i.e., the "(4 × 4)" structure. Initially, for $\theta_{Cu} < 0.5$ ML, a disordering of Pt(111)-Br structure is caused due to a random replacement of Br adatoms from platinum sites by Cu adatoms , and consequently Cu adatoms are in contact with both bromide adsorbed on platinum (first layer) and bromide adsorbed on the top of Cu adatoms (second layer). In this case, a hexagonal close packed monolayer of bromide does not exist neither in the first layer nor in the second layer, although the surface coverage by bromide is almost the same as at high potentials (0.6 V), where the surface is free of Cu adatoms. For a critical value of the θ_{Cu} all bromide adatoms will be entirely displaced by Cu adatoms and the ordered hexagonal "(4 × 4)" structure is formed, which ideally has a one to one ratio between Cu and Br adatoms, see Figure 7. A fundamental difficulty with the formation of the "(4 × 4)" structure is explaining how it can form and exist over such a wide range of stoichiometries, *e.g.*, the in-plane peaks appear to lock-in at a particular wavevector over the whole potential range in between two Cu UPD peaks (≈ 70 mV). It is difficult to propose any physical model which unambiguously can explain such behavior. It should be note, however, that in the structural model we derived for the Pt(111)-Cu-Br, it was necessary to include the large static Debye-Waller factor for Cu adatoms. This might imply a large disorder in the Cu adlayer, perhaps due to the presence of defects in the Cu layer, that would allow a variable Cu coverage in the bilayer "(4 × 4)" phase. Another possibility is the coexistence of disordered Cu-Br domains with the "(4 × 4)" phase, so that an ordered Pt(111) Cu-Br surface structure persists even during a continued Cu deposition into a disordered phase. Nevertheless, a further deposition of Cu ($\theta_{Cu} > 0.55$ ML) was causing a (4 × 4) in-plane peak to disappear, forming a pseudomorphic (1 × 1) Cu monolayer with a disordered Br overlayer.

Summary.

We review previous studies on the adsorption of Br^- on Pt(111), and present some new data for the effect of Br^- on the UPD of Cu onto the Pt(111) surface. The Pt-(111) Cu-Br system was examined by utilizing *ex-situ* UHV and *in-situ* surface x-ray scattering spectroscopic techniques (SXS) in combination with the rotating ring-disk measurements. In copper free solution, we observe at 0.2 V a hexagonal, close-packed bromide monolayer that is incommensurate, but aligned with the underlying platinum lattice; between 0.2 and 0.7 V the nearest-neighbor distance is a continuous function of the electrode potential. Absolute coverages of Br as a function of potential were determined from ion flux measurements using the RRDE technique; bromide adsorption begins at ca. -0.15 V and reaches the coverage of $\approx 0.42 \pm 5\%$ ML at 0.5 V. Cu UPD on Pt(111) in the presence of bromide is a two-step process, with the total amount of Cu deposited at underpotentials $\approx 0.95 \pm 5\%$ ML. The RRDE analysis unambiguously shows that Br coverage changes relatively little (< 0.05 ML) over the entire Cu UPD region. We propose a model wherein the first stage of deposition occurs by displacement of the close-packed Br adlayer by Cu adatoms through a "turn-over" process in which Cu is sandwiched between the Pt surface and the Br overlayer, leading to the formation of an ordered Pt(111)-Cu-Br bilayer intermediate phase that closely resembles the (111) planes of the Cu(I)Br crystal. Due to the strong Cu-Br interaction, it seems likely that the mobility of Cu-Br species is greatly enhanced over the single Br and/or Cu adatoms, thus allowing the formation of a structure with the large unit cell, the "(4 × 4)" phase. The coverage of both Cu and Br in this intermediate phase is ≈ 0.5 ML. The second stage is the filling-in of the Cu plane of the bilayer to form a pseudomorphic (1 × 1) Cu monolayer and a disordered Br adlayer with a coverage of $\approx 0.4 \pm 5\%$ ML.

Acknowledgment

We would like to thank Lee Johnson and Frank Zucca for their technical assistance, and also the staff and user administration at the Stanford Synchrotron Radiation Laboratory and at the NSLS, Brookhaven National Laboratory for their support on the synchrotron beamlines. This work was supported by the Office of Energy Research, Basic Energy Science, Materials Science Division of the U.S. Department of Energy under Contract No. DE-AC03-76SF00098.

REFERENCES

[1]   Stickney, J. L.; Rosasco, S. D; Hubbard A. T. *J. Electroch. Soc.* **1984,** *131,* 260.
[2]   Andricacos, P.; Ross, P. N. *J. Electroanal. Chem.,* **1984,** *167,* 301.
[3]   Marković, N.; Ross, P. N.; *Langmuir* **1993,** *9,* 580.
[4]   Michaelis, R.; Zei, M. S.; Zhai, R.S.; Kolb, D. M. *J. Electroanal. Chem.,* **1992,** *339,* 299.

[5] Leung, L-W.; Gregg, T. W.; Goodman, D. W. *Chem. Phys. Lett.*, **1992**, *188*, 467.

[6] Varga, K.; Zelenay, P; Wieckowki, A. *J. Electroanal. Chem.*, **1992**, *330*, 453.

[7] Sashikata, K.; Furuya, N.; Itaya, K. *J. Electroanal. Chem.*, **1991**, *316*, 361.

[8] White, J.; Abruña, H. D. *J. Phys. Chem.*, **1990**, *94*, 894.

[9] Gómez, R.; Yee,H. S.; Bommarito, G. M.; Felui, J. M.; Abruña, H. D. *Surf. Sci.*, **1995**, *335*, 101.

[10] Shingaya, Y.; Matsumoto, H.; Ogasawara, H.; Ito, M. *Surf. Sci.*, **1995**, *335*, 23.

[11] Marković, N. M.; Gasteiger, H. A.; Lucas, C. A.; Tidswell, I. M.; Ross, P. *Surf. Sci.*, **1995**, *335*, 91.

[12] Tidswell, I. M.; Lucas, C. A.; Marković, N. M.; Ross, P. *Phys. Review B*, **1995**, *51*, 10205.

[13] Marković, N. M.; Gasteiger H. A.; Ross, P. N. *Langmuir*, **1995**, *11*, 4098.

[14] Marković, N. M.; Gasteiger H. A.; Ross, P. N. *J. Phys. Chem.*, **1995**, *99*, 3411.

[15] Salaita, G. N.; Stern, D. A.; Lu, F.; Baltrushat, H.; Shard, B. C.; Stickney, J. L.; Soriaga, P. M.;. Frank, D. G; Hubbard, A.T. *Langmuir* **1986**, *2*, 828.

[16] Orts, J. M.; Gómez, R.; Feliu, J. M.; Aldaz, A.; Clavilier, J. *Electrochim. Acta* **1995**, *39*, 1519.

[17] Gasteiger H. A.; Marković, N. M.; Ross, P. N., *Langmuir*, in press.

[18] Sandy, A. R. et al. *Phys. Rev. B.* **1991**, *43*, 4667.

[19] *Standard Potentials in Aqueous Solution*; Bard, A. J.; Parsons, R., Jordan, J., Eds.; Marcel Dekker, Inc.: New York, 1985.

[20] Clavilier, J. *J. Electroanal. Chem.*, **1980**, *107*, 211.

[21] Molina, F. V.; Parsons, R. *J. Chim. Phys.*, **1991**, *88*, 1339.

[22] Lynch, M. L.; Barner, B. J.; Corn, R. M. *J. Electroanal. Chem.*, **1991**, *300*, 447.

[23] Marković, N.; Ross, P. N. (Jr.) *J. Electroanal. Chem.*, **1992**, *330*, 499.

[24] Melroy, O. R.; Toney, M. F.; Borges, G. L.; Samant, M. G.; Kortright, J. B.; Ross, P.N.; Blum, L. *J. Electroanal. Chem.*, **1989**, *258*, 403.

[25] Wang, J.; Ocko, B. M.; Davenport, A. J.; Isaacs, H. S. *Phys. Rev. B.* **1992**, *46*, 10321.

[26] Tidswell, I. M.; Lucas, C. A.; Marković, N. M.; Ross, P. N. *Phys. Review B.* **1993**, *47*, 16542.

[27] Lucas, C. A.; Marković, N. M.; Ross, P. N. *Surf. Sci.*, **1995**, *340*, L949.

[28] Bittner, A. M.; Wintterlin, J.; Berran, B.; Ertl, G. *Surf. Sci.*, **1995**, *335*, 291.

[29] Tanaka, S.; Lin, S-L.; Itaya, K. *J. Electroanal. Chem.*, **1995**, *396*, 125.

[30] Bondy, A. *J. Phys. Chem.*, **1964**, *68*, 441.

[31] Bauer, E. *Surf. Sci.*, **1967**, *7*, 351.

[32] Lucas, C. A.; Marković, N. M.; Ross, P. N. *Phys. Review B.* submitted.

Chapter 8

Electron Spectroscopy Studies of Acidified Water Surfaces

D. Howard Fairbrother[1], H. S. Johnston[2], and G. A. Somorjai[2,3]

[1]Lawrence Berkeley National Laboratory, University of California,
1 Cyclotron Road, Berkeley, CA 94720
[2]Department of Chemistry #1460, University of California,
Berkeley, CA 94720

Acidified water systems, composed principally of nitric and/or sulfuric acid, are catalytically active in many significant heterogenous atmospheric processes (1). Since the initial step in any of these reactions involves adsorption from the gas phase, the surface composition in these systems is important in governing the accommodation coefficient and consequently the overall rate and ultimate reaction pathway. Using modified surface sensitive spectroscopic probes it is possible to monitor the composition of one of these catalytically significant surfaces, notably the H_2SO_4/H_2O binary system, under stratospherically relevant conditions, notably $P_{H2O} \approx 10^{-6}$ - 10^{-4} Torr and T = 180-240K (2). Analysis reveals that under these conditions the chemical composition of the bulk and near surface region are, within experimental error, identical. This result has been interpreted on the basis that in this system the surface composition is much more strongly controlled by complex compound formation than by segregation induced by surface tension differences between the chemical constituents.

A custom designed ultra-high vacuum chamber incorporating differentially pumped Auger electron (AES) and X-ray photoelectron spectrometers (XPS) has been designed to perform under partial pressures extending as high as 10^{-4} Torr. Both surface sensitive spectroscopies have been calibrated on pure sulfuric acid solutions to illustrate their ability to provide accurate atomic surface stoichiometries. In the case of AES this ability has also been extended for the study of the surface composition in the H_2SO_4/H_2O binary system under equilibrium conditions. Compositional analysis of this acidified surface indicates that under these experimental conditions the chemical composition of the bulk and near surface region are equal within experimental error (10-15%). The attainment of equilibrium in this system has also been verified using exchange between D_2O and H_2SO_4 detected by mass spectrometry.

[3]Corresponding author

Stratospheric Significance of the Sulfuric Acid/Water Surface

Research on the chemistry associated with stratospheric aerosol particles has intensified since the realization that these systems play a catalytic role in many stratospheric processes including those that contribute to the formation of the "ozone hole" above antarctica (3). Specifically, stratospheric sulfuric acid aerosols (SSAs), composed of 60-80 wt % sulfuric acid in the temperature range 205-240K, can affect the overall ozone concentration in these regions by acting as catalysts, for example, in the reaction between N_2O_5 and water to form two molecules of nitric acid (4). As such the composition of SSAs is expected to play a key role in determining their ability to promote heterogeneous reactions. For example, the reaction probabilities of $ClONO_2$ and N_2O_5 on sulfuric acid surfaces has been shown to vary significantly as a function of acid composition in the 40-75 wt % range (5). In addition the solubilities of other acids such as HCl and HNO_3 are found to increase with decreasing sulfuric acid content (6). Since any stratospheric process is initiated by adsorption of gas phase molecules at the surface the composition at the solid/liquid - gas interface is expected to be an important parameter in determining the overall catalytic efficiency of these systems. Although the bulk composition in many of these acidified water systems is well established, thermodynamic effects, such as surface tension, often result in a surface composition which differs markedly from that of the bulk (7).

Surface Sensitive Measurements Under Stratospheric Conditions

The essence of the experimental apparatus consists of a quadrupole mass spectrometer (QMS) for residual gas analysis as well as facilities for Auger electron and X-ray photoelectron spectroscopic measurements. This arrangement is shown schematically in Figure 1. A significant obstacle in obtaining surface specific information under equilibrium conditions prevalent in the stratosphere relates to the relatively high partial pressures ($< 10^{-4}$ Torr) compared to ultra-high vacuum (UHV) conditions typically used for surface analysis ($< 10^{-9}$ Torr). Thus in order to maintain the operational viability of both AES and XPS techniques under these "high" pressure conditions it is necessary to modify these spectroscopic probes.

AES measurements are performed using a double pass cylindrical mirror analyzer mounted at 60° from the surface normal. To reduce the partial pressure at the multiplier under experimental conditions a differential pumping stage is incorporated into the design in order to reduce the local pressure at the electron multiplier. Two sets of viton O-rings located between the shield and (a) the outer cylinder, and (b) the walls of the chamber serve to isolate the electron multiplier from the main chamber. Differential pumping is accomplished through two 1.25" diameter holes located at the base of the outer cylinder with a twin set of turbo molecular pumps. Under experimental conditions this arrangement maintains a differential pressure gradient of between one and two orders of magnitude between the main chamber and the electron multiplier. Complementary XPS measurements are carried out using a X-ray gun, mounted on a linear translation stage, inclined at 45° to the surface normal. In addition the whole XPS unit is also equipped with facilities for differential pumping with a diffusion pump attached to the X-ray source through a UHV gate valve (see Figure 1).

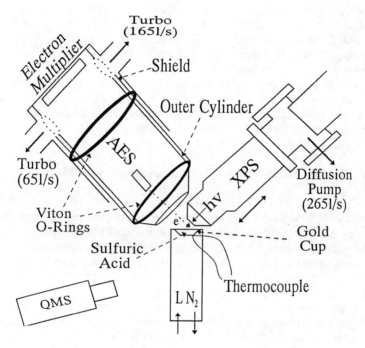

Figure 1. Schematic of experimental apparatus

Electron Based Spectroscopies on Liquid Samples. Typical AES on solid surfaces are recorded with approximately 2mA of beam current in conjunction with analogue detection. However, the use of such a large beam current on liquid samples usually results in a measurable temperature rise accompanied by an increase in the background pressure, the later arising from either thermal and/or electron stimulated desorption. In addition, sample charging is typically observed upon solidification. These undesirable effects have been avoided using incident beam currents of between 50-150nA coupled with pulse counting detection.

In Situ Preparation of Sulfuric Acid

Sulfuric acid samples (typically 1-3 cm^3) were prepared at room temperature and pressure by adding the liquid directly to a gold cup using a glass applicator (see Figure 1). The low vapor pressure of H_2SO_4, 1.4 x 10^{-5} Torr at S.T.P, (8) allows the system to be pumped down to a base pressure of $\approx$ 5 x 10^{-5} Torr, without bakeout. In order to prevent evaporation of sulfuric acid, which under these conditions is rapid at $\approx$ 320K (9), the sample temperature is maintained at $\approx$ 290K by maintaining a constant flow of cooled dry nitrogen through the manipulator. This procedure ensures that liquid sulfuric acid samples can be maintained for 3-4 days before a substantial fraction of the liquid sample is pumped away. By using a gold cup to hold the acid the potentially corrosive effects of the acidified system upon the stainless steel chamber and surroundings are also minimized. However, prolonged use of H_2SO_4 over the course of several months produces discoloration on various surfaces within the chamber due to the formation of a stable sulphate or oxide phase.

Experimental Calibration on Pure Sulfuric Acid. To ensure the validity of the surface science techniques employed it is important to establish a reliable stoichiometric calibration. Figure 2, which shows AES and XPS spectra of sulfuric acid under conditions of $P_{H2O} < 1$ x 10^{-5} Torr and T = 290K refers to a liquid which is > 99.99% pure sulfuric acid (10). Consequently the measured S:O ratio, $(S/O)_{exp}$, should reflect the inherent 1:4 S:O stoichiometry, $(S/O)_{stoich.}$, modified by the influence of the relevant AES and XPS sensitivities (s_S and s_O) (11) in addition to the mean free paths (λ_S and λ_O) associated with the secondary electrons of these two chemical elements (12). This can be expressed in the form:

$$\left(\frac{S}{O}\right)_{exp} = \left(\frac{S}{O}\right)_{stoich.} \left(\frac{s_O}{s_S}\right) \frac{1 - (\exp(-1 / \lambda_S \sin \theta))}{1 - (\exp(-1 / \lambda_O \sin \theta))} \tag{1}$$

where the $\exp(-1/(\lambda \sin \theta))$ term represents the relative contribution from species below the surface. The $\sin \theta$ term accounts for the fact that the take-off angle (θ) for both AES and XPS is off-normal ($< 90°$) so as to enhance the inherent surface sensitivity of both techniques. Using eqn. (1), the S:O ratio as determined from the

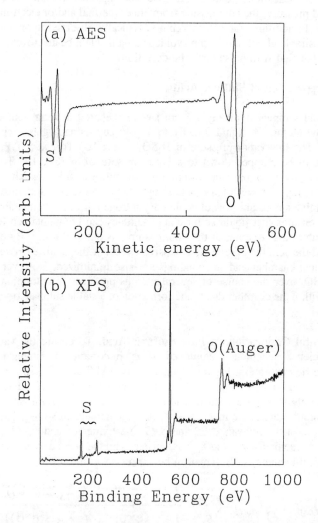

Figure 2. Auger and X-ray photoelectron spectra of pure sulfuric acid taken at room temperature and a total pressure of $\approx 1 \times 10^5$ Torr. AES was acquired with $E_{primary} = 3$KeV and beam current of ≈ 100nA; XPS was recorded using Mg(Kα) radiation (1253.6eV) with a pass energy of 100eV.

AES data ($\theta=30°$) shown in Figure 2(a), is calculated to be 1:3.8, very close to the stoichiometric value of 1:4. An analogous determination based on the XPS data ($\theta=45°$) in Figure 2(b) yields a S:O stoichiometry of 1:4.2. In addition the pressure invariant AES S:O signal observed up to $P_{H2O} = 1 \times 10^{-4}$ Torr indicates that none of the AES measurements are influenced by a contribution from background gas phase water molecules. Taken in conjunction, the good agreement ($\pm 5\%$) between these experimentally determined and theoretically predicted S:O stoichiometry values indicates that both AES and XPS are capable of providing reliable S:O values (Fairbrother, D. H.; Johnston, H. S.; Somorjai, G. A. to be published *J. Phys. Chem.*).

Degree of Surface Sensitivity. Analysis of the $(1 - \exp(-1/(\lambda \sin \theta)))$ term reveals that for AES both S and O over 80% of the observed signal originates from the first three layers below the surface (this calculation is based upon the equation, $S(n) = \exp(-n/ \lambda_{o,s} \sin 30°)$, where $S(n)$ represents the relative signal contribution from species n layers below the surface ($n=0$) and $\lambda_{o,s} = 4.745$ and 2.612 from (*12*)).

Surface Composition Under Stratospheric Conditions

As the sulfuric acid (H_2SO_4) sample is cooled below 250K in the presence of a constant partial pressure of water ($P_{H2O} = 1 \times 10^{-6} - 1 \times 10^{-4}$ Torr) the uptake of water (H_2O) into the sulfuric acid,

$$H_2O_{(g)} \rightleftharpoons H_2O_{(s/l)} \qquad (2)$$

results in a reduction in the relative mole fraction of S present at the interface, and hence a decrease in the measured S:O AES ratio, $(S:O)_{exp}$. Using eqn. (1) the $(S:O)_{exp}$ values can be converted into their corresponding $(S:O)_{stoich}$ values. This in turn allows the mole fraction of acid at the interface, $x(H_2SO_{4(surf)})$ to be determined from the stoichiometric S/O ratios, $(S/O)_{stoich.}$, utilizing the different chemical composition of the two components, H_2O and H_2SO_4, thus;

$$(\frac{S}{O})_{stoich.} = \frac{x(H_2SO_{4(surf)})}{4x(H_2SO_{4(surf)})+(1-x(H_2SO_{4(surf)}))} \qquad (3)$$

Figure 3 shows a plot of the variation in the measured AES S:O peak-to-peak ratio as a function of surface temperature measured at constant water partial pressures of (a) 1×10^{-4} (solid circles) and (b) 3×10^{-6} Torr (open circles).

Verification of Dynamic Equilibrium. To ensure that the measurement outlined in the previous paragraph was taken under conditions of dynamic equilibrium, mass spectrometry can be employed to monitor exchange between D_2O and H_2SO_4. This process has the general form;

$$D_2O_{(g)} + H_2SO_{4(l/s)} \rightleftharpoons HDO_{(g)} + HDSO_{4(l/s)} \qquad (4)$$

A comparison of the H_2SO_4/D_2SO_4 cracking patterns (*13*) in the 75-85 a.m.u region shown in Figure 4 indicates that exchange can be monitored by following the peak ratio at 81 (HSO_3^+) and 82 (DSO_3^+), specifically using D_2O incorporation into

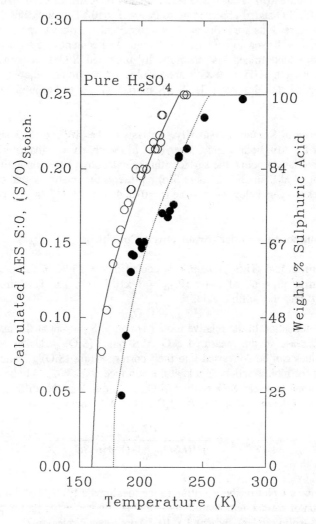

Figure 3. Calculated S:O ratio (left hand axis), as well as weight percentage of sulfuric acid (right hand axis), for the H_2SO_4/H_2O system measured as a function of sample temperature, with a constant water partial pressure of 1 x 10^{-4} and 3 x 10^{-6} Torr, represented by solid and open circles respectively. Also shown are the theoretically predicted bulk values for both, 1 x 10^{-4} Torr (solid line) and, 3 x 10^{-6} Torr (dotted line) calculated from (10).

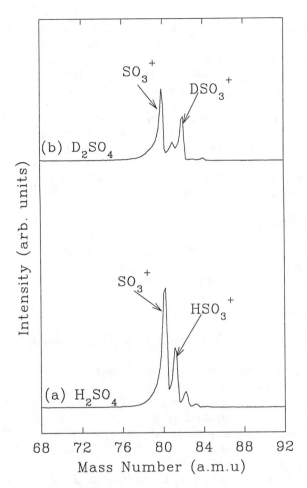

Figure 4. Mass spectrum of (a) H_2SO_4 and (b) D_2SO_4 (75-85 amu) recorded using an impact energy of 70eV.

H_2SO_4. Experiments carried out with $P_{D2O} = 1 \times 10^{-4}$ Torr at H_2SO_4 sample temperatures of 200, 210 and 220K revealed that after 30 minutes of equilibration there was a concomitant increase in the mass 66: mass 65 ratio. This is taken as evidence for exchange between D_2O and H_2SO_4 as the result of an equilibration of the general form shown in eqn. (4). The establishment of equilibrium can also be verified by confirming that the S:O ratio can be increased or decreased by raising or lowering the sample temperature respectively, under conditions of constant water partial pressures.

Surface vs Bulk Composition

Bulk compositional values can be determined as a function of temperature and water partial pressure from (*10*). These results are also shown as solid and dotted lines in

Figure 3 and, when compared to the experimental data, indicate that over the range of temperatures and pressures employed in this study, the composition of the near surface region is virtually identical to that of the bulk. This result is in contrast to that predicted purely on the basis of surface tension of the pure components, which in this case would give rise to considerable surface segregation where one component is in considerable ($> \approx 70\%$) excess. Consequently this result implies that, at least in this system, surface tension effects are not the dominant thermodynamic driving force controlling surface composition. Indeed a recent thermodynamic study (*14*) has shown that the surface concentration of water as free H_2O is much more strongly controlled by complex formation rather than any surface tension effects. Consequently, the surface composition can be predicted in this system, at least to a first approximation, from the bulk phase diagram.

Acknowledgments

The authors would like to acknowledge support of this work by the National Science Foundation under Grant No. ATM-9412445 and the U.S. Department of Energy under Contract No. DE-AC03-76SF00098.

Literature Cited

(*1*):Solomon, S.; Garcia, R. R.; Rowland, F. S.; Wuebbles, D. J. *Nature* **1986**, *321*, 755.
(*2*):Steele, H. M.; Hamill, P. *J. Aerosol Sci.* **1981**, *12*, 517.
(*3*):Solomon, S. *Rev. Geophys.* **1988**, *26*, 131.
(*4*):Rodriguez, J. M.; Ko, M. K. W.; Sze, N. D. *Nature* **1991**, *352*, 134.
(*5*):Hanson, D. R.; Ravishankara, A. R. *J. Geophys. Res.* **1991**, *96*, 17,307.
(*6*):Zhang, R.; Woolridge, P. J.; Molina, M. J. *J. Phys. Chem.* **1993**, *97*, 8541.
(*7*):Somorjai, G. A. *An Introduction to Surface Chemistry and Catalysis, Wiley-Interscience*, New York, NY, 1994.
(*8*):Jaecker-Voirol, A.; Ponche, J. L.; Mirabel, P. *J. Geophys. Res.* **1990**, *95*, 11857.
(*9*):Guldan, E. D., Schindler, L. R.; Roberts, J. T. *J. Phys. Chem.* **1995**, *99*, 16059.
(*10*):Zhang, R. J.; Wooldridge, P. J.; Abbatt, J. P. D.; Molina, M. J. *J. Phys. Chem.* **1993**, *97*, 7351.
(*11*):ESCA Operator's Reference Manual. ESCA Version 4.0 and Multi-Technique Version 2.0, Perkin-Elmer Corporation, Physical Electronics Division, Eden Prairie, MN, 1998.
(*12*):Riviere, J. C. *Surface Analytical Techniques*, Oxford University Press, Oxford, 1990.
(*13*):Snow, K. B.; Thomas, T. F. *Int. J. Mass Spectrom. Ion Phys.* **1990**, *96*, 49.
(*14*):Phillips, L.F. *Aust. J. Chem.* **1994**, *47*, 91.

Chapter 9

Electrochemical Digital Etching: Atomic Level Studies of CdTe(100)

T. A. Sorenson, B. K. Wilmer, and J. L. Stickney[1]

Department of Chemistry, University of Georgia, Athens, GA 30602

Atomic level control in the etching of CdTe(100) is being investigated, in an attempt to develop an electrochemical digital etching procedure. In principle, surface atoms on the crystal should show higher reactivity then those contained on the interior, due to their decreased coordination. Electrochemical oxidation in 50 mM K_2SO_4, resulted in removal of the surface Cd atoms and a tellurium enriched surface, as observed by Auger electron spectroscopy. Subsequent reduction at -1.8 V reduced the surface excess of Te, and returned the surface composition to stoichiometric. Selection of an appropriate potential for the oxidation of surface cadmium atoms was complicated by the observation that bulk CdTe is oxidized at potentials close to that used to oxidize the surface cadmium atoms.

Etching of CdTe single crystals can be categorized as dry etching and wet etching. Dry etching is performed using techniques such as argon ion sputtering in ultrahigh vacuum (UHV) (1,2) and plasma etching using CH_4/H_2 (3,4). Wet etching has been demonstrated using a variety of solutions including: $HNO_3/K_2Cr_2O_7/Ag^+$ (5), Ce^{4+} (6,7), $S_2O_4/NaOH$ (8), 15% $HNO_3(8)$, $Cr_2O_7^{2-}/HNO_3(1,8)$ and Br_2/CH_3OH (1,9). Interest in CdTe for solar energy conversion has led to a number of studies of the CdTe/electrolyte interface (8,10-12), and development of photoelectrochemical etching (13-15). In general, the above studies focused on macroscopic etching as a microfabrication process or surface cleaning technique for CdTe.

As the dimensions of semiconductor structures approach the atomic scale, atomic level control becomes increasingly important. With regards to etching, a technique known as digital etching has begun to be studied (16-22). Digital etching makes use of surface limited reactions to etch a material one atomic or molecular layer at a time. Gas phase digital etching generally involves an initial surface limited adsorption step, of a precursor gas. The resulting surface is then subjected to a beam of energetic particles: photons, electrons, ions, etc. The yield of volatile species resulting from the bombardment is then limited to a monolayer by the limited amount

---

[1]Corresponding author

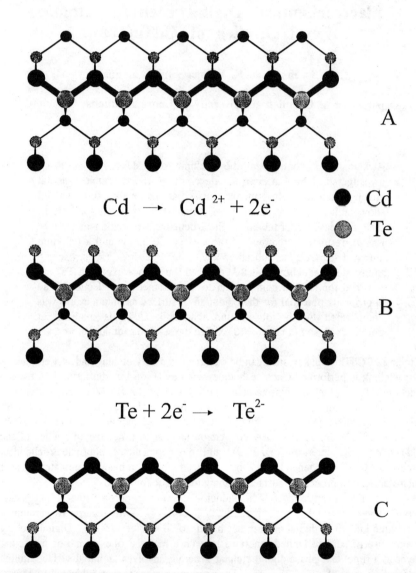

$$Cd \rightarrow Cd^{2+} + 2e^-$$

● Cd
◉ Te

$$Te + 2e^- \rightarrow Te^{2-}$$

Figure 1.   Schematic illustrating the electrochemical digital etching process on CdTe(100).

of preadsorbed gas. The process is repeated, in a cyclic manner, until the desired etch depth is achieved. Digital etching has been demonstrated on silicon and gallium arsenide using fluorine, chlorine, selenide, and hydrogen as gaseous adsorbates and Ar^+ ions, electrons, and photons (from XeF and KrF lasers) as energetic particles.

Research in our laboratory involves studies of the electrochemical atomic layer processing of compound semiconductors. A major project involves atomic layer epitaxy (ALE). ALE is the formation of a compound one atomic layer at a time using surface limited reactions, and the electrochemical analog of ALE (ECALE) is presently under development in this group. This study concerns development of the electrochemical analog of digital etching, where a compound is dissolved one atomic layer at a time (*23,24*). The present article is an atomic level study of electrochemical digital etching.

A schematic illustration of electrochemical digital etching is shown in Figure 1. The principle is that potentials are used for which only the top atomic layer of a particular element is electrochemically removed, in a surface limited reaction. Removal of just the top atomic layer is facilitated by those atoms being in a lower coordination state on the surface, than corresponding atoms in the bulk of the compound. If the surface is terminated by a layer of cadmium atoms (Figure 1A), careful selection of an oxidative potential should result in the anodic dissolution of the top cadmium atoms, while leaving a tellurium terminated surface (Figure 1B). Interior Cd atoms are not removed as they are tetrahedrally coordinated. The resulting tellurium terminated surface can then be reductively etched by selection of an appropriate reduction potential, leaving a cadmium terminated surface and completing one digital etching cycle (Figure 1C). The cycle is then repeated until the desired etch depth is achieved. The use of an electrochemical etching cycle, with both oxidation and reduction components has been well demonstrated by Kohl et. al., in the etching of InP (*25*). The present studies, however, focus on the development of a cycle based on removal of a single monolayer per cycle, using surface limited reactions.

Experimental

Undoped cadmium telluride single crystals, cut and oriented to the (100) plane, were obtained from II-IV Incorporated. Before use, the CdTe single crystals were chemically etched in 1% Br_2/CH_3OH. An ohmic contact was established by soldering an indium strip on the top of the crystal. The crystal was held between two molybdenum plates, tightened by 0-80 bolts. Al foil was sometimes wrapped around the In strip to help with conduction and prevent the crystal from falling out. Tungsten wires were connected to the molybdenum bars to facilitate electrical contact to the CdTe(100) crystal and allow thermal annealing in the UHV surface analysis instrument.

All experiments in this paper were performed in an UHV surface analysis instrument equipped with a Phi model 3017 Auger electron spectroscopy subsystem and Phi model 15-120 low energy electron diffraction (LEED) optics. The instrument was also equipped with an electrochemical antechamber used to perform the etching experiments (*1,26,27*). A standard, in-house built, three electrode

Figure 2. LEED image showing (1X1) pattern of CdTe(100) that resulted from ion bombarding with 0.5 keV Ar$^+$ ions for 20 minutes followed by annealling for 5 minutes. Electron beam energy 52.9 eV.

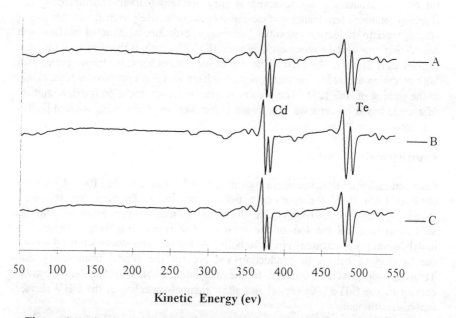

Kinetic Energy (ev)

Figure 3.   Auger spectra of CdTe(100) taken at various points in the electrochemical digital etching cycle.   A) Argon ion bombarded, annealed surface. Cd/Te = 1.32. B) Surface resulting from oxidation at +0.25V for 5 minutes. Cd/Te = 1.08. C) Surface resulting from reduction at -1.8V for 2 minutes. Cd/Te = 1.34.

potentiostat was used for all electrochemical experiments. All potentials are reported versus a Ag/AgCl(3M KCl) reference electrode obtained from Bioanalytical Systems, Inc. During electrochemical experiments, solution levels were kept at least 2 mm below the In strip and Mo plates. Problems with the electrolyte solution contacting the In strip and Mo plates were encountered but were easily recognized by a dramatic increase in the observed current. The electrolyte solution used was 0.050 M K_2SO_4 for all electrochemical experiments followed by a rinse in 1mM H_2SO_4, to remove excess electrolyte prior to transfer to vacuum. All solutions were made with Nanopure water and were purged with argon for at least fifteen minutes. All chemicals were high purity (Fisher) and used without further purification. The CdTe electrode was illuminated during oxidations using a Fiber-Lite Model 190 (Dolan-Jenner Industries, Inc.), with a W-halide bulb on high intensity, focused on the electrode through a viewport. All reductions were performed in the dark.

Results and Discussion

The surface structure of CdTe(100) has been the focus of a number of recent studies. A cadmium terminated CdTe(100) surface, for instance, displays a c(2X2) + (2X1) reconstruction (*28-34*), while a tellurium terminated surface displays a (2X1) reconstruction (*29-32*). Prior to each electrochemical study, the CdTe(100) crystal was ion bombarded and annealed, resulting in an unreconstructed (1X1) LEED pattern (Figure 2). Why a (1X1) is observed in the present studies, as opposed to a c(2X2) or a (2X1) is not clear, but may result from variations in the pretreatment procedures. The corresponding Auger spectrum (Figure 3A) shows a Cd/Te ratio of 1.32, calculated from the heights of the cadmium peak at 376 eV and tellurium peak at 483 eV. Figures 4A and B show voltammetry for the electrochemical reduction and oxidation, respectively, of this surface in 50 mM K_2SO_4. When the potential was scanned negative of open circuit in the dark, only a slight reductive current due to hydrogen evolution was observed, with no detectable change in the Cd/Te ratio. Scanning the potential positively, however, with the surface illuminated, resulted in an oxidation feature consisting of a small peak at about +0.1 V, on top of an exponentially increasing background. When the resulting oxidized surface was immersed in fresh electrolyte and scanned negatively in the dark, a surface limited peak, at approximately -1.65 V, was observed on a small background of hydrogen evolution (Figure 4C).

To characterize the oxidation process a series of experiments were performed in which the ion bombarded and annealed CdTe(100) crystal was immersed at increasingly positive potentials, for five minutes each time. Examples of the chronoamperograms obtained are shown in Figure 5 and are very similar in shape to those observed in previous studies of photoelectrochemical etching (*14*). Two features are evident, a steady state current that increases with increasing potential, and a transient current that decays to the steady state current after about 60 seconds, for potentials positive of -0.25 V. Auger spectra taken after each immersion indicated that Cd was preferentially dissolved, leaving a Te enriched surface. Figure 3B shows the Auger spectrum recorded following oxidation at +0.25 V, and a Cd/Te ratio of 1.08.

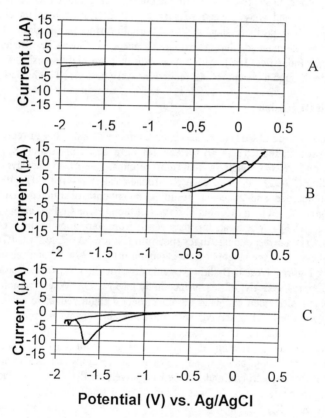

Figure 4. Voltammograms on an argon ion bombarded, annealed CdTe(100) surface in 50 mM K$_2$SO$_4$, pH = 5.6.   A) Reduction from the open circuit potential to -2.0V. B) Oxidation from the open circuit potential to +0.30V and reversing to -0.55V under illuminated conditions. C) Reduction following B) from open circuit potential to -1.8V and reversing to -0.50V.

Subsequent reduction of the Te enriched surface was accomplished by immersing the oxidized CdTe(100) single crystal in a fresh aliquot of the 50mM K_2SO_4 solution at -1.8 V. The resulting chronoamperogram is shown in Figure 6. Similar behavior was observed for Te reduction following the oxidation steps. That is, an initial reduction peak was observed, which decayed to background after 1 minute. Auger spectra obtained after these reduction steps all indicated that the initial stiochiometry of the CdTe(100) surface had been restored, that Cd/Te ratios were equivalent to those observed for the ion bombarded/annealled surface (Figure 3C).

Figure 7 shows the dependence of the charge passed, converted to monolayers of the corresponding elements, as a function of the potential used during the oxidation step. The total oxidation charge (triangles pointing down) increased dramatically between -0.25 V and -0.12 V. In addition, the charges for the transient (squares) and constant background (circles) features in Figure 5, are plotted in Figure 7. For potentials more positive than -0.25 V, charge due to the transient actually decreased as the potential increases, while the charge due to the constant background oxidation increased markedly. Charge passed in the subsequent Te reduction step at -1.8 V (triangles pointing up) followed the transient oxidation charge.

Given the data presented above, it appears that at potentials between open circuit (-0.5 V) and -0.25 V, cadmium is oxidized in a surface limited manner, with a 0.5 monolayer removed at -0.25 V. Reduction of the resulting Te layer was consistent with the amount of Cd removed, and restores the original surface. However, at more positive oxidation potentials, a second oxidation process began, the constant background oxidation, which appeared to involve oxidation of both Cd and Te. The potential for oxidative stripping of bulk tellurium from a gold electrode has been shown to begin around 0.2 V (*27,35*), possibly explaining the steep increase in charge at that potential in Figure 7. Why the total charge increases abruptly at -0.12 V is not clear, given a formal potential for Te of 0.2 V. On the other hand, the formal potential for Cd is close to -0.7 V, and though it is clear that Cd is stabilized by bonding with Te to form CdTe, by -0.125V a sufficient over potential should exist to strip Cd atoms away from Te atoms in the lattice, leaving the Te under-coordinated and destabilized. The resulting Te atoms should be suceptible to underpotential oxidation, explaining the loss of both Cd and Te.

Creation of a pitted morphology after several cycles of etching is indicated by examination of the LEED patterns after each step. No change in the symmetry of the LEED patterns was observed after either the oxidative or reductive steps, but each step resulted in an increase in diffuse intensity, indicating surface disordering or roughening. In addition, the mirror like finish, initially obtained with the Br_2/CH_3OH etch, was significantly less reflective after three or four experiments. It is quite possible that the co-dissolution of Cd and Te, at potentials above -0.25, is occurring more extensively at specific sites, rather than homogeneously across the surface.

Conclusion

An electrochemical method for digitally etching CdTe(100) is being investigated. In principle, two potentials are needed, one where surface atoms of Cd are oxidatively

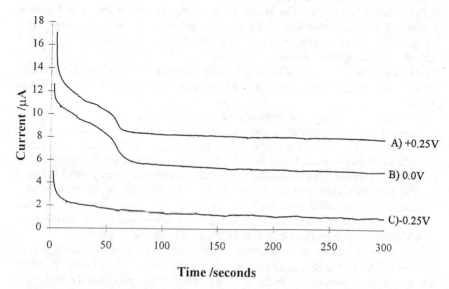

Figure 5. Chronoamperograms of ion bombarded, annealed CdTe(100) in 50 mM K_2SO_4 at different oxidative potentials for 5 minutes. A) +0.25V. B) 0.0V. C) -0.25V.

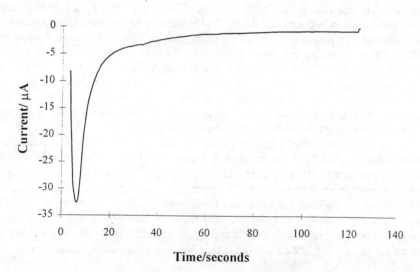

Figure 6. Chronoamperogram showing the reduction of the previously oxidized CdTe(100) surface in 50 mM K_2SO_4 at -1.8V.

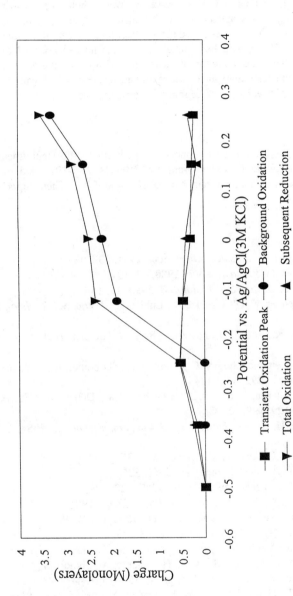

Figure 7. Graph showing the charge passed, converted to monolayers, as a function of potential used for oxidation. Total oxidative charge has been separated into two components: transient oxidation and background oxidation. In addition, the subsequent reduction charge for Te is listed as well.

removed in a surface limited reaction, and one where surface atoms of Te are reductively removed in a surface limited reaction. Finding an optimal potential for cadmium oxidation is complicated by the observation that some tellurium is oxidatively etched at potentials 0.3 V negative of where bulk Te is stripped (underpotential stripping). This appears to be due to destabilization of the Te by stripping the Cd atoms to which it was coordinated. Results from this study indicate that a potential near -0.25V would be optimal for surface limited removal of Cd without removal of the Te. The reduction of tellurium at -1.80V, on the other hand, has been shown to be an easily controlled surface limited reaction, which can restore the stiochiometry of the CdTe(100) surface following an oxidative step.

Acknowledgments

Acknowledgment is made for financial support of this work to the Department of Navy, Office of Naval Research, under Grant No. N00014-91-J-1919 and to the National Science Foundation under Grant No. DMR-94000570. Their support is gratefully acknowledged.

Literature Cited

1. Villegas, I.; Stickney, J. L. *J. Electrochem. Soc.* **1991**, 138, 1310.
2. Solzbach, U.; Richter, H. J. *Surface Sci.* **1980**, 97, 191.
3. Pearton, S. J.; Ren, F. *J. Vac. Sci. Technol. B* **1993**, 11, 15.
4. Neswal, M.; Gresslehner, K. H.; Lischka, K.; Lubke, K. *J. Vac. Sci. Technol. B* **1993**, 11, 551.
5. Lu, Y. C.; Route, R. K.; Elwell, D.; Feigelson, R. S. *J. Vac. Sci. Technol. A* **1985**, 3, 264.
6. Iranzo Marin, F.; Debeimme-Chouvy, C.; Vigneron, J.; Triboulet, R.; Etcheberry, A. *Jpn. J. Appl. Phys.* **1995**, 34, L1344.
7. Iranzo Marin, F.; Vigneron, J.; Lincot, D.; Etcheberry, A.; Debeimme-Chouvy, C. *J. Phys. Chem.* **1995**, 99, 15198.
8. Ricco, A. J.; White, H. S.; Wrighton, M. S. *J. Vac. Sci. Technol. A* **1984**, 2, 910.
9. Lincot, D.; Vedel, J. *J. Electroanal. Chem.* **1984**, 175, 207.
10. Lincot, D.; Vedel, J. *J. Cryst. Growth* **1985**, 72, 526.
11. Lincot. D.; Vedel, J. *J. Electroanal. Chem.* **1987**, 220, 179.
12. Tanaka, S.; Bruce, J. A.; Wrighton, M. S. *J. Phys. Chem.* **1981**, 85, 3778.
13. Tenne, R; Marcu, V.; Yellin, N. *Appl. Phys. Lett.* **1984,** 45, 1219.
14. Muller, N.; Tenne. R. *Appl. Phys. Lett.* **1981**, 39, 283.
15. Marcu, V.; Tenne, R.; Yellin, N. *Appl. Surf. Sci.* **1987**, 28, 429.
16. Horilke, Y.; Tanaka, T.; Nakano, S.; Iseda, H.; Sakaue, A.; Shindo, S.; Miyazaki, S.; Hirose, M. *J. Vac. Sci. Technol. A* **1990**, 8, 1844.
17. Ishii, M.; Meguro, T.; Gamo, K.; Sugano, T.; Aoyagi, Y. *Jpn. J. Appl. Phys.* **1993**, 32, 6178.
18. Bourne, O. L.; Hart, D.; Rayner, D. M.; Hackett, P. A. *J. Vac. Sci. Technol. B* **1993**, 11, 556.

19. Takatani, S.; Kikawa, T. *Appl. Phys. Lett.* **1994**, 65, 2585.
20. Meguro, T.; Hamagaki, M.; Modaressi, S.; Hara, T.; Aoyagi, Y.; Ishii, M.; Yamamoto, Y. *Appl. Phys. Lett.* **1990**, 56, 1556.
21. Luviksson, A.; Xu, M.; Martin, R. M. *Surface Sci.* **1992**, 277, 282.
22. Aoyagi, Y.; Shinmura, K.; Kawasaki, K.; Nakamoto, I.; Gamo, K.; Namba, S. *Thin Solid Films* **1993**, 225, 120.
23. Lei, Q. P.; Stickney, J. L. *Mat. Res. Soc. Symp. Proc.* **1992**, 237, 335.
24. Willmer, B. K. B.; *ECALE Processing of Compound Semiconductors and Scanning Probe Microscopy of Mg(0001)* Masters Dissertation, University of Georgia, 1995, pp. 1-50.
25. Kohl, P. A.; Harris, D. B.; Winnick, J. *J. Electrochem. Soc.* **1990**, 137, 3315.
26. Soriaga, M. P.; Harrington, D. A.; Stickney, J. L.; Wieckowski, A. In *Ultrahigh-Vacuum Surface Analytical Methods in Electrochemical Studies of Single Crystal Surfaces*; Conway, B. E.; Bockris, J. O'M.; White, R. E., Eds. Modern Aspects of Electrochemistry, Number 28; Pluenum Press, New York, 1995; pp.1- 60.
27. Suggs, D. W.; Stickney, J. L. *Surface Sci.* **1993**, 290, 362.
28. Veron, M. B.; Sauvage-Simkin, M.; Etgens, V.H.; Tatarenko, S.; Van Der Vegt, H. A.; Ferrer, S. *Appl. Phys. Lett.* **1995**, 67, 3957.
29. Tatarenko, S.; Bassani, F.; Klein, J. C.; Saminadayar, K.; Cibert, J.; Etgens, V. H. *J. Vac. Sci. Technol. A* **1994**, 12, 140.
30. Tatarenko, S.; Daudin, B.; Brun, D.; Etgens, V. H.; Veron, M. B. *Phys. Rev. B* **1994**, 50, 18479.
31. Wu. Y. S.; Becker, C. R.; Waag, A.; Bicknell-Tassius, R. N.; Landwehr, G. *J. Appl. Phys.* **1991**, 69, 268.
32. Wu, Y. S.; Becker, C. R.; Waag, A.; Kraus, M. M.; Bicknell-Tassius, R. N.; Landwehr, G. *Phys. Rev. B* **1991**, 44, 8904.
33. Seehofer, L.; Etgens, V. H.; Falkenberg, G.; Veron, M. B.; Brun, D.; Daudin, B.; Taterenko, S.; Johnson, R. L. *Surface Sci.* **1996**, L55.
34. Seehofer, L.; Falkenberg, G.; Johnson, R. L.; Etgens, V. H.; Tatarenko, S.; Brun, D.; Daudlin, B. *Appl. Phys. Lett.* **1995**, 67, 1680.
35. Gregory, B. W.; Suggs, D. W.; Stickney, J. L. *J. Electrochem. Soc.* **1991**, 138, 1279.

Chapter 10

Adsorption of Bisulfate Anion on the Au(111), Pt(111), and Rh(111) Surfaces

A Comparative Study

S. Thomas[1], Y.-E. Sung[1,3], and Andrzej Wieckowski[1,2,4]

[1]Department of Chemistry and [2]Frederick Seitz Materials Research Laboratory, University of Illinois at Urbana-Champaign, 600 South Mathews Avenue, Urbana, IL 61801

We report on adsorption of bisulfate anion on the Au(111), Pt(111), and Rh(111) electrodes in sulfuric acid media using electrochemistry, Auger electron spectroscopy, low energy electron diffraction and core-level electron energy loss spectroscopy. The key observations for the bisulfate adsorbate on all surfaces studied are: (i) the oxygen-to-sulfur ratio is 4, (ii) the S(LMM) Auger electron transitions and S2p energy loss spectra show signatures characteristic of the sulfate anion, (iii) the spectral data are typical of the S^{6+} valency in the adsorbate. These results indicate that no decomposition or dehydration of the adsorbed anion occur in UHV. On Pt(111) and Rh(111) the maximum anion coverage is very close to 1/3 of the monolayer (ML). This is in an excellent agreement with the Me(111)($\sqrt{3}$ x $\sqrt{3}$)R30° from electron diffraction. For Au(111), the maximum coverage is 1/5 ML while the diffraction data indicate a diffuse Au($\sqrt{3}$ x $\sqrt{3}$)R30° structure. Whereas in the specific case of the Au(111) electrode the low energy electron diffraction data provide only an indirect basis for structural assignment of the sulfate adsorbate, we may conclude from these data that the structure of bisulfate on gold is different from that on Pt(111) and Rh(111). The core level energy loss spectra show that the electron density on surface sulfur is higher than in the salt, evidently due to backdonation of electrons from the substrate to the adsorbate. To complete the description of the anion-surface chemical bond requires the assessment of the electron donation contribution to the bond. This will be interrogated in a separate study.

Anions, together with cations and solvent molecules, are building blocks of the boundary layer that develops in the interface between metal and electrolytic solution. When specifically adsorbed, they alter the charge distribution at the interface and the

[3]Current address: Department of Chemistry and Biochemistry, University of Texas, Austin, TX 78731
[4]Corresponding author

electronic structure of the substrate, thereby influence the electrochemical processes at the surface (*1-7*). More specifically, they have a major influence on underpotential deposition processes as, for instance, in copper deposition on gold where anion coadsorption affects both reaction kinetics and the growth mode (*5*). Also, some anions are mild corrosion inhibitors in metal/alloy corrosion (*8*). More recently it has been shown that oxidation rates of methanol on Pt(111) vary by an order of magnitude between perchloric acid and sulfuric acid solutions (*7*). Therefore, a better understanding of the anion interactions with metal electrodes is imperative in making further inroads in interfacial electrochemistry as well as in the fields of electrocatalysis and corrosion.

Advancements made in recent years in the development of both in-situ and ex-situ techniques of electrode surface characterization have led to better understanding of the adsorbate structure and the nature of adsorbate/substrate interactions. Due to the unique role platinum (*9-27*) and rhodium (*28-37*) play in electrocatalysis, electrodes made of these metals have received a considerable attention. Gold has also been actively investigated as a model substrate for studying weakly bonded, electrode-adsorbate systems. In this article, we present a comparative study of bisulfate adsorption on the (111) single crystal faces of gold, platinum and rhodium. Even though we are dealing with a single anion as the adsorbate precursor, bisulfate, many of the findings in this study may be generalized to other "specifically" adsorbed anions. We used Auger Electron Spectroscopy (AES), Low Energy Electron Diffraction (LEED) and Core-Level Electron Energy Loss Spectroscopy (CEELS). The first two techniques provided mainly a combined compositional and structural insight. With CEELS, the technique that has received much attention in our studies, we focused on the loss spectra from S2p and Pt4f levels, and on the effect of electrode potential on the loss spectra. In the energy loss spectroscopy a monochromatic beam of electrons interact with the inner shell electrons of the target atoms, and lifts the core level electron to a final (empty) state above the Fermi level (*38-44*). The loss energy, which is the difference between the primary electron energy and the energy of the outgoing electrons, can then be monitored to obtain information on the electronic environment and the chemical state of the atoms involved. By keeping the energy of the primary electrons low, 500 eV in our studies, one increases the cross-section for the loss events and hence, surface sensitivity. Since one probes only the difference in the energy levels between the core-level states and the empty electronic states in the excited adsorbate, the results are not affected by charging effects and work function changes. As a result, loss energy shifts provide genuine information on surface/adsorbate chemical states, and on the effect of electrode potential on such states. Therefore, on top of our structural considerations presented below, we use CEELS data to comment on the nature of electronic-level interactions between surface anions and the substrate surfaces.

Experimental

All surfaces were mechanically polished by a diamond paste (Buehler) to 0.25 μm, and the (111) orientation was confirmed by X-ray back reflection to be correct within ± 0.5°. The crystals were then positioned in a vacuum-electrochemistry transfer system,

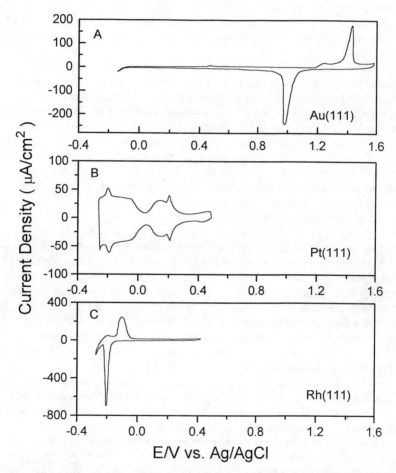

Figure 1. Cyclic Voltammograms of UHV prepared (A) Au(111) (B) Pt(111) and (C) Rh(111) electrodes in 50 mM H_2SO_4 solution. Scan rate is 50 mV/s.

as reported earlier (*45*). The Au sample was cleaned by 340 eV Ar^+ ion bombardment and annealing at 650 K, until the Auger electron spectrum and LEED showed a clean and ordered sample. Rh and Pt were cleaned by repeated Ar^+ ion bombardment (IBB) at 1 keV, 6 μA ion current at 5 x 10^{-5} Torr pressure for 15 minutes. The samples were then annealed at 800 °C for 20 minutes in a 5 x 10^{-8} Torr oxygen atmosphere, and for another 5 minutes in UHV without oxygen. The IBB/annealing cycle was repeated until the appropriate order and cleanliness of the surface was confirmed by LEED and Auger electron spectroscopy. After obtaining a clean and ordered surface, the electrodes were transferred to an air isolated electrochemical chamber (*44*), the chamber was brought to atmospheric pressure with ultra-pure argon, and -- using a meniscus-type configuration of the cell (*1*) -- the crystal surface was connected to electrolytic solution for voltammetric characterization and adsorption studies. Following electrochemistry, the chamber was isolated, pumped down first with a sorption pump, and then with a liquid helium cryopump.

AES measurements were carried out in a differentiated mode with a 2 eV modulation amplitude, at either 3 or 0.5 keV of primary electron energy, and 0.5 μA sample current, using a Perkin Elmer (PHI) 10-155 cylindrical mirror analyzer (CMA). The analysis was conducted at a low beam current to minimize electron beam damage. Spectra were acquired using a digital data acquisition system, and smoothed one time using a 11 point averaging technique..

Quantitative AES analysis was carried out using a standardization technique developed in our laboratory (*38*). Namely, a dry, thin layer of sulfate deposit was obtained on the Me(111) template from a 0.3 M Na_2SO_4 solution and subsequent water evaporation. This Na_2SO_4 covered Me(111) was used as a standard for work with monolayer (bi)sulfate adlattices. The procedure involved a comparison of the peak-to-peak (p/p) intensities of sulfur and oxygen at 131 eV and 516 eV, respectively, relative to the Me p/p (Au 69 eV, Rh 302 eV, Pt 64 eV) intensity of the clean Me(111) sample. As mentioned above, the chemical state of the adsorbate was interrogated by the Core Level Energy Loss Spectroscopy. The loss energies reported here were measured relative to the electron elastic peak of 500 eV electrons.

The electrochemical measurements were carried out at room temperature using a conventional three-electrode cell. An EG&G PAR 362 potentiostat was used along with a BAS X-Y recorder for the voltammetric data acquisition and potentiostatic procedures. The potentials are given with respect to a silver/silver chloride in reference with [Cl$^-$] = 1 M. Working solutions were made of Millipore water (18 MΩ·cm) and ultra-pure grade sulfuric acid (Ultrex from VWR). Solutions were deaerated and blanketed with nitrogen (Linde, Oxygen Free, 99.99%). The Na_2SO_4 solution was made of anhydrous Fisher Scientific ACS certified reagent and the NaHS solution with Johnson Matthey anhydrous reagent.

Results

Voltammetry. Cyclic voltammetric curves for the three surfaces investigated are shown in Figure 1. They are in a very good agreement with earlier data (*4, 46-53*). As already agreed upon (*54-56*), the voltammetric activity on the Pt(111) electrode in the potential ranges from -0.25 to 0.05V and from 0.05 to 0.25V corresponds to

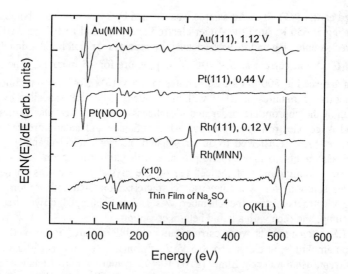

Figure 2. Auger Electron Spectra (AES) of the (bi)sulfate covered Au(111), Pt(111) and Rh(111) electrodes. Also shown is the Auger spectrum from a thin film of Na_2SO_4 deposit formed on Pt(111). Primary beam energy was 3 keV. Major Auger transitions from Au (MNN), Pt (NOO), Rh (MNN), S (LMM) and O (KLL) are shown in the spectra.

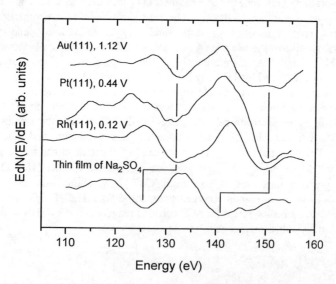

Figure 3. Spectral AES S (LMM) region of the samples shown in Figure 2. Notice similarities between the spectra of the (bi)sulfate adlattice and the thin film of Na_2SO_4.

hydrogen and bisulfate adsorption, respectively. The hydrogen adsorption range does not overlap the HSO_4^- adsorption range; except at the highest concentrations of sulfuric acid investigated, where it overlaps to a small degree (*1, 57*). This is to a clear contrast to what we observe with Rh(111), Figure 1C, where bisulfate and hydrogen adsorption overlap in the hydrogen range of the electrode potentials completely (*58-61*).

AES Studies. Figure 2 shows the 50 - 550 eV region of the Auger electron spectra for the three electrodes emersed from the 5×10^{-4} M H_2SO_4 solution, along with the spectrum of the thin layer of Na_2SO_4. In addition to the spectral features characteristic of the clean metal substrate, three Auger transition peaks are: peak (1) at ~122 eV, peak (2) at ~132 eV, and peak (3) at ~152 eV, overlapped with the Au 150 eV peak (Figure 3). The spectra shown in Figure 3, in essence S(LMM) sulfur fingerprints, are very similar to each other, and are in a good agreement with the spectrum from the thin Na_2SO_4 film (*62, 63*). The shift toward higher kinetic energy in the surface sulfate vs. that measured with the Na_2SO_4 deposit is due to the change in extramolecular relaxation energy, as reported earlier (*64*). Therefore, we may clearly assign the spectral morphology obtained with the emersed electrodes to an adsorbed sulfate (or bisulfate) anion, and conclude that the oxidation state of sulfur in the adsorbate is +6 (see Discussion section).

For Au(111), like for the remaining two (111) surfaces, the AES measurements with the adsorbate emersed below the electrode potential of 1.00 V show that the oxygen-to-sulfur ratio is equal to 4. Additional surface oxygen on gold was found in the potential range of 1.00 to 1.12 V. The adsorption of the excessive oxygen does not lead to sulfate desorption, as inferred from the AES sulfate sulfur data (*4*). The oxygen clearly occupies interstitial surface sites, that is, those not used by adsorbed anion, confirming earlier conclusions by Angerstein-Kozlowska et al. (*49*). The oxygen adatoms not chemically associated with sulfate attain as much as 0.20 ML coverage before sulfate desorption. Obviously, the O/S ratio further increases when the sulfate anion begins to desorb.

The Pt(111) sample emersed from sulfuric acid solutions (or from perchloric acid containing sulfuric acid solutions) exhibits an already reported Auger electron spectrum associated with bisulfate adsorption, Figures 2 and 3 (*1, 38*). Again, the Auger spectral morphology from the adsorbate is in excellent agreement with that obtained from the thin film of Na_2SO_4.

A characteristic Auger electron spectrum for the Rh(111) sample emersed from 50 mM H_2SO_4 is also shown in Figure 2. Three (bi)sulfate AES peaks characteristic of the S(LMM) region of the bisulfate adsorbate (Figure 3) are seen, as for Au and Pt. Apart from the shift in the AES peak positions (due to the extramolecular relaxation energy) (*64*), the spectral morphology of the electrode emersed from the H_2SO_4 solution compares very well with that of the Na_2SO_4 film.

For both Pt and Rh, the maximum sulfate coverage obtained by the quantitative AES is almost at 1/3 ML (0.32 ± 0.02 ML) that closely corresponds to the coverage data deduced from the ($\sqrt{3} \times \sqrt{3}$)R30° LEED pattern found in this study. For Au(111), the quantitative Auger analysis yields a maximum SO_4^{2-} coverage of 1/5 ML (0.20 ± 0.02 ML) which is in agreement with previous *in situ* radiochemical and

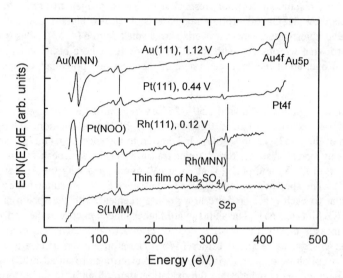

Figure 4. Auger spectrum from (bi)sulfate covered Au(111), Pt(111) and Rh(111) as well as that from a thin film of Na_2SO_4, measured from 40 - 450 eV using 500 eV primary beam electrons. In addition to the Auger transitions, electron energy loss peaks from S2p, Pt4f, Au4f, etc. can be seen.

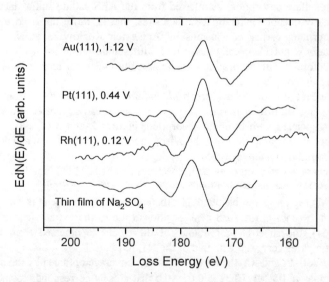

Figure 5. The S2p electron energy loss spectra of the (bi)sulfate adlattice for the samples shown in Figure 4. Note the similarity in spectral morphology between the loss peaks from the sulfate adsorbate and the thin film of Na_2SO_4.

chronocoulometric coverage (*46*). Besides structural significance of these data, the comparison indicates that there is no SO_4^{2-} loss upon transfer to UHV.

LEED and CEELS Data. The Au(111) electrode prepared in UHV showed a broad, integral (1 x 1) spots, indicating a reconstruction of the surface into the Au(111)-(1 x 23) structure (neither Pt(111) nor Rh(111) are subject to surface reconstruction in a broad temperature range) (*47, 48, 50-53*). After bisulfate adsorption on the Au(111) electrode and the electrode emersion of to UHV the measurements showed a weak Au(111)($\sqrt{3}$ x $\sqrt{3}$)R30° LEED pattern (*46*). This pattern was observed only at the maximum adsorbate coverage. Equivalent LEED measurements for both Pt(111) and Rh(111) produced sharp ($\sqrt{3}$ x $\sqrt{3}$)R30° LEED patterns. The corresponding, highly ordered ($\sqrt{3}$ x $\sqrt{3}$)R30° surface structures were obtained in a broad bulk concentration range of sulfuric acid.

The core level spectra obtained using 500 eV primary electron beam energy are shown in Figure 4 (together with the AES transitions). These loss peaks are derived from the S2p, Au4f and Pt4f core levels. Figure 5 shows the S2p loss regions of the adsorbate on the three metal surfaces along with S2p loss data from the thin layer of Na_2SO_4. The sulfate adlattices and the thick Na_2SO_4 deposit exhibit very similar spectral features. The loss peaks appear at 191 eV, 183.0 eV and 175.0 eV and are caused by the excitation of the S2p core electrons into empty states close to the vacuum level (*64, 65*). The principal sulfate loss, ~175.0 eV, can be compared to the 168.8 eV binding energy of the S2p core level measured for bulk Na_2SO_4 (with respect to the Fermi level) (*66*). The difference between S2s and S2p binding energies is 59.5 eV from XPS (*66*) and ~63 eV from CEELS. The S2p loss energy is by ~9 eV higher than the principal loss energy for NaHS, which is similar to the 8 eV shift seen with XPS. This agrees with the conclusions of Cazaux et al (*67*) as to an extent of the chemical shifts involved in electron energy loss spectroscopy.

Discussion

The adsorbed (bi)sulfate and the thin layer of sodium sulfate gave electron spectra with almost identical loss energy and Auger electron features. The electron spectra from (bi)sulfate monolayers are very close to each other, and to those from the thin Na_2SO_4 film, including spacing between the energy transitions (Figures 2 - 5). Therefore, we conclude that the oxidation state of sulfur in the (bi)sulfate adsorbate is +6, as in Na_2SO_4. The quantitative AES treatment shows that the oxygen-to-sulfur ratio remains at 4 until the substrate metal surface starts to oxidize. As potential is further increasing, the O/S ratio increases due to the incorporation of oxygen atoms into the surface and the adsorbate desorption.

The Au(111) surface gave the cyclic voltammetric curve (Figure 1A) with a sharp capacitive peak at 0.44 V associated with a structural transition of the reconstructed Au(111) surface into deconstructed (1x1) phase (*47, 48, 50-52*). In a narrow electrode potential range bracketing the sulfate adsorption maximum a poorly defined Au(111) ($\sqrt{3}$ x $\sqrt{3}$) R30° LEED pattern was found. Since the sulfate coverage is only around 2/5 ML vs. 1/3 ML expected for a regular ($\sqrt{3}$ x $\sqrt{3}$) R30° surface structure, we conclude that the structure we observe in UHV is not the same as that in electrochemical cell. The UHV diffraction results occur, most likely, from two-dimensional islands of a local ($\sqrt{3}$ x $\sqrt{3}$) surface structure. The formation of such islands would require a lateral attraction of the sulfate adsorbate whose charge is neutralized be the surface charge screening.

In order to appreciate the difficulties in the structural assignment one recalls that Magnussen et al. (47) and Edens et al. (48), using scanning tunneling microscopy (STM), found a $\left(\begin{smallmatrix} 2 & 1 \\ -1 & 2 \end{smallmatrix}\right)$ structure in the same narrow potential range as we identified the weak ($\sqrt{3}$ x $\sqrt{3}$) R30° pattern in UHV. According to the authors (47), a phase transition from a disordered, low coverage sulfate adlayer occurs into the $\left(\begin{smallmatrix} 2 & 1 \\ -1 & 2 \end{smallmatrix}\right)$ anion structure, with all of the STM maxima assignable to bisulfate adsorbate placed in *bridge* sites. This would require the coverage of 0.40 ML, not supported by the previous radiochemical, chronocoulometric and the present UHV data (46). However, Edens et al. (48) using combined STM and infra-red absorption spectroscopy) measurements concluded that sulfate anions were adsorbed in *on top* positions only. To reconcile the difference the authors (48) postulated that sulfate anion was coadsorbed with hydronium cations in the 1:1 ratio, and that the cation also appeared in the STM image. Since the 5th oxygen (from H_3O^+) was not found in UHV, we conclude that the hydronium ion cation, or water molecules engaged in the proton hydration, desorbs upon electrode emersion to UHV. This leaves the Au(111)($\sqrt{3}$ x $\sqrt{3}$) R30° islands available for the structural assay. The H_3O^+ desorption may be considered likely in view of the fact that the work function of the emersed electrodes becomes more positive due to the dehydration occurring upon system evacuation (68).

For bisulfate adsorption on the Pt(111) and Rh(111) surfaces, we have the ($\sqrt{3}$ x $\sqrt{3}$) R30° LEED patterns with sharp spots and coverages that are close to 1/3 ML. Therefore, we believe that the structure is Me(111)($\sqrt{3}$ x $\sqrt{3}$) R30° both in UHV and in the electrochemical cell. Since STM measurements conducted by other investigators gave the same $\left(\begin{smallmatrix} 2 & 1 \\ -1 & 2 \end{smallmatrix}\right)$ structure for Pt(111) (26) and Rh(111) (69, 70) as for Au(111), (4) we report that there is a conflict between LEED and STM data. Synchrotron X-ray structure determinations are needed to reconcile the disagreement.

All Auger electron spectra from the surface anion originate from S(LMM) Auger cross transitions (65, 71) and appear in the order of increasing energy: $L_{II}M_\sigma M_O < L_{II}M_\sigma M_\sigma < L_{II}M_\sigma M_n$. Here, M_n stands for non-bonding orbitals in SO_4^{2-}, predominantly localized on the oxygen atoms, M_σ, for bonding orbitals (between the sulfur and the oxygen atoms), and M_O, for O(2s) molecular orbitals (65, 72). We have already mentioned that the Auger electron energies of the adsorbate are ~ 9 eV higher than those in the Na_2SO_4 deposit. The extramolecular relaxation energy that accounts for this 9 eV difference is most likely due to effective screening of the electron holes formed in SO_4^{2-} during the Auger electron process by conduction electrons from the metallic substrate, as proposed before (1, 2, 4).

We have recently reported -- using the Pt(111) electrode -- that at low coverage of bisulfate the bisulfate energy loss is significantly lower than that from the sulfate anion in the salt matrix. The lowering of the S2p loss energy is due to a higher electron density on the sulfate sulfur as referenced to that in the salt. We believe that the decrease in the energy was mainly due to electron backdonation from the substrate atoms into the unfilled orbitals of the bisulfate adsorbate (*1*). The most probable arrangement is that the electron density from Pt 5d is donated to the empty orbitals of bisulfate (LUMO), which are mainly S3s and S3p in nature (*73*). Following the Pt5d-HSO$_4^-$ LUMO orbital interactions, the increased electron density around the S atom adds to the intramolecular electronic repulsion in the bisulfate, hence lowers the electron binding energies. There is also a σ donation from the HOMO of SO$_4^{2-}$ to Pt5d. (*74*). Recently, Sellers et al. (*73*) have employed *ab initio* calculations to understand the nature of the sulfate-surface bonding from a theoretical standpoint. Even though this study mainly focused on the Ag(111) and Au(111) surface, the general conclusions can be extended to other similar systems. In their conclusion, the electronic interaction between the substrate and the SO$_4^{2-}$ involved a charge transfer from the anion to the substrate and a backdonation of electrons from the substrate to the adsorbate. This results in an increased electron density around sulfur as shown by electron density difference maps. Ab-initio molecular orbital calculations conducted by Ito et al (*75*) indicate that the HOMO is mainly localized on the oxygen lone-pairs. These lone-pairs exhibit antibonding character in the O-S bond and are favored to form the bond with the metal surface. Therefore, the donation mechanism is quite likely. This is in agreement with Attard et al. (*76*) who have shown that the σ donation from the HOMO level in HSO$_4^-$ to the vacant Pt5d orbitals is supported by theoretical coupled-Hartree-Fock calculations.

We have also found that the S2p core loss was strongly influenced by the electrode potential (*1*). First, in the range from 0.0 to 0.40 V, a downshift in the loss energy was found. As the adsorption potential was increased, the loss energy stabilizes at 174.5 eV until the beginning of surface oxidation (0.6 V). Ultimately, the loss energy increased with potential and reached the value of 175.5 eV at 1.0 V. The Pt4f$_{7/2}$ loss energy spectra are also affected by the change in the electrode potential. Here, three distinctively different trends were found (*1*), a) no change in loss energy (73.2 eV) in the potential range from -0.20 to 0.40 V (in dilute sulfuric acid solutions), b) a gradual change where the Pt4f$_{7/2}$ loss energy increases from 73.2 to 73.5 eV at 0.80 V, and c) a sharp increase in the energy above 0.80 V, with the final value at 74.3 eV. Since there is no spectral change in the electrode potential range where bisulfate adsorption commences and develops we conclude that bisulfate adsorption does not change the 4f$_{7/2}$ energy level. Between 0.40 to 0.80 V, the surface electronic environment around a platinum atom is progressively changing due to OH adsorption, which is reflected by the increase in the Pt4f$_{7/2}$ loss energy.

Conclusions

Our AES, LEED and CEELS data indicate that bisulfate anion is undergoing specific adsorption on these three (111) surfaces, but the coverage and the adsorption potential vary considerably among these substrates. For Au(111), the maximum is only 1/5 ML, both from the quantitative AES, chronocoulometric and radiochemical studies. For Pt(111) and Rh(111), the coverage is close to 1/3 ML. The latter coverage is typical of the Me(111)($\sqrt{3} \times \sqrt{3}$) R30° surface structure which we confirmed by electron diffraction. On gold, the ($\sqrt{3} \times \sqrt{3}$)-type structure is poorly developed and we conclude that the electron diffraction originates from remnants of the $\left(\begin{smallmatrix} 2 & 1 \\ -1 & 2 \end{smallmatrix} \right)$ unit cell

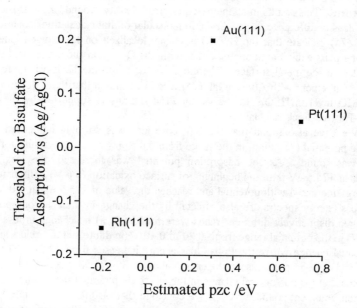

Figure 6.   Plot of the adsorption threshold of (bi)sulfate adsorption against the estimated pzc (in aqueous solution) of Au(111), Pt(111) and Rh(111).

formed in the electrochemical cell.   That is, we conclude that sulfate structure on Au(111) is different that on Pt(111) and Rh(111).   The apparent reason here is that the adsorbate on Au(111) is *sulfate* (*47*) in contrast to Pt(111) (*1*) and, most probably Rh(111), where the adsorbate is *bisulfate*.

While there is no atomistic view why we produce different adsorbates from the same bisulfate precursor in solution, we point out that the outer electronic configurations of Pt and Rh are $5d^9 6s^1$ and $4d^8 5s^1$, respectively, while that of Au is $5d^{10} 6s^1$.   Pt and Rh are the *d*-band metals while Au is a *sd* metal, whose behavior is in between that of *sp* and *d* metals.   The *d* band metals are hydrophilic and water molecules are expected to orient with the oxygen atom towards the surface (*77*, *78*).   Electroreflectance spectra, for instance, have shown that water molecules are chemisorbed on *d*-band metals with the partial injection of the oxygen lone pair electrons into the *d*-band of the substrate (*79*).   Such chemisorbed water, deprived of a fraction of its electron density, may have a weaker affinity for abstracting protons from bisulfate anion upon its primary interaction with the surface.   Therefore, on platinum (*1*), the bisulfate anion remains intact in the adsorbed state.   On the contrary, physisorbed water on Au(111) may have a potential for accepting protons and creating the SO_4^{2-}/H_3O^+ surface complex, as postulated by Edens et al. (*48*).   These preliminary conclusions are far from complete, they only indicate a direction of thought.   In fact, there are two competing pieces of experimental evidence as to the bisulfate surface structure on Pt(111) that need to be reconciled before further conceptual progress can be made, namely, the possibility of a coadsorption of H_3O^+ cation with sulfate anions also on Pt(111), as postulated by Faguy (*80*), or to the hypothetical recombination of bisulfate with hydronium cation to produce a neutral sulfuric acid molecule on Pt(111), as concluded by Ito (*81*).

The following data are of interest to this project.   The work function of the Au(111), Pt(111) and Rh(111) faces are 5.3 ($\pm$ ~0.2) (82-84), 6.1 eV ($\pm$0.06) (*85*) and 5.2 ($\pm$ ~0.2) eV (86-89), respectively.   Recall also that from among these three surfaces only Au(111) has no vacancies in *d* band.   The interaction of Au(111) with water will therefore be weak and the reduction in work function due to water adsorption small, much smaller that with Pt(111) and, most likely with Rh(111).   The corresponding literature data are -1.02 eV and -0.60 eV for Pt(111) and Au(111), respectively (*91*).   The difference in the potentials of zero charge (pzc) between Pt(111) and Au(111) is lower than the work function difference due to the electronic and chemical origin.   For Au(111) the experimentally determined pzc is 0.30 V vs. Ag/AgCl (*46*).   For Pt(111), Weaver et al. also derived 0.30V (vs Ag/AgCl) using the work function of 5.7 eV (*35*).   If the updated value for the work function is used, 6.1 eV (*85*), the previous Weaver's estimate would give the pzc of 0.70 V vs Ag/AgCl. The estimated pzc for Rh(111), assuming the same work function effect from water adsorption as with platinum, would be -0.2 V (vs. Ag/AgCl).   The starting potentials of bisulfate adsorption for Au(111), Pt(111) and Rh(111) are 0.20, 0.05, -0.15 V respectively (Figure 1, refs. 1, 2, 46).   Plotting formally the adsorption threshold as a function of the pzc gives data shown in Figure 6.   If, for the sake of argument, a linear response is expected, and the adsorption threshold is a clearly observed quantity, the pzc of Pt(111) should be much lower than presently known (*35*) and/or the pzc of Au(111) much higher than known (*46*).   This does not seem likely, especially in view of the recent work in $HClO_4$ medium (*90*) where the pzc as high as 0.9 V vs Ag/AgCl was concluded. The lack of linearity may demonstrate that the electron charge transfer component in the anion bonding is the dominating factor in adsorption. While in agreement with this article spirit, more work is needed, and the pzc of Rh(111) better known, to consider this latter statement as the grand conclusion of this paper. However, the paper summarizes new experimental material and offers some considerations for further progress in the anion adsorption sub area of electrochemistry.

Acknowledgments

This work is supported by the Department of Energy, under Grant DE-AC02-76ER01198, administrated by the Frederick Seitz Research Laboratory at the University of Illinois, and by the National Science Foundation under Grant CHE 94-11184.

References

(1) Thomas, S.; Y. -E. Sung, Wieckowski, A. *J. Phys. Chem.*, in press
(2) Sung, Y.-E.; Thomas, S.; Wieckowski, A. *J. Phys. Chem.* **1994**, *99*, 13513.
(3) Zhang, J.; Sung, Y.-E.; Rikvold, P.R.; Wieckowski, A. *J. Chem. Phys.* **1996**, *104*, 5699.
(4) Mrozek, P.; Sung, Y.-E.; Wieckowski, A. *Surf. Sci.* **1995**, *335*, 44.
(5) Chen, C.-H.; Vesecky, S.M.; Gewirth, A.A. *J. Am. Chem. Soc.* **1992**, *114*, 451
(6) Mrozek, P.; Sung, Y. -E.; Han, M.; Gamboa-Aldeco, M.; Wieckowski, A; Chen, C. -H.; Gewirth, A.A. *Electrochim. Acta* **1995**, *40*, 17.
(7) Herrero, E.; Franaszczuk, K.; Wieckowski, A. *J. Phys. Chem.* **1994**, *98*, 5074.
(8) Thomas, A. PhD. Thesis, Univ. of Illinois at Urbana-Champaign, May, 1996.
(9) Scharifker, B.R.; Chandrasekaran, K.; Gamboa-Aldeco, M.E.; Zelenay, P.; Bockris, J.O'M. *Electrochim. Acta* **1988**, *33*, 159.
(10) Kunimatsu, K.; Samant, M.G.; Seki, H.; Philpott, M.R. *J. Electroanal. Chem.* **1988**, *243*, 203.
(11) Kunimatsu, K.; Samant, M.G; Seki, H. *J. Electroanal. Chem.* **1989**, *258*, 163.
(12) Nart, F.C.; Iwasita, T. *J. Electroanal. Chem* **1991**, *308*, 277.
(13) Nart, F.C.; Iwasita, T. *J. Electroanal. Chem.* **1992**, *322*, 289.
(14) Paulissen, V.B; Korzeniewski, C. *J. Electroanal. Chem.* **1993**, *351*, 329.
(15) Kunimatsu, K.; Samant, M.G.; Seki, H. *J. Electroanal. Chem.* **1989**, *272*, 185.
(16) Faguy, F.C.; Markovic, M.; Azdic, R.R; Fierro, L.A.; Yeager, E.B. *J. Electroanal. Chem.* **1990**, *289*, 245.
(17) Ugasawava, H., Sawatari, Y.; Inuki, J.; Ito, M. *J. Electroanal. Chem.* **1993**, *358*, 337.
(18) Llorca, M.J.; Feliu, J.M.; Aldez, A.; Clavilier, J. *J. Electroanal. Chem.* **1993**, *351*, 299.
(19) Gamboa-Aldeco, M.E.; Herrero, E.; Zelenay, P.S.; Wieckowski, A. *J. Electroanal. Chem.* **1993**, *348*, 451.
(20) Nart, F.C.; Iwasita, T.; Weber, M. *Electrochim. Acta* **1994**, *39*, 961.
(21) Faguy, P.W.; Markovic, N.; Ross, P.N. *J. Electrochem. Soc.* **1993**, *140*, 1683.

(22) Kita, H.; Gao, Y.; Ohnishi, K. *Chemistry Letters* **1994**, 73.

(23) Shingaya, Y.; Ito, M. *J. Electroanal. Chem.* **1994**, *372*, 283.

(24) Ugasawava, H.; Inuki, J.; Ito, M. *Surf. Sci.* **1994**, *311*, L665.

(25) Feliu, J.M.; Orts, J.M.; Gomez, R.; Aldez, A.; Clavilier, J. *J. Electroanal. Chem.* **1994**, *372*, 265.

(26) Funtikov, A.M.; Link, U.; Stimming, U.; Vogel, R. *Surf. Sci.* **1995**, *324*, L343.

(27) Savich, W.; Sun, S.G.; Lipkowski, J.; Wieckowski, A. *J. Electroanal. Chem.* **1995**, *388*, 233.

(28) Rhee, C. K.; Wasberg, M.; Zelenay, P.; Wieckowski, A. *Catal. Lett.* **1991**, *10*, 149.

(29) Rhee, C. K.; Wasberg, M.; Hourani, G.; Wieckowski, A. *J. Electroanal. Chem.* **1990**, *291*, 281.

(30) Clavilier, J.; Wasberg, M.; Petit, M.; Klein, L. H. *J. Electroanal. Chem.* **1994**, *374*, 123.

(31) Wasberg, M.; Horanyi, G. *J. Electroanal. Chem.* **1995**, *381*, 151.

(32) Zurawski, D.; Rice, L.; Hourani M.; Wieckowski, A. *J. Electroanal. Chem.* **1987**, *230*, 221.

(33) Wasberg, M.; Hourani, M.; Wieckowski, A. *J. Electroanal. Chem.* **1990**, *278*, 425.

(34) Hourani, M.; Wasberg, M.; Rhee, C.; Wieckowski, A. *Croat. Chem. Acta* **1990**, *63*, 373.

(35) Leung, L. H.; Chang, S.; Weaver, M. J. *J. Chem. Phys.* **1989**, *90*, 7426

(36) Chang, S.-C.; Weaver, M. J. *Surf. Sci.* **1990**, *238*, 142.

(37) Yau, S.-L.; Gao, X.; Chang, S.-C.; Schardt, B. C.; Weaver, M. J. *J. Am. Chem. Soc.* **1991**, *113*, 6049.

(38) Mrozek, P.; Han, M.; Sung, Y.-E.; Wieckowski, A. *Surf. Sci.* **1994**, *319*, 21.

(39) Cazaux, J.; Coliiex, C. *J. Electron. Spectros. Rel. Phenom.* **1990**, *52*, 837.

(40) Cazaux, J.; Jbara, O.; Kim, K.H. *Surf. Sci.* **1991**, *247*, 360.

(41) Sharma, J.K.N.; Chakraborty, B.R.; Shivaprasad, S.M.; Cazaux, J. *Surf. Sci.* **1988**, *193*, L58.

(42) Strasser, G.; Rosina, G.; Matthew, J.A.D.; Netser, F.P. *J. Phys. F: Met. Phys.* **1988**, *15*, 739

(43) Ludeke, R; Koma, A. *Phys. Rev. Letters* **1975**, *34*, 817

(44) Hitchcock, A.P.; Tronc, M.; Modelli, A. *J. Phys. Chem.* **1989**, *93*, 3069.

(45) Wasberg, M.; Palaikis, L.; Wallen, L.; Kamrath, M.; Wieckowski, A. *J. Electroanal. Chem.* **1988**, *256*, 51.

(46) Shi, J.; Lipkowski, M.; Gamboa, M.; Zelenay, P.; Wieckowski, A. *J. Electroanal. Chem.* **1994**, *114*, 451.

(47) Magnussen, O.M.; Hagebock, J.; Behm., R.J. *Faraday Disc.* **1992**, *94*, 329.

(48) Edens, G.J; Gao, X.; Weaver, M. J. *Electroanal. Chem.* in press (1994)

(49) Angerstein-Kozlowska, H; Conway, B.E.; Hamelin, A.; Stoicovicu, L. *J. Electroanal. Chem.* **1987**, *228*, 429.

(50) Nakai, Y.; Zei, Y.; Kolb, D.M.; Lehmpful, G. *Ber. Bunsenges. Phys. Chem.* **1987**, *88*, 340.

(51) Kolb, D.M; Schneider, J. *Elecrochim. Acta* **1986**, *31*, 929.

(52) Zei, M.S.; Scherson, D.; Lehmpfuhl, G; Kolb, D.M. *J. Electroanal. Chem.* **1987**, *229*, 99.

(53) Wang, J.; Ocko, B.M.; Davenport, A.J; Isaacs, H.S. *Phys. Rev. B*, **1992**, *46*, 10321.

(54) Scherson, D.A.; Kolb, D.M. *J. Electroanal. Chem.* **1984**, *176*, 353.

(55) Al Jaaf-Golze, K.; Kolb, D.M.; Scherson, D. *J. Electroanal. Chem.* **1986**, *200*, 353.

(56) Herrero, E.; Feliu, J.M.; Wieckowski, A.; Clavilier, J. *Surf. Sci.* **1995**, *325*, 131.

(57) Gamboa-Aldeco, M.; Rhee, C.K.; Nahle, A.; Wang, Q.; Zhang, J.; Richards, H.L.; Rickvold, P.A.; Wieckowski, A. *The Electrochemistry Society Proceedings Volume* **94-21**, p. 184.

(58) Rikvold, P.A.; Gamboa,-Aldeco, M.; Zhang, Z.; Han, M.; Wang, Q.; Richards, H.L; Wieckowski., A. *Surf. Sci.* **1995**, *335*, 389.

(59) Clavilier, J.; Rhodes, A.; El Achi, K.; Zamkhchari, M.A. *J. Chim. Phys.* **1991**, *88*, 1291.

(60) Gamboa-Aldeco, M.; Wieckowski, A unpublished, **1994**.

(61) Zelenay, P.; Horanyi, G.; Rhee, C. K.; Wieckowski, A. *J. Electroanal. Chem.*, **1991**, *300*, 499.

(62) Bernett, M.M.; Murday, J.S.; Turner, N.H. *J. Electron Spectrosc. Rel. Phenom.* **1975**, *12*, 375.

(63) Turner, N.H.; Murday, J.S.; Ramaker, D.E. *Anal. Chem.* **1980**, *52*, 84.

(64) Briggs, D; Rivere, J.C. in *Practical Surface Analysis*, 2nd Ed., Vol. 1, Auger and X-ray Photoelectron Spectroscopy, Briggs, D.; Seah, M.P. Editors, (John Wiley & Sons, New York, **1990**), p 85.

(65) Farrell, H.H *Surf. Sci* **1973**, *43*, 465.

(66) Moulder, J.F.; Stickle, W.F.; Sobol, P.E.; Bomben, K.D. *Handbook of X-ray Photoelectron Spectroscopy*, Chastain, J. Editor, (Physical Electronics Industries, Inc., Eden Prairie, Minnesota, **1992**).

(67) Cazaux, J.; Ibara, O.; Kim, K.H. *Surf. Sci.* **1991**, *247*, 360.

(68) Chang, S.-C.; Jeung, L.-W.H.; Weaver, M.J *J. Phys. Chem.* **1989**, *93*, 5311.

(69) Wan, L.-J.; Yau, S. -L.; Swain, G.M.; Itaya, K. *J. Electroanal. Chem.* **1995**, 381, 105.

(70) Wan, L.-J.; Yau, S. -L.; Itaya, K. *J. Phys. Chem.* **1995**, *99*, 9507.

(71) Weissmann, R.; Muller, K. *Surf. Sci. Rep.* **1981**, *1*, 251.

(72) Manne, R. *J. Chem. Phys.* **1967**, *46*, 4645.

(73) Patrito, E.M; Olivera P.P; Sellers, H. under publication

(74) Howard, S.T.; Attard, G.A.; Liebermann, H.F. *Chem. Phys. Lett.* in press.

(75) Sawatari, Y.; Inukai, J.; Ito, M.; *J. Electron. Spectrosc. Relat. Phenom.* **1993**, *64/65*, 515.

(76) Howard, S.T.; Attard, G.A.; Liebermann, H.F. *Chem. Phys. Lett.* in press

(77) Trasatti, S. in *Modern Aspects of Electrochemistry* (Edited by Conway B.E. and Bockris J.O'M) Vol. 13, p 81, Plenum Press, New York (1979).

(78) Trasatti, S. *J. Electroanal. Chem.*, **1975**, *65*, 815.

(79) Kolotyrkin, Ya. M.; Lazorenko-Manevich, R.M.; Plotnikov, V.G; Sokolova, L.A. *Electrokhimiya* **1977**, *13*, 695.

(80) Faguy, P.W.; Marinkovic, N.S.; Adzic, R.R. *Langmuir* **1996**, 12, 243.

(81) Shingaya, Y.; Ito, M., submitted for publication, courtesy by Ito, M.

(82) Potter, H.C.; Blakeley, J.M. *J. Vac, Sci Technol.*, **1975**, *12*, 635.
(83) *Handbook of Chemistry and Physics*, David R. Lide Editor in Chief, 76 th Edn, p 12-122, 1995-96.
(84) Lecoeur, J.; Bellier, J.P.; Koehler, C. *Electrochim. Acta*, **1990**, *35*, 1383.
(85) Derry G.N.; Ji-Zhong Z., *Phys. Rev. B.*, **1989**, *39*, 1940.
(86) Bertel, E.; Rosina, G.; Netzer, F.P. *Surf. Sci.* **1986**, *172*, L515.
(87) Nieuwenhuys, B.E.; Bouwman, R.; Sachtler, W.M.H. *Thin Solid Films* **1974**, *21*, 51.
(88) Kalis, T.; Belyaeva, M.E.; Sergeev, S.I. *Sov. Electrochem.* **1987**, *23*, 112.
(89) Mate, C.M.; Kao, C.-T.; Somorjai, G.A. *Surf. Sci.* **1988**, *206*, 145.
(90) Hamm, U.W.; Kramer, D.; Zhai, R.S. ; Kolb, D.M. *J. Electroanal. Chem.*, accepted for publication.
(91) Trasatti, S. *Electrochim. Acta*, **1983**, *28*, 1083.

Chapter 11

Ex Situ Low-Energy Electron Diffraction and Auger Electron Spectroscopy and Electrochemical Studies of the Underpotential Deposition of Lead on Cu(100) and Cu(111)

Gessie M. Brisard[1], Entissar Zenati[1], Hubert A. Gasteiger[2,3], Nenad M. Markovic[2], and Philip N. Ross, Jr.[2]

[1]Département de chimie, Université de Sherbrooke, Sherbrooke, Québec J1K 2R1, Canada
[2]Lawrence Berkeley National Laboratory, University of California, 1 Cyclotron Road, Berkeley, CA 94720

The underpotential deposition of Pb was studied on Cu(100) and Cu(111) single crystal surfaces prepared both by sputtering/annealing in ultrahigh vacuum and by a novel electropolishing procedure. Identical results were found with both methods. A rotating ring disk assembly is used for the unambiguous determination of the lead adsorption isotherm on both Cu(100) and (111). The LEED pattern after emersion experiments shows that Pb atoms are deposited underpotentially on Cu(111) into a compact non-rotated hexagonal overlayer while on Cu(100) Pb UPD deposits form a $(\sqrt{2}\times\sqrt{2})R45°$.

The presence of Cl^- in the supporting electrolyte has a strong effect on the potential region where deposition/stripping of Pb occurs and on the reversibility of the reaction.

It has been known for a long time that the catalytic activity of electrodes strongly depends on various modifications of the surface properties (1). This stimulated many studies aimed towards the better understanding of electrocatalytical reactions at the atomic level (2). The Underpotential Deposition (UPD) phenomena exhibits a characteristic dependence on the crystal structure of the substrate (i.e. electrode) surface. The use of well ordered, single crystal electrodes has become a meaningful approach to study the crystallographic dependence of various deposition reactions. Two methods for the preparation of clean and well-ordered single crystal surfaces prior to their transfer into an electrochemical cell have been applied: flame-annealing (3) and sputter/anneal cycles in ultra high vacuum (UHV) (4). The advantage of UHV preparation method lies in its ability to utilize Auger electron spectroscopy (AES) and LEED to assess both the cleanliness of the electrode surface and the

[3]Current address: Institute of Surface Chemistry and Catalysis, Ulm University, D–89069 Ulm, Germany

structural symmetry of its outermost surface layers prior to electrochemical experiments. Copper single crystals, however, as well as many other materials cannot be flame-annealed and the only alternative to UHV preparation is electropolishing. Electrochemical studies of copper in solution are much more demanding than the investigations of the two other group 1B metals Ag and Au, chiefly due to the difficulties to produce oxide free, well defined copper surfaces in the electrochemical environment. Using a tree-step electropolishing method, we have showed that the chemically prepared surface gives identical results for Pb UPD than a surface prepared by sputtering/annealing in ultrahigh vacuum (5). Furthermore, resolving the question regarding the *ex-vacuo* cleanliness of copper single crystal, reproducible studies of copper electrodes in solution at the submonolayers levels are possible and offers exciting new perspectives in understanding interfacial properties and kinetics of UPD processes which are relevant to the electrochemistry of copper.

This paper presents a concise review of our study on Pb UPD on Cu single crystals combining *ex-situ* UHV techniques and electrochemical measurements by comparing the effect of halide on the Pb UPD process on Cu(100) and Cu(111) surfaces. Specific interactions of halides with metal substrates are fundamental interfacial aspects in the understanding of the mechanisms governing metal underpotential deposition reactions since the structure of the metal overlayers is affecting the catalytic activity of the electrode material. The adsorption of anions can impair or enhance the UPD kinetics of the metal monolayers. Moreover, concomitant anion adsorption at the electrode/solution interface plays an important role in the ordering and growth of the UPD metal. In the case of Pb on Cu single crystal in the presence of Cl^- it is possible to measure an apparent electrovalency greater than the true charge of the Pb^{2+} cation when concomitant adsorption or desorption processes of the anion occur (6,7). A knowledge of the adsorption isotherm of the metal is crucial for the elucidation of the Pb/Cu UPD process. We are presenting a rotating ring Cu single crystal disk electrode, *RRD$_{Cu(hkl)}$E*, used for the determination of the charge associated with the Pb UPD on Cu(*hkl*). The structure of the Pb UPD overlayers is examined by *ex-situ* LEED and AES experiments, and the measured Pb coverage compared with the adsorption isotherm established by RRD$_{Cu(hkl)}$E measurements.

Experimental

Electrochemical Measurements. Cu(*hkl*) single crystal (Monocrystals Company) was mechanically polished with emery paper and mirror-finished with 0.25 μm diamond paste (Buehler). With the perimeter being wrapped with Teflon tape, the 6 mm OD crystal (0.283 cm^2) was mounted in a Kel-F collet for the three-step electropolishing procedure (5). Before the electrochemical measurements, the crystal was protected by a drop of water from air-borne contamination and it was mounted into a rotating ring Cu-single-crystal-disk electrode assembly (RRD$_{Cu(hkl)}$E). All experiments on electropolished samples were carried out in a standard three-compartment electrochemical cell, with a calomel reference electrode separated from the solution by a bridge.

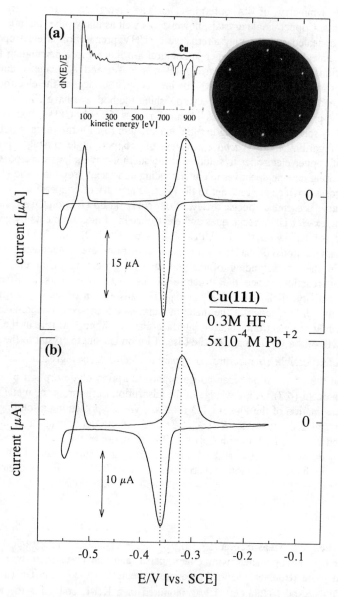

Figure 1. Comparison of lead UPD on Cu(111): **(a)** AES (3 keV incident beam) and LEED (60 eV) of a crystal prepared in UHV by sputter/anneal cycles, **(b)** crystal prepared by the electropolishing procedure. 0.1 M HF, 10 mV/s. The dotted line marks the peak potentials in the cyclic voltammograms. (Reproduced with permission from reference5. Copyright 1995, ACS)

The 0.01 M HClO$_4$ electrolyte (Baker, *Ultrex*) solution was prepared with triply pyrolytically distilled water and /or Millipore UV-Q distilled water. Lead was added as PbClO$_4 \times 3\,$H$_2$O (EM Science) to the desired concentration and the addition of Cl$^-$ was in form of NaCl (Aldrich, 99.99%). The electropolished Cu(*hkl*) surface was immersed at a potential of -0.2 V. A Pine Instruments bi-potentiostat (model AFRDE4) was used for the potentiodynamic measurements; data were acquired digitally on an IBM PC using LabView for Windows. A PAR EG&G Model 173 potentiostat was used for the electropolishing experiments.

UHV Measurements. Cu(*hkl*) single crystal was prepared by cycles of sputtering (0.5 keV Ar$^+$) and annealing (550°C in UHV) until AES and LEED indicated a clean and well-ordered surface, respectively. Afterwards, the crystal was transferred into a thin-layer electrochemical cell with a Pd/H reference electrode and immersed under potential control where no Pb adsorption occurs. 0.3 M HF (Baker, *Ultrex*) was chosen as supporting electrolyte for these measurements as it affords emersion experiments without the interference from non-volatile molecules (*e.g.*, HClO$_4$). The electrochemical experiments were done at room temperature. After electrochemical characterization in the presence of lead, the crystal was emersed from the electrolyte and returned to the UHV environment for postelectrochemical analysis. AES spectra were recorded for each emersed surfaces and the LEED pictures were taking at 46 eV for Cu(100) and 60 eV for Cu(111).

Results and Discussion

Comparison between UHV and Electropolished Prepared Cu(*hkl*). An important experimental feature is the preparation of the copper single crystal for quantitatively reproducible results. This was done by assessing the cleanliness of our electropolishing method with a vacuum prepared surface. The validity of the electrochemical preparation of the single crystal face was determined for both Cu(111) and Cu(100) and both experiments showed that the less elaborate electropolishing procedure produces a clean and well-ordered Cu(*hkl*) surface comparable in quality to a UHV prepared sample. Cu(111) is given as an example in Figure 1. Figure 1 represents the cyclic voltammetry curves of Pb UPD on UHV and chemically prepared Cu(111) surfaces. In figure 1a, the AES spectrum of a clean Cu(111) surface and the sharp LEED pattern recorded for this surface indicates the well-ordered bulk termination of the (111) plane of the Cu crystal. After surface characterization by AES and LEED the crystal was transferred under inert atmosphere into an electrochemical cell and immersed at ≈-0.2 V into 0.1 M HF with 5·10^{-4} M PbClO$_4$. In this solution the voltammogram taken at 10 mV/s exhibits a UPD lead deposition peak at ≈ -0.36 V and a Pb stripping peak at ≈ -0.32V. The similarity of the Cu(111), especially in terms of the peak potentials, obtained with both chemically (Fig. 1b) and UHV (Fig. 1a) prepared surfaces confirmed that the electropolished surface is comparable with the UHV prepared surface. The slight difference in the peak width of the UPD peaks exhibited in this figure may be rationalized by a reduced terrace size (*8*).

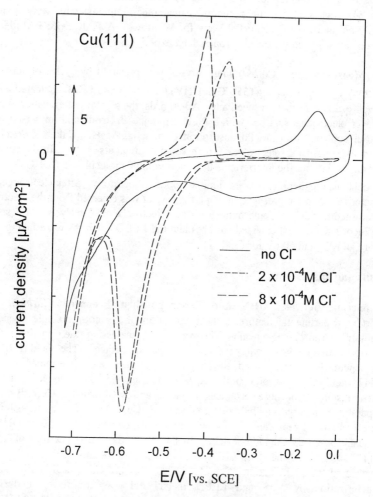

Figure 2. Cyclic voltammetry of electropolished Cu(111) in 0.01 M HClO$_4$ supporting electrolyte with and without chloride at 10 mV/s. (Reproduced with permission from reference 5. Copyright 1995, ACS)

Behavior of Cu(111) and Cu(100) in presence chloride. In figure 2 and 3 the interaction of Cu(111) and Cu(100) with chloride are compared respectively. In a earlier work we have reported the specific adsorption of chloride on Cu(111) (5). The different voltammetry response put in perspective the difference in the desorption potential of Cl⁻ on the two faces. On Cu(111) (Fig. 2) the adsorption/desorption phenomena of Cl⁻ are obvious in the potential range between hydrogen evolution and copper dissolution. On the contrary, Cu(100) in presence of chloride is featureless (Fig. 3). The featureless cyclic voltammetry on Cu(100) face comes from the fact that hydrogen evolution reaction occurs at potential positive to the desorption of Cl⁻, hence suppressing the stripping of the halide from the surface and consequently the submonolayer readsorption of chloride on the positive going sweep. In the potential region shown in Fig. 3, Cl⁻ stays on the surface and neither the adsorption nor the desorption peaks are apparent as it is for Cu(111). In due course, at very negative potentials chloride would be expected to desorb from the Cu(100) surface. This behavior may be rationalized in term of the difference in potential of zero charge (pzc) of the two faces. It has been reported that the pcz of Cu(100) is more negative than Cu(111) (*10*). The more negative pzc implies that the onset of chloride anion adsorption is shifted to more negative value and its desorption would be masked by the hydrogen evolution reaction. However, the evidence of the presence of chloride on the surface was indeed reported by Ehlers *et al* (*9*) using AES analysis of Cu(100) emersed from HCl and will be discuss in a later section. Nonetheless the interaction of Cu(111) and chloride can be formulated in terms of a two-dimensional (2D) CuCl(111) bilayer:

$$(CuCl)_{2D} + e^- \leftrightarrow Cu + Cl^-; \quad E_{Cu/Cl} \tag{1}$$

where $E_{Cu/Cl}$ is the reversible potential of the above 2D- process, different from the equilibrium potential of the half cell reaction for bulk CuCl which is at approximately -0.13 V on the SCE reference potential scale (*11*). Accordingly, the anodic/cathodic wave in Figure 2 (in the presence of Cl⁻) corresponds to the adsorption/desorption of Cl⁻ and $E_{Cu/Cl}$ is approximated by the mean of the anodic/cathodic voltammetric peak potentials, yielding a value of »-0.5 V.

Chloride effect on Pb UPD. On both Cu(111) and Cu(100), the presence of Cl⁻ in the supporting electrolyte has a strong effect on the potential region where deposition/stripping of Pb occurs and on the reversibility of the reaction. Table I summarizes the lead voltammetric responses on Cu(100) in HClO₄ solution of pH 2 at different sweep rates. On the negative going sweep deposition of Pb⁺² takes place, the stripping process is observed on the positive going sweep. In absence of Cl ion in the solution the deposition occurs at a more positive potential as the sweep rate decreases showing a more bonded state as the process is allow to take place more reversibly.

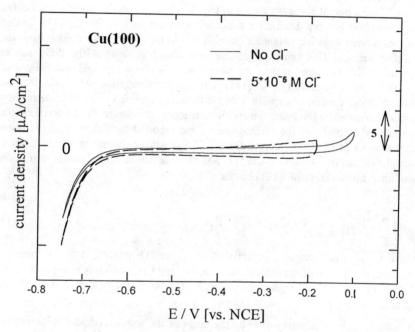

Figure 3. Cyclic voltammetry of electropolished Cu(100) in 0.01 M HClO$_4$ supporting electrolyte with and without chloride at 10 mV/s.

Table I. Difference in peak potential between deposition and stripping of Pb on Cu(100) in presence and absence of chloride ion at various sweep rate

ΔE_P (mV)	20 mV/s	10mV/s	5mV/s
without Cl⁻	245	190	150
with Cl⁻	115	100	85

A broad deposition peaks is obtained in absence of Cl⁻ and indicates a slow deposition process on Cu(100). The Pb UPD process occurs with a slow kinetics on Cu(100). This behavior is also apparent on the 111 face (5). Pb UPD on Cu(111) in chloride-free electrolyte resulted in a strongly bonded UPD structure of Pb characterized by, however, rather slow kinetics. The effect of specifically adsorbed chloride ion on the UPD process of lead is shown for Cu(100) in Table I. In presence of Cl⁻, Pb deposition occurs in a single sharp peak and a single stripping peak. The voltammetric response showed that the kinetics of the anodic process (Pb stripping) are faster than the UPD deposition of Pb, revealed by the stronger sweep rate dependency of the cathodic process compared to the anodic process (see also Figure 3b in reference 5). The peak splitting is considerably smaller at a sweep rate of 5 mV/s. The stability of Pb UPD is reduced considerably by the presence of chloride on the surface. The shift in the Pb UPD peak to more negative potentials is the result of the competition between chloride and Pb for the surface adsorption sites. In presence of chloride, Pb has to displace Cl⁻ from the electrode surface in order to adsorb and consequently adsorption take place at more negative potentials.

However, from potentiodynamic experiments on the disk, the charge contribution of Cl⁻ on the UPD process during the competitive process is indistinct. Assuming that chloride desorption/adsorption contributes to the coulombic charge under the lead deposition/stripping peak, an evaluation of the lead coverage as a function of potential from simple coulometry is not meaningful, resulting in an apparent UPD lead electrosorption valency in excess of 2 (6,7). Independent charge measurements with a rotating ring Cu-single-crystal disk electrode ($RRD_{Cu(hkl)}E$) demonstrates that concomitant adsorption of Pb and desorption of Cl⁻ can not be neglected during Pb UPD on Cu single crystal.

Evaluation of Pb monolayer coverage. In this section we will summarize the potentiodynamic measurements conducted with a rotating ring disk assembly which is a valuable method to assess accurately the UPD lead coverage of Cu single crystal. Experimental data collected with a RRCu(100)E are given in Figure 4. This figure shows the Cu(100) disk and ring electrodes voltammetry in HClO₄ solution of pH 2 in the presence of $5 \cdot 10^{-5} M$ Pb⁺² and $1 \cdot 10^{-4} M$ Cl⁻. The top section of that figure shows the cyclic voltammetry for Pb UPD on a Cu(100) disk electrode and the lower part represents the current on the ring electrode (which has been transformed into a Pb ring electrode (5)). While sweeping the potential on the Cu disk, the ring electrode was potentiostated at -0.68 V, a potential at which the reduction of Pb⁺² on the ring electrode is described accurately by a two-electron process :

$$Pb^{+2} + 2e^- \rightarrow Pb^0 \tag{2}$$

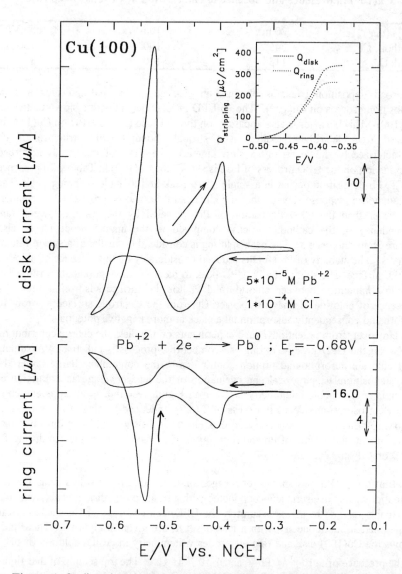

Figure 4. Cyclic voltammogram on an electropolished Cu(100) disk electrode in a RRDE assembly at 900 rpm in 0.01 M HClO$_4$ in the presence of 5·10^{-5}M Pb^{+2} and Cl$^-$. **Top:** Pb UPD on Cu(100) at 20 mV/s. **Bottom:** Ring electrode currents recorded with the ring being potentiostated at -0.68 V.

At potential $\approx$ -0.7V, reaction (2) occurs at a diffusion controlled rate and the ring unshielded value is given by equation (3) :

$$I_R^\infty = 0.62 \beta^{2/3} \, n \, F \, A \, c_0 \, D^{2/3} \, \nu^{-1/6} \, \omega^{1/2} \tag{3}$$

where β is a constant based on the ring and disk radii (β=0.89) (12), A denotes the disk area (0.283 cm^2), D is the diffusivity of Pb^{+2} (D = $0.925 \cdot 10^{-5}$ cm^2 / s) (13), ν is the viscosity of the electrolyte ($\nu = 8.86 \cdot 10^{-3}$ cm^2 / s) (14) and ω is the rotation rate in radians per second. The ring disk assembly was rotated at 900 rpm to ensure a diffusional transport of lead to the electrode during the UPD process. This condition was sufficient to allow multilayer deposition of lead (E > -0.6V) and moreover as the rotation rate was increased above 900 rpm the deposition of Pb was still determined by the diffusion limited transport of Pb to the Cu disk and the bulk stripping peak remains essentially unchanged. The theoretical diffusion limited current on the ring electrode at 900 rpm is evaluated as $I_R^\infty = -14.7\mu A$ (Equation 3), thus $\approx$9 % smaller (within experimental error) than the unshielded experimental ring current of -16.0 μA (lower part of Figure 4). As the electrode potential is scanned negatively from -0.25 V, the cathodic voltammetric peak on the Cu(100) disk electrode (top part of Figure 4) is followed by a reduction in the ring shielding current (bottom of Figure4), indicating the deposition of Pb on the disk. At $\approx$-0.54 V the formation of a Pb UPD layer is completed and the ring current returns to its unshielded value. The onset of multilayer Pb deposition near the negative potential limit again brings about a reduction of the ring current from its unshielded value until, in the reverse sweep, the onset of multilayer Pb stripping at the Nernst potential is reached, at which point the disk current increases above zero and the ring current grows above its unshielded value due to the increase of the lead concentration in the vicinity of the ring electrode. After completion of bulk lead stripping at $\approx$-0.5 V no Pb^{+2} is released by the disk electrode until the UPD Pb stripping peak commences at $\approx$-0.48 V; consequently, the ring electrode current in this potential range is at its unshielded value, above which it rises as UPD Pb is stripped off the disk (up to -0.34 V). The potentiodynamic experiment with the RRD$_{Cu(100)}$E gives a voltammogram of Pb UPD on the Cu(100) disk electrode that is mirrored perfectly by the currents on the ring electrode, establishing qualitatively that the voltammetric features on the disk electrode are essentially due to the deposition/stripping of Pb. Hence the quantitative determination of Pb without the possible interference of Cl⁻ discharge with the coulometry on the Cu(100) disk electrode is done *via* the lead deposition currents on the ring electrode knowing its collection efficiency, N. N is evaluated from the ratio of the bulk ring current over the bulk disk current and the value lies generally between 0.18 and 0.20. Knowing the collection efficiency it is possible to evaluate θ_{Pb} by integrating the ring current in excess of its unshielded value during the stripping of UPD Pb in the positive-going sweep according to:

$$\theta_{Pb} = \frac{1}{247\mu C / cm^2} \frac{Q_R}{A\,n\,N} = \frac{1}{247\mu C / cm^2} \frac{\frac{1}{\nu}\int(I_R - I_R^\infty)\,dE}{A\,n\,N} \tag{4}$$

where v is the sweep rate, Q_R is the coulombic charge on the ring electrode, and 247 μC/cm^2 corresponds to the ideal one-electron surface charge density of Cu(100) (based on the atomic density,ρ, of the Cu (100) plane, $\rho = 1.54 \cdot 10^{15}$ atoms/cm^2 (15)). The potentiodynamic experiments gives a value of 265 μC/cm^2 for the charge calculated from the ring current. Integration of the disk stripping peak gives a value of 343 μC/cm^2. The insert of Figure 4 (lower dashed line) gives the numerical evaluation of Equation 4 in terms of the lead stripping charge, $Q_{ring}=Q_R/N$. Integration of the disk currents in the same potential window (insert of Figure 4, upper dashed line) gives the charge associated with the UPD Pb process and chloride desorption. As establish earlier on Cu(111) the Cl- adsorption/desorption process contributes significantly to the charge of the Pb UPD and under similar conditions we have found that the UPD lead deposition peak is much more negative on Cu(100) than on Cu(111) (5). The charge contribution for Cl$^-$ desorption as a complete Pb monolayer is being formed can only be quantified by the difference between the coulombic charges of the disk and the ring electrode at negative potential. According to the ring current, the stripping of the entire UPD Pb layer indicates a coverage of $\Theta = 0.53$ ML. The theoretical saturation coverage for a UPD Pb on Cu (100) is 0.53 ML which corresponds to 262 μC/cm^2 from the discharge of lead according to equation (1). This charge is referenced to the maximum atomic density of the fcc (100) lead plane ($\rho = 0.82 \times 10^{15}$ atoms/cm^2 (15)) undergoing a complete 2 electrons transfer. An independent confirmation of these results as well as structural information on the Pb UPD layer were attained from *ex-situ* experiments.

Ex-Situ **LEED and AES Measurements.** For both Cu(111) and Cu(100) surfaces the results of the electrochemical, AES and LEED measurements were taken in an UHV system. At a potential where the Pb UPD layer formation is completed, the electrode was emersed and transferred in the UHV chamber for postelectrochemical analysis. For both Cu surfaces the resulting Auger spectra is compared with the spectrum of the clean UHV-prepared surface showing the presence of lead on the surface. AES of the emersed Cu(111) electrode indicates a surface nearly free of oxygen and chloride, the major features being the Auger transitions of Cu form the substrate and of Pb in the overlayer. This result clearly indicates the complete replacement process of Cl ion by Pb and supports our hypothesis, developed in the above discussion of our *in-situ* electrochemical experiments, that the deposition of Pb on Cu(111) is concomitant with the desorption of chloride from the copper surface and *vice versa*. In contrast, the emersion of Cu(100) with a saturated UPD layer of Cu in the same electrolyte shows in addition to the Auger transitions of Cu and Pb, a small amount of chloride on the surface. However, the presence of the trace amount of chloride on the surface may explain, the disordered structure of Pb/Cu(100) which was found at room temperature, given an unclear LEED picture and only the Cu principal spots were visible. Nonetheless, upon heating, a visible commensurate $(\sqrt{2} \times \sqrt{2})R45°$ lead superstructure of Pb on Cu(100) has been observed and subsequent annealing at 150°C for 10 minutes did not produce any changes in the AES spectrum. As for Cu(111), a clear LEED pattern was observed showing a lead

(a) LEED Schematic of Pb/Cu (100)

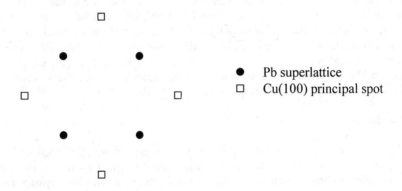

● Pb superlattice
□ Cu(100) principal spot

(b) LEED Schematic of Pb/Cu(111)

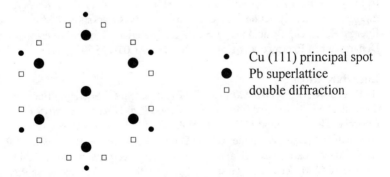

● Cu (111) principal spot
● Pb superlattice
□ double diffraction

Figure 5. Schematic representation of the LEED pattern after emersion at the saturated Pb overlayer with assignment of all the observed LEED spots. (a) Pb on Cu(100), (b) Pb on Cu(111).

overlayer as an expanded, non-rotated hexagonal structure with respect to the Cu substrate (5). The schematic representation of the LEED pattern for Cu(111) and Cu(100) is given in Fig. 5. The LEED-derived coverage of Pb based on the respective Cu(hkl) substrate was evaluated to be 0.53ML according to the following equation:

$$\theta_{Pb} = \left(\vec{k}_{Cu} / \vec{k}_{Pb}\right)^{-2} \tag{5}$$

and are in excellent agreement with the values assessed from the $RRD_{Cu(hkl)}E$ measurements. Those results also showed that chloride must be desorb from the surface in order to allow a full monolayer of Pb on the surface, a phenomena which was difficult to distinguished simply on the voltammogram of Cu(100) since Cl⁻ desorption peak was not noticeable. However, the rotating ring disk experiments support the hypothesis of a coulombic charge associated with the discharge of chloride at potential where Pb is adsorbed on the surface.

Conclusions

The dependence of adsorption isotherm on the crystallographic orientation for Cu(hkl) has been studied with ($RRD_{Cu(hkl)}E$) and confirmed by the ex-situ UHV measurements. From independent charge and coverage measurements we showed that the concomitant adsorption of Pb and desorption of Cl⁻ can be distinguished.

Acknowledgments

We would like to thank Denis Poulin (Sherbrooke) , Frank Zucca and Lee Johnson (LBNL) for their invaluable help in building many parts and in polishing the single crystals for the experimental set-up. E. Zenati acknowledges her fellowship from the Canadian International Development Agency, ACDI Marocco. The financial support from Fonds pour la Formation de Chercheurs et l'Aide à la Recherche, FCAR Equipe, is gratefully acknowledged. This work was also supported by the Office of Energy Research, Basic Energy Sciences, Materials Science Division of the U.S. Department of Energy under Contract No. DE-AC03-76SF00098.

Literature Cited

1- Brieter, M.W.,Electrocatalysis, The Electrochemical Society, Princeton, NJ, 1974
2- Kolb D.M., In Advances in Electrochemistry and Electrochemical Engineering, Gerischer H. and Tobias C.W., Eds.,Wiley, NY ,1978 vol. 11, pp. 125.
3- Clavilier J.,Faure R.,Guinet G. and Durant R.;J.Electroanal.Chem.;1980,107, 205
4- Stickney J.L., Rosasco S. and Hubbard A.; J. Electrochem. Soc.; 1984, 131, 260.
5- Brisard G.M., Zenati E., Gasteiger H.A., Markovic N.M., and Ross P.N. (Jr), Langmuir 1995, 11, 2221.
6- Siegenthaler H. and Jüttner K., J. Electroanal. Chem., 1984, 163, 327.
7- Vilche J.R. and Jüttner K., Electrochim. Acta. 1987, 32, 1567.
8- Ross P.N., J. Vac. Sci. Technol. 1987, A, 5, 948.
9- Ehlers C.B., Villegas I., and Stickney J.L., J.Electroanal. Chem., 1990, 284, 403.
10- Lecoeur J. and Bellier J.P., Electrochim. Acta, 1985, 30, 1027.

11- Bertocci, U. and Wagman, D.D. In *Standard Potential in Aqueous Solution;* Bard, A.J., Parsons, R., Jordan., Eds; Marcel Dekker: New-York, **1985**, 292.

12- Albery W.J. and Hitchman M.L., *Ring-Disc Electrodes*, Oxford University Press, New York, **1971**.

13- Mills R. and Lobo V.M.M., *Self-diffusion in Electrolyte Solutions*, Elsevier, New York, **1989**.

14- *CRC Handbook of Chemistry and Physics*, 65th edition, Ed. Weast, RC., CRC Press, Boca Raton, Florida, **1985**.

15- Kittel, C. *Introduction to Solid State Physics*, 6th ed. John Wiely & Sons, New York, **1986**.

Chapter 12

Anion Adsorption and Charge Transfer on Single-Crystal Electrodes

Roberto Gómez, José M. Orts, and Juan M. Feliu[1]

Departamento de Quimica Fisica, Universidad d'Alacant, Apartat 99, E–03080 Alacant, Spain

Anion adsorption on single crystal electrode surfaces plays a key role in the understanding of surface electrochemical processes. A detailed knowledge about the charge involved in the adsorption process as well as the surface coverage and charge state of the species forming the adlayer is required. In order to do so, a coupling of cyclic voltammetry, charge displacement and scanning tunneling microscopy is exploited. The necessity of charge correction for voltammetric stripping processes, derived from hydrogen and anion adsorption, is put in evidence.

Since their first publication in 1980 (*1*), the voltammetric profiles of clean, ordered platinum single crystals have been a subject of controversy, especially in the case of Pt(111). At first, the main discussion point was related to the surface state of the sample after the flame treatment. Did the obtained voltammetric profile correspond to the clean, atomically ordered surface expected for the crystallographic basal plane? or was the ordered surface obtained only after some voltammetric cycles involving oxygen adsorption (*2*)? The answer to these questions was gotten by means of well-established Surface Science techniques, such as ex-situ LEED, either with clean samples (*3*) or with those prepared by I-CO replacement (*4*), and by in-situ STM (*5*). These results proved that the flame treatment and subsequent water protection led to clean, well-ordered Pt(111) surfaces, which gave a voltammetric response never observed before 1980.

The following discussion point dealt with the interpretation of the voltammetric profiles. It is well known that two groups of adsorption states can be distinguished in the voltammogram of Pt(111) in sulfuric acid media in spite of the existence of a single family of adsorption sites on this orientation. The one appearing at lower potentials could be already discerned in papers published in the late 70s and early 80s dealing with voltammetric profiles obtained after the transfer of UHV-characterized Pt(111) samples to the electrochemical cell (*6,7*). The

[1]Corresponding author

differences in the measured charge density as well as in the voltammetric profiles among these studies were linked to surface contamination during transfer and to the so-called "electrochemical activation technique", which cleans, but also disorders, the surface. The second group of adsorption states, appearing at higher potentials, were termed "unusual adsorption states", since they had never been observed before the introduction of the flame treatment.

On the other hand, these unusual states were observed with several working solutions containing different anions (7). The potential range in which these states appeared was found to depend on the nature and concentration of the electrolyte in a significant way. The charge densities measured for the unusual states were not strongly dependent on the nature of the anions in the test solution.

Different processes were proposed as responsible for this voltammetric charge transfer, but they can be simply divided into two groups. In the first one (7), hydrogen is considered to be at the origin of the electronic transfer. Above 0.5 V in 0.5 M sulfuric acid (all potentials referred to the RHE scale), the Pt(111) voltammogram presents a current density corresponding to a double layer capacity of 70 $\mu F.cm^{-2}$, including the contributions of specifically adsorbed anions; as the potential becomes less positive, the (bi)sulfate anions desorb competitively with hydrogen adsorption, the latter being responsible for the charge transfer recorded in the voltammogram. In perchloric acid, where specific adsorption of anions is expected to be absent, hydrogen adsorption would take place at higher potentials, involving energies unprecedented under UHV conditions, where water and the applied potential are absent.

Another explanation (8) consisted in interpreting the voltammetric profile by considering that in 0.5 M sulfuric acid the bisulfate adlayer begins to desorb at 0.5 V in a process that involves the observed charge transfer in a relatively narrow potential range, thus giving rise to abnormally high current densities. In perchloric and hydrofluoric acids where specific adsorption of both anions is generally discarded, the species responsible for the unusual states was assumed to be -OH.

In summary, the unusual states are due to an extremely reversible process involving either reductive adsorption/oxidative desorption of a cation (H^+) or reductive desorption/oxidative adsorption of an anion, cyclic voltammetry being unable to distinguish between both of them. Different experimental results were published and interpreted favoring either the first or the second way of understanding the voltammograms, but none of them could unequivocally discriminate between both hypotheses on the adspecies undergoing the charge transfer.

In this paper we would like to emphasize a different experimental approach (described in references 9-13) for investigating the nature and the amount of the species responsible for the charge transfer observed during voltammetric scans with platinum electrodes. It is based on the fact that CO is strongly adsorbed on platinum in a wide potential range. It also takes advantage of the fact that CO adsorption is electroneutral. Indeed, the strong adsorption of CO is effective in displacing from the surface most of the common adsorbates. The desorption of these species will supply a charge, released through the circuit, corresponding to a process opposite to that taking place during their adsorption. Insofar as in this technique only desorption processes account for the measured value and sign of the

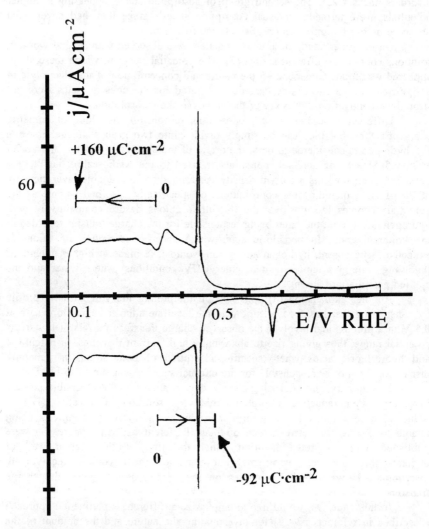

Figure 1. Voltammetric profile of Pt(111) in 0.5M H_2SO_4. Sweep rate: 50 mV·s⁻¹. Numerical values correspond to the displaced charge density at the highest and lowest potentials in charge displacement experiments.

displaced charge at a given potential, reductive desorption becomes distinguishable from reductive adsorption. This is a key point in the interpretation of the voltammogram.

Experimental

In order to illustrate the experimental procedure (*9,12*), we can start with the well-known voltammetric profile of the Pt(111) electrode in 0.5 M sulfuric acid (Figure 1). The usual states range from 0.35 V down to the hydrogen evolution threshold, whereas the characteristic unusual states are observed at higher potentials, up to 0.50 V. The voltammetric charges involved in the usual-state potential range can be easily measured in this voltammogram, including or not the apparent double layer contribution. If we correct the apparent double layer charging current, the charge density determined for the usual states is around 160 $\mu C/cm^2$.

Let us describe briefly the charge displacement technique. The potential is fixed at a selected value, e.g. 0.08 V, where the background current is almost zero. Then, CO is introduced into the gaseous atmosphere of the cell near the meniscus. CO molecules get into the solution phase, particularly in the meniscus and reach the electrode surface. During the subsequent adsorption process, an oxidation transient current flows during a certain time interval, decreases and finally attains an almost zero value again. This indicates the end of the displacement of the previously adsorbed species. The displaced charge can be easily evaluated. The excess of CO molecules is removed from the solution by Ar bubbling for several minutes. Then the current corresponding to the fully blocked surface can be voltammetrically recorded and subsequently the involved capacitive charge be measured. Next, the potential window is opened, CO is stripped from the surface in a single sweep and the initial voltammetric profile is strictly recovered. The cleanliness of the working solution is thus verified, as well as the stability of the electrode surface structure during the overall experiment. After a new flame treatment the displacement of charge can be carried out at another potential, i.e. 0.5 V. In this case, a reduction transient current is recorded. The sum of the absolute values of both charges amounts, within the experimental error (±6 $\mu C cm^{-2}$), to the charge measured in the voltammogram between both displacement potentials. At 0.32 V the total displaced charge is nearly zero. This potential can be taken as the Potential of Zero Total Charge (PZTC) (*14*)

The experiment can be carried out at any potential between 0.08 and 0.50 V. It is worth pointing out that the following relationship is held:

$$Q_t(E_1) - Q_t(E_2) = \int_{E_1}^{E_2} \frac{j}{v} dE$$

where $Q_t(E_1)$ and $Q_t(E_2)$ are the charge densities displaced at two different adsorption potentials, j is the voltammetric current density and v is the sweep rate.

Results and Discussion

In the light of the previous interpretations, and taking into account the new information coming from the charge displacement results for Pt(111) in contact with H_2SO_4 0.5 M, the voltammetric profile (Figure 1) can be read in the following way: at 0.5 V, i.e. more positive than any significant voltammetric feature, anions are adsorbed on the Pt(111) surface. At 0.4 V the amount of adsorbed anions present on the surface is lower and, finally, at 0.32 V there is not a significant amount of anions adsorbed on the Pt(111) electrode. At potentials lower than 0.32 V, hydrogen begins to adsorb, its coverage being governed by the imposed potential. The displaced charge now increases as the displacement potential becomes more negative. In any case, the maximum charge density displaced prior to the hydrogen evolution onset amounts to around 160 μCcm^{-2}, which means that the full hydrogen monolayer is not completed before hydrogen evolution. At this point the hydrogen coverage is about 2/3 of the full monolayer. Both displacement processes could be summarized as:

$$Pt(111)\text{-}HSO_4 + CO + e^- \rightarrow Pt(111)\text{-}CO + HSO_4^-$$
$$Pt(111)\text{-}H + CO \rightarrow Pt(111)\text{-}CO + H^+ + e^-$$

the first one occurring at E>0.32 V, and the second one at E<0.32 V.

In this scheme, we are assuming that the charge transfer only involves the displaced adsorbate. That is, no significant charge release comes from the adsorption of CO. In order to test if CO is such a neutral probe we can use it to displace an adsorbed species with well-known coverage and oxidation state on Pt(111). Iodine is adsorbed on Pt(111) as neutral iodine atoms with well-known adlayer structures (15). Its coverage is of 0.44 for the fully blocked surface, thus implying a negative charge density of -104 μCcm^{-2} exchanged during the whole desorption of I atoms as I^- anions. In fact, this is the value obtained in the displacement experiment, irrespective of the potential fixed for displacement of the iodine adlayer (10). Therefore, this experiment proves that CO is a neutral probe in the charge displacement experiment and allows us to interpret the charge densities of the transients in terms of coverage of preadsorbed species.

The case of Pt(111) electrodes in contact with perchloric acid solutions deserves more attention. Conventional experiments were performed up to 0.50 V. It was confirmed that the usual states were due to a cationic adsorption and, indirectly, that the unusual states in this case were not due to hydrogen adsorption, but to an anionic adsorption. A direct confirmation of the anionic nature of the unusual states was not possible since displacement at potentials as positive as 0.90 V were compulsory and in this potential range CO oxidizes when dosed. A new probe was necessary, being stable as an adsorbate at positive potentials. Molecular iodine was used succesfully, because of its neutral, strong adsorption on platinum (16). Nevertheless, the experiment cannot be as controlled as in the case of CO, given that dissolved iodine cannot be removed by bubbling Ar and consequently the final voltammetric profile cannot be compared to the initial one. In any case, the displaced charges also confirm that the unusual states are due to oxidative adsorption (OH_{ads} is always a possibility in aqueous solutions). All these experiments are self-consistent and confirm the possibility of using neutral probes to investigate the amount of electroactive adsorbed species on platinum electrodes.

So far, we have limited ourselves to expound results obtained with the Pt(111) electrode in contact with sulfuric and perchloric acid solutions. The technique can be extended to other electrolytic solutions. These studies are significant since they can provide insight into the voltammetric behavior of platinum samples as well as reasonable values of the maximum hydrogen and anion coverages attained in electrolytes containing bromide, chloride, acetate, oxalate,... (*11,17*). The two-fold nature of the whole voltammetric behavior has been put in evidence in all cases.

Given the neutral character of the CO adsorption process, the displacement technique can be advantageously used to assess the coverage of different species adsorbed on Pt electrodes, especially anions. At potentials in the double layer region, where the anions are possibly adsorbed, the displaced charge can be related to the anion coverage. In order to do so, we implicitly accept that there is a monoelectronic charge transfer per adspecies during both anion adsorption and desorption. This assumption should be checked experimentally. If this is the case, the equation used for calculate the coverage is:

$$\theta_A = \frac{Q_t}{Q_{Pt(hkl)}}$$

where Q_t is measured in the double layer region and $Q_{Pt(hkl)}$ is the nominal atomic density of the surface expressed in electric units.

The evaluation of anion coverages is important to double layer studies. More specifically, it is a piece of information that can nicely complement the data furnished by other in situ techniques, such as STM and Infrared Spectroscopy. As an example of the synergetic nature of the STM and Charge Displacement combination, we briefly comment on the study of bromine adlayers on Pt(111) (*18*). The charge density displaced at 0.45 V in a 0.1M $HClO_4$ solution containing 10^{-4}-10^{-2} M KBr amounts to -106 μCcm^{-2}. This value is almost the same as that displaced in a 0.1 M $HClO_4$ solution at the same potential for Pt(111) electrodes that, after the flame treatment, had been put in contact either with Br_2 vapor atmospheres (-116 μCcm^{-2}) or bromide containing solutions (-111 μCcm^{-2}). Hence, it is reasonable to accept that bromine adsorbates are discharged. The displaced charge densities imply bromide coverages between 0.44 and 0.48 Br/Pt. These data can help us to interpret complex STM images of the Br adlayer adsorbed on Pt(111) after similar sample pretreatments. The Br adlayer on Pt(111) corresponds to a hexagonal close-packed layer of bromine adatoms separated by a distance close to twice their van der Waals radius (Figure 2). However, in most cases, the STM images obtained with this system depicted different patterns of tunnel current maxima, not always showing a hexagonal symmetry. Most likely, the different images resulted from the coincidence between the almost close-packed bromine adlayer and the (1x1)-Pt(111) substrate. The proposed structures (Figure 3) were deduced by making compatible the results coming from charge displacement measurements and the symmetry of the adlattice in the corresponding STM images. Similar results have been obtained by other authors (*19-21*).

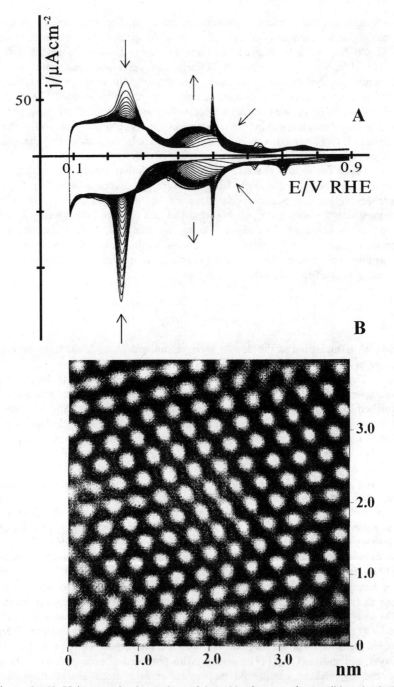

Figure 2. A) Voltammetric desorption of bromide from an irreversibly adsorbed Pt(111)-Br layer, in 0.1M H_2SO_4. Arrows indicate the evolution of the voltammetric profile. Sweep rate: 50 mV·s⁻¹. B) Constant current STM image of the close-packed bromine adlayer on Pt(111).

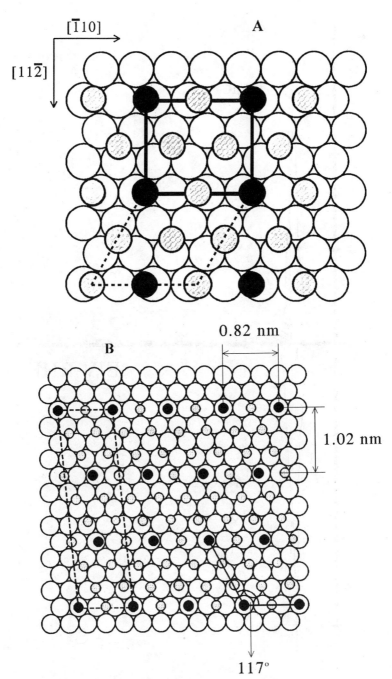

Figure 3. Ball models for adlayer structures of Pt(111)-Br imaged by STM.

$A)\ (3x\dfrac{\sqrt{3}}{2})rect, \theta_{Br} = 0.44;\quad B)\begin{pmatrix} 3 & 0 \\ 12 & -5 \end{pmatrix}, \theta_{Br} = 0.50.$

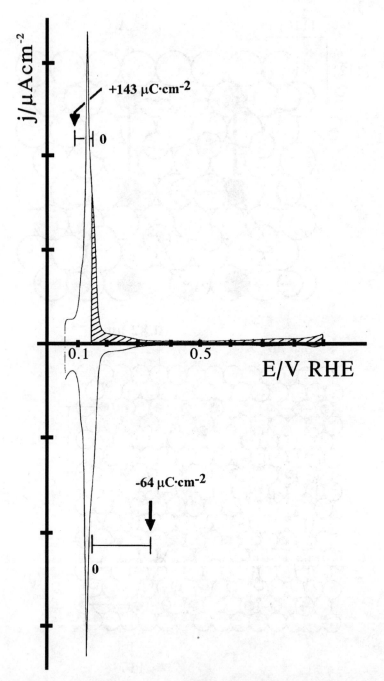

Figure 4. Voltammetric profiles of a Pt(110) electrode, same conditions as in figure 1. Shadowed zone corresponds to the negative displaced charge density.

Is cyclic voltammetry capable of giving indications on the state (reconstructed or not) of the electrode surface? In order to answer this question let us put our attention on the Pt(110) electrode, for which the interpretation of its voltammetric profile through the hydrogen adsorption hypothesis never was a subject of controversy. Figure 4 shows the voltammetric profile for this orientation after cooling the sample in a H_2+Ar atmosphere. When we perform charge displacement experiments, the displaced charge at 0.10 V is 143 μCcm^{-2} and that displaced at 0.34 V is -64 μCcm^{-2}. The overall voltammetric charge, 206 μCcm^{-2} between the same potential limits, was interpreted as an indication of the existence of a full hydrogen monolayer on a missing-row (1x2) reconstruction of the Pt(110) surface. This is the structure of the clean surface under UHV conditions. This hypothesis is no longer necessary after charge displacement measurements, given that the reported values would agree with the possibility of a full hydrogen monolayer on the Pt(110)-(1x1) surface, the other charge contribution being due to anodically adsorbed species. However, this result does not provide a direct evidence for the existence of a Pt(110)-(1x1) surface under electrochemical conditions. The data are also compatible with a Pt(110)-(1x2) surface provided that the hydrogen coverage is lower than that corresponding to the full monolayer on the reconstructed surface.

Besides Pt(111) and Pt(110) electrodes, we have applied the displacement technique to Pt(100) surfaces. We have obtained results akin to those presented with the Pt(110) electrode *(13)*; the voltammetric charge density also consists of two contributions. The charge density displaced prior to the onset of the hydrogen evolution is compatible with almost a full hydrogen monolayer on a Pt(100) substrate. In such a way, it is no longer necessary to invoke surface reconstructions to explain the experimental voltammetric charge found in this case (240 μCcm^{-2}), which is greater than that corresponding to a full monolayer on Pt(100)-(1x1) (209 μCcm^{-2}). The apparent excess of charge would be attributed to the reductive desorption of anionic species. It is noteworthy that the amount of the latter is dependent on the long-range order of the surface, being larger on (100) surfaces presenting steps and ramdomly-distributed defects *(13)*.

A more systematic approach to the investigation of the influence of defects on bisulfate adsorption can be achieved employing well-defined stepped monocrystalline platinum electrodes. We have focused on surfaces having terraces with (111) symmetry. Both (110)- and (100)-symmetry steps produce distinct voltammetric features in the usual-state potential range (Figure 5). We take advantage of the fact that the charge displaced from (111) terraces is zero at 0.32 V. Thus, the possible charge coming from the displacement at 0.32 V can be directy related to the adsorption on the steps of species formed anodically. It has been observed that from 1/3 to 1/2 of the voltammetric charge involved in the step feature is related to the adsoption of anions *(17)*.

Another point that has been investigated is the interpretation of the well-known voltammetric profile of a polycrystalline electrode "activated" by electrochemical cycling between the oxygen and hydrogen evolution onsets. The characteristic reversible processes immediately before hydrogen evolution have been classically considered as due to the adsorption/desorption of strongly and weakly adsorbed hydrogen. At present, it seems evident to assign the weakly

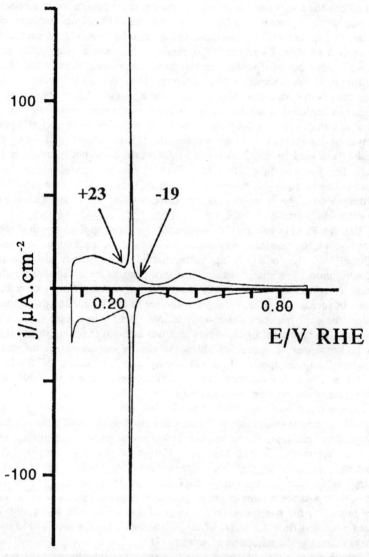

Figure 5. Voltammetric profile of a Pt(755) electrode in 0.5M H₂SO₄. Sweep rate = 20 mV·s⁻¹. Charge density values are given in μC·cm⁻².

bonded state to (110) contributions and the strongly bonded state to (100) contributions from the voltammograms of the monocrystalline samples. The state at 0.11 V is present in the Pt(110) electrode profile. In the same way, the contribution around 0.27 V always comes from stepped surfaces having (100) rows. There is, however, a question that arises when the voltammetric profile of a cycled polyoriented electrode is analyzed: why the voltammetric peaks, particularly those at 0.27 V, are wider than those obtained with stepped surfaces?. To understand this, we should remind the STM results of platinum basal plane electrodes subjected to this electrochemical "activation" treatment: the long range order of the terraces and steps is broken and the surface appears to be roughened by the electrochemical treatment. It seems that the breakdown of the surface order, particularly at monodimensional level, leads to a widening of the narrow peaks at 0.27 V observed with stepped surfaces containing (100) sites, which are monodimensionally ordered.

To confirm this assumption we have cut and polished a platinum electrode surface with the (531) orientation, which is at the center of the stereographic triangle (*22*). The hard sphere model for this highly kinked surface has the same surface concentration of each of the three types of sites ((111), (100) and (110)). In the model surface, sites of the same symmetry are never in adjacent positions; each type of site being separated from another one of the same symmetry by one site of different symmetry. Hence, if we consider domains composed of a single site symmetry, it is nominally the most disordered platinum surface that can be prepared. The voltammetric profiles of such an electrode show wide peaks and an almost complete absence of terrace contributions, as expected (Figure 6). Moreover, the charge measured from the voltammogram, after a classical double layer correction, agrees with the theoretical value of 210 μCcm^{-2} expected for this surface. This value coincides with that generally used to estimate the real surface, or surface roughness, of electrochemically "activated" polycrystalline surfaces. It is likely that faceting could occur to some extent on such an open structure. This explains the small differences observed between the voltammetric profiles of samples cooled in air and in a reductive atmosphere.

The results derived from CO displacement experiments with Pt(531) samples show that only a fraction of its voltammetric charge is due to hydrogen adsorption. The same can be stated with respect to the polyoriented surface, cycled or not in the oxygen adsorption/desorption region (30% of the whole voltammetric charge is linked to anion adsorption). This stresses that the hydrogen adsorption interpretation is no longer valid for platinum surfaces, including polycrystalline activated electrodes, for which this interpretation has never been questioned as yet.

The knowledge of the charge density due to hydrogen and anion adsorption is a key point as to calculate the net coulometric charges involved in well-defined stripping reactions occurring at electrode surfaces. What is meant by this term is that only a single sweep is required to strip off all the adsorbed molecules. The substrate adsorption properties and state of charge of the electrode are fully recovered once the stripping has been carried out. The most studied case concerns CO adlayers at full coverage. It is necessary to subtract the anion charge contribution (*9*) to obtain coverage values that agree with those determined by means of other techniques. On the other hand, at potentials negative to the PZTC,

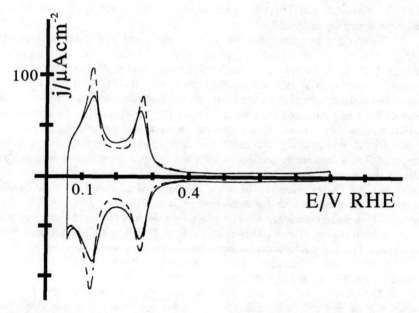

Figure 6. Voltammetric profiles of a Pt(531) electrode. Full line: cooled in a H_2+Ar atmosphere. Dashed line: cooled in air.

it is necessary to correct for the hydrogen contribution only, instead of taking all the voltammetric charge density of the blank in the low potential range. For instance, we have applied this correction when evaluating the coverage of nitric oxide on Pt(100) from its reductive stripping to give NH_4^+ *(23)*.

Conclusions

We have highlighted a new type of experiment which aims at understanding the origin of the voltammetric features of platinum electrodes. The results obtained so far strongly support the existence of important contributions to the overall voltammetric charge transfer due to anion adsorption. Although recent results suggest that hydrogen is responsible for the whole voltammetric charge transfer on Pt *(24)*, we believe that displacement data correspond to well defined properties of the platinum metal surfaces and should be incorporated into the overall interpretation of their adsorption properties.

Charge displacement in the upper potential range of the voltammogram can serve to estimate the amount of adsorbed anions in some cases. This helps to interpretate complex STM images in which only a fraction of the atoms in the adlayer can be easily imaged. Charge corrections are especially important in the comprehension of the stripping processes.

It should be stated, however, that this technique lacks chemical specificity. The only information that we obtain is related to the overall charge transfer, but not to the detailed nature of the process. In the same way, CO displacement experiments cannot distinguish between formally different pathways like:

$$A^- \rightleftharpoons A + e^-$$
$$Pt_n + A \rightleftharpoons Pt_n\text{-}A$$

that certainly holds for halide anions, or:

$$Pt_n \rightleftharpoons Pt_n^+ + e^-$$
$$Pt_n^+ + A^- \rightleftharpoons Pt_n\text{-}A$$

Other possibilities can also be envisaged for the same charge displacement. More experiments are required to get a full understanding of the surface properties of this fascinating material.

Acknowledgments

Financial support from CICYT through projects PB93-0944 and UE94-0031 is greatly acknowledged. We are also grateful to the Generalitat Valenciana for providing the funds as to purchase the STM equipment.

Literature cited

(1) Clavilier, J. *J. Electroanal. Chem.*, **1980**, *107*, 211.
(2) Scortichini, C.L.; Reilley, C.N. *J. Electroanal. Chem.*, **1982**, *139*, 247.

(*3*) Aberdam, D.; Durand, R.; Faure, R.; El Omar, F. *Surf. Sci.*, **1986**, *171*, 303.

(*4*) Zurawski, D.; Rice, L.; Hourani, M.; Wieckowski, A. *J. Electroanal. Chem.*, **1987**, *230*, 221.

(*5*) Itaya, K.; Sugawara, S.; Sashikata, K.; Furuya, N. *J.Vac.Sci.Technol.A*, **1990**, *8*, 5145.

(*6*) Motoo, S.; Furuya, N. *J. Electroanal. Chem.*, **1984**, *172*, 339 and references therein.

(*7*) Clavilier, J.; Rodes, A.; El Achi, K.; Zamakhchari, M.A., *J. Chim. Phys.*, **1991**, *88*, 1291 and references therein.

(*8*) Ross, P.N., *J. Chim. Phys.*, **1991**, *88*, 1353 and references therein.

(*9*) Clavilier, J.; Albalat, R.; Gómez, R.; Orts, J.M.; Feliu, J.M.; Aldaz, A. *J. Electroanal. Chem.*, **1992**, *330*, 489.

(*10*) Clavilier, J.; Albalat, R.; Gómez, R.; Orts, J.M.; Feliu, J.M. *J. Electroanal. Chem.*, **1993**, *200*, 353.

(*11*) Orts, J.M.; Gómez, R.; Feliu, J.M.; Aldaz, A.; Clavilier, J. *Electrochim. Acta*, **1994**, *39*, 1519.

(*12*) Feliu, J.M.; Orts, J.M.; Gómez, R.; Aldaz, A.; Clavilier, J. *J. Electroanal. Chem.*, **1994**, *372*, 1519.

(*13*) Clavilier, J.; Orts, J.M.; Gómez, R,; Feliu, J.M.; Aldaz, A. *Proc. Electrochem. Soc.*, **1994**, *94-21*, 167.

(*14*) Frumkin, A.; Petry, B.; Damaskin, B., *J. Electroanal. Chem.*, **1970**, *27*, 81.

(*15*) Schardt, B.C.; Yau S.-L.; Rinaldi, F. *Science*, **1989**, *243*, 1050.

(*16*) Herrero, E.; Feliu, J.M.; Wieckowski, A.; Clavilier, J. *Surf. Sci.*, **1995**, *325*, 131.

(*17*) Gómez, R. *Tesi Doctoral;* Universitat d'Alacant: Alacant, Spain, **1994**.

(*18*) Orts, J.M.; Gómez, R.; Feliu, J.M.; Aldaz, A.; Clavilier, J. *J. Phys. Chem.*, **1996**, *100*, 2334.

(*19*) Bittner, A.; Wintterlin, J.; Beran, B.; Ertl, G. *Surf. Sci.*, **1995**, *335*, 291.

(*20*) Tanaka, S.; Yau, S.-L.; Itaya, K. *J. Electroanal. Chem.*, **1995**, *396*, 125.

(*21*) Gasteiger, H.A.; Markovic, N.M.; Ross, P.N. *Langmuir*, **1996**, *12*, 1414.

(*22*) Clavilier, J.; Orts, J.M.; Gómez, R,; Feliu, J.M.; Aldaz, A. *J. Electroanal. Chem.*, **1996**, *404*, 281.

(*23*) Rodes, A.; Gómez, R.; Orts, J.M., Feliu, J.M.; Aldaz, A. *J. Electroanal. Chem.*, **1993**, *359*, 315.

(*24*) Tadjeddine, A.; Peremans, A. *J.Electroanal. Chem.*, **1996**, *409*, 115.

Chapter 13

In Situ Scanning Tunneling Microscopy of Organic Molecules Adsorbed on Iodine-Modified Au(111), Ag(111), and Pt(111) Electrodes

K. Itaya, N. Batina[1], M. Kunitake, K. Ogaki, Y.-G. Kim, L.-J. Wan, and T. Yamada

Department of Applied Chemistry, Faculty of Engineering, Tohoku University, Sendai 980–77, Japan

Adlayer structures of iodine on Au(111) and Ag(111) single crystal electrodes are briefly described based mainly on our recent in situ STM and ex situ LEED studies. It is concluded that complimentary use of these two techniques is a powerful method for characterizing the iodine adalyers with an atomic scale. It is also shown that the iodine-modified electrodes are ideal substrates for the investigation of adsorption processes of various organic molecules on the electrode surface in solution. The internal molecular structure, orientation, and packing arrangement of ordered molecular layers can be determined with near-atomic resolution under electrochemical conditions.

The adsorption of iodide on single crystal electrodes such as Au, Ag, Pt, Rh, and Pd is one of the most intensively studied electrochemical processes not only by traditional voltammetry but also by recently developed surface sensitive methods (1,2). It is well known that the iodine adlayers with the commensurate (3x3) and ($\sqrt{7}$x$\sqrt{7}$)R19.1° structures are formed on the Pt(111)-(1x1) surface in the double layer potential range, depending on the electrode potential and pH of the solution. These structures were previously confirmed by ex situ low-energy electron diffraction (LEED) (3) and more recently by scanning tunneling microscopy (STM) (4-6). It has also been concluded using both techniques that only the commensurate ($\sqrt{3}$x$\sqrt{3}$)R30° structure can be formed either on Rh(111) (7,8) or on Pd(111) (9).

However, it has recently been recognized that the iodine adlayer structures are more complicated on other substrates such as Au and Ag. Particular attention has recently been paid on the structure of iodine on Au(111) because of a specific variation in its adlayer structure. In situ surface X-ray scattering (SXS) studies by Ocko et al. (10) revealed two different phases of the iodine adlattice on Au(111) designated as the rectangular (px$\sqrt{3}$) phase and the rotated hexagonal (rot-hex) phase. It was clearly demonstrated by using SXS that the adlattice constants vary continuously in each of the two phases with the electrode potential (10). We have recently confirmed that complimentary use of in situ STM and ex situ LEED is a powerful technique for characterizing the atomic structure and the phase transition of the iodine adlayers on Au(111) (11,12) as well as on Ag(111) (13). Our results for the iodine adlayers on Au(111) (11) agreed very well with Ocko et al.'s SXS results (10).

[1]Current address: Departamento de Quimica, Universidad Automa Metropolitana-Iztapalapa, Apartado Postal 55–534, 09340 Mexico

On the other hand, the adsorption of organic molecules on electrode surfaces in electrolyte solutions has long been an important subject in electrochemistry for elucidating roles of adsorbed molecules in various electrode reactions (14). In contrast to a large number of successful efforts made for imaging organic molecules in air and ultrahigh vaccum (UHV) (15), there has not been much success in electrolyte solutions. Only a few pioneering papers recently reported images of molecules such as DNA bases (adenine, guanine, and cytosine) adsorbed on highly ordered pyrolytic graphite (HOPG) and Au(111) in solution (16). More recently, xanthine and its oxidized form (17), and porphyrines (18) were found to form ordered adlayers on HOPG. It is also noteworthy that an interesting order-disorder phase transition in the monolayer of 2,2'-bipyridine on Au(111) has been discerned by in situ STM as a function of electrode potential (19).

HOPG and similar layered crystals such as MoS_2 have almost exclusively been used for recent STM studies of adsorbed molecules in many scientific fields (15) because of the ease of preparing clean and atomically flat surfaces. However, HOPG usually exhibits fairly large corrugation amplitudes for the individual carbon atoms in STM images, which sometimes prohibit distinction of small corrugations from adsorbed organic molecules. Well-defined metal surfaces were also used as the substrate for the study of adsorption of organic molecules in solution because of relatively strong tendencies of metal surfaces to immobilize organic molecules (14). Hubbard and coworkers have intensively investigated the adsorption of various small aromatic molecules mainly on Pt(111) in solutions and determined the adlayer structures with LEED in UHV (1,2). In gas phase, there are voluminous reports describing the study of the adsorption of small aromatics such as benzene on various metal surfaces using surface sensitive techniques such as LEED, Auger electron spectroscopy (AES), and electron energy-loss spectroscopy (EELS), which were performed to evaluate the role of adsorbed molecules in catalytic reactions (20). Although we have very recently succeeded in imagionging benzene molecules adsorbed on well-defined Rh(111) and Pt(111) in HF solutions (21), it was found in our preliminary studies with in situ STM that relatively large molecules such as porphyrines did not form ordered adlayers on well-defined Au(111) nor Pt(111) in solutions, probably because of low surface mobility of the adsorbed molecules.

We have long been interested in finding more appropriate substrates to investigate the adsorption of organic molecules in solution. Recently, we disclosed a novel property of iodine-modified electrodes for the adsorption of organic molecules (22-25). It was surprisingly found, for the first time, that a water soluble porphyrine, 5, 10, 15, 20-tetrakis(N-methylpyridinium-4-yl)-21H, 23H-porphine (TMPyP), formed highly-ordered molecular arrays via self-ordering on the iodine-modified Au(111) electrode in $HClO_4$ solution (22). The orientation, packing arrangement, and even internal molecular structure were clearly visualized with near-atomic resolution by in situ STM. The iodine-modified Au(111) electrode (I-Au) was established as one of the most promising substrates for the investigation of the adsorption of organic molecules in solution. Indeed, we discovered this electrode with great generality to be suitable on which to form highly-ordered adlayers not only TMPyP but also various other molecules such as cystal violet (23). It was also described in our recent paper that various iodine-modified metal electrodes such as Ag, Rh, and Pt can also be employed as a substrate on which to investigate the adsorption of organic molecules (24,25). It was clarified that the iodine adlayer on these metals played a crucial role in the formation of highly-ordered molecular arrays.

In this chapter, we first briefly summarize the structures of iodine adlayers on Au(111) and Ag(111) based on our recent results obtained by using in situ STM and ex situ LEED (11-13), and then in situ STM of organic molecules adsorbed on the iodine-modified Au(111), Ag(111), and Pt(111) (22-25), demonstrating the extremely important new aspect of iodine-modified electrodes.

Structures of Iodine Adlayers

On Au(111). Although iodine adlayers on Au(111) in solutions with and without iodide ions have previously been characterized by many authors employing different techniques, several discrepancies are found in the literature. Bravo et al., using ex situ LEED, found that the iodine adlayer formed upon emersion from CsI solution possesses a $(\sqrt{3}x\sqrt{3})R30°$ lattice at a low iodine coverage (*26*). A $(5x\sqrt{3})$ structure was also found at more positive potentials (high coverage). On the other hand, using STM in air, McCarley and Bard reported only the $(\sqrt{3}x\sqrt{3})R30°$ structure (*27*). Haiss et al. reported several structures such as $(\sqrt{3}x\sqrt{3})R30°$, $(5x\sqrt{3})$, $(7x7)R21.8°$ in air and in a nonaqueous solvent (*28*). Gao and Weaver reported on the first in situ STM study on Au(111) in a KI solution under potential control, revealing several potential-dependent structures such as $(5x\sqrt{3})$ and $(7x7)R21.8°$ (*29*). Under a very similar set of experimental conditions, Tao and Lindsay found a potential-dependent phase transition from $(\sqrt{3}x\sqrt{3})R30°$ to $(3x3)$ (*30*). We have also reported $(5x\sqrt{3})$ and $(7x7)R21.8°$ structures found on an I-Au(111) electrode in pure $HClO_4$ solution in the absence of KI (*31*).

Apart from electrochemical environment, Cochran and Farrell first investigated the dissociative adsorption of I_2 molecules onto Au(111) in UHV (*32*). They reported a series of LEED patterns that changed from $(\sqrt{3}x\sqrt{3})R30°$, via patterns with the $\sqrt{3}$ spots splitting into "triads", and finally to "rosette" patterns with increasing I_2 exposure. The recent in situ SXS studies by Ocko et al. (*10*) revealed a series of adlattices of I-Au(111), which is in good agreement with Cochran and Farrell's LEED observation (*32*). The 2-dimensional phases designated by Ocko et al. (*10*), the rectangular $(px\sqrt{3})$ phase and the rotated hexagonal (rot-hex) phase, correspond to Cochran and Farrell's triad phase and rosette phase (*32*), respectively. As the electrode potential was shifted in the positive direction, the nearest I-I distance in the adlattices was shortened in each phase. This phenomenon is called electrocompression (*10*), which was originally observed in the case of underpotential deposition of metals by Toney et al. (*33,34*).

The above mentioned results suggested us that complimentary use of LEED and STM might be capable of determining continuous structural changes in the iodine adlattices at electrode-electrolyte interfaces. Our first study using ex situ LEED and in situ STM was carried out to establish the structures of iodine adlayers on the Au(111) surface in greater detail (*11*). Figure 1 shows a cyclic voltammogram for a well-defined Au(111) in 1 mM KI at a scan rate of 5 mV/s, in which the peaks are sharper than those obtained at 20 mV/s as reported in our previous paper (*11*). The peaks observed in the potential range between 0V and -0.4 V vs. Ag/AgI correspond to the adsorption-desorption reactions of iodide on Au(111).

Typical LEED patterns shown in Figure 2 were recorded at two different emersion potentials shown by the arrows in Figure 1 (*11*). A genuine $(\sqrt{3}x\sqrt{3})R30°$ structure was observed upon emersion at -0.2 V as shown in Figure 2a. At more positive emersion potentials, the LEED patterns underwent mainly splitting into "triads" and shifting of the original $(\sqrt{3}x\sqrt{3})R30°$ spots. As the potential was increased, the subspots moved away further from the center, and the distances between the split spots increased. The LEED pattern shown in Figure 2b was acquired at 0.3 V. Quantitative analysis of these LEED patterns was performed to obtain the structure and the adlattice parameters of the iodine adlayer. It was concluded that the LEED patterns shown in Figure 2b can be explained by the rectangular $(px\sqrt{3})$ structure, which can be designated as $c(px\sqrt{3}R-30°)$ according to Wood's notation. The vaules of p continuously decreased from 3 to 2.5 with increasing positive potential (*11*). Figure 3a shows the observed LEED patterns reproduced from Figure 2b with fundamental reciprocal unit vectors a_1^*, a_2^* of Au(111)-(1x1) and reciprocal unit vectors b_1^*, b_2^*

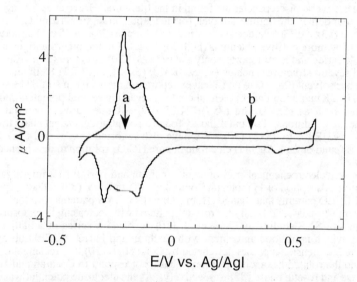

Figure 1. Cyclic voltammogram for Au(111) in 1 mM KI. Scan rate = 5 mV/s.   Arrows **a** and **b** indicate the potentials of emersion after which the LEED data in Figure 2 were recorded.

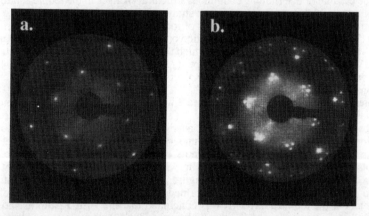

Figure 2. Two typical LEED patterns obtained at the emersion potentials shown in Figure 1. (Reproduced with permission from ref. 11. Copyright 1995 American Chem. Soc.)

of a centered iodine adlattice. To verify c($px\sqrt{3}$-R30°) as the proper structure, the LEED patterns were regenerated as reciprocal lattice plots for various p values by taking into consideration a 3-domain symmetric structure and double scattering via the 6 fundamental spots nearest to the center spot. Figure 3b shows one of these plots when the value of p equals to 2.5 (*11*). The series of such plots completely reproduced the series of experimental LEED patterns. The LEED data revealed that the parameter p varied continuously from 3 to 2.49 in the broad range of emersion potential from -0.2 V to 0.7 V. Our LEED results (*11*) were in good agreement with the in situ SXS results reported by Ocko et al. (*10*).

In situ STM also revealed real space atomic images of the iodine adlayers in both phases. In the c($px\sqrt{3}$R-30°) phase, all iodine atoms appeared in STM images with equal corrugation heights, suggesting that all iodine atoms are situated on physically equivalent positions. The registry of iodine atoms in the c($px\sqrt{3}$-R30°) will be described in a later section (see Figure 5b). Although our LEED patterns obtained in the potential range for the rot-hex phase did not show a rosette pattern, probably because of a partial desorption of iodine atoms in UHV, in situ STM clearly discerned the rot-hex phase with characteristic Moire patterns (*11,12*). It is reported in our recent paper that such Moire patterns in STM images can be quantitatively analyzed to evaluate the lattice parameters in the rot-hex phase (*13*).

On Ag(111). Several recent reports describ adlayer structures of I on Ag(111) determined by using STM. Hossick Schott and White presented STM images of a flat adlayer of I-Ag(111) obtained in air, which indicated proximity to the so-called (5x$\sqrt{3}$) (*35-37*). Kawasaki and Ishii also performed STM and x-ray photoelectron spectroscopy (XPS) on halogen-covered, sputtered Ag(111) films exposed to air (*38*). They found a contracted Ag(111)-($\sqrt{3}$x$\sqrt{3}$)R30°-I adlattice.

However, prior to these findings, Salaita et al. (*39*) had presented two LEED patterns obtained by using UHV-electrochemical system for Ag(111) emersed from 0.1 mM HI under more controlled conditions, one with triangular splitting of subspots in the negative potential range, and the other with hexagonal splitting in the positive range. Stickney et al. also studied the underpotential deposition of Ag on an I-Pt(111) electrode (*40*) and described similar triangular and hexagonal splittings in the LEED pattern. We expected that the two LEED patterns observed on Ag(111) (*39*) should be similar to the "triad" and "rosette" patterns observed on Au(111) as described above. In view of our LEED results for the iodine adlayers on Au(111), we anticipated that I-Ag(111), which might possess the same structure as I-Au(111), would also exhibit the electrocompression in both phases of adlattice structure (*13*).

Figure 4 shows cyclic voltammograms of Ag(111) obtained in an acidic HI solution and a buffered 0.1 mM KI solution at pH 10, respectively (*13*). In 0.1 mM HI solution, a small reversible pair 2-2' appears at potentials slightly negative than the potential for the bulk formation of AgI. The sharp LEED pattern of ($\sqrt{3}$x$\sqrt{3}$)R30° was obtained at the emersion potential of -0.68 V vs. Ag/AgI. The $\sqrt{3}$ subspots split into "triads" at more positive potentials. It was also found that the splitting distance increased with increasing electrode potential. This triad type of patterns can be designated as c($px\sqrt{3}$R-30°), which is very similar to the iodine adlayer on Au(111) as described above. Based on the same analysis as that illustrated in Figure 3, the value of p was obtained as a function of emersion potential. It was found that the c($px\sqrt{3}$R-30°) structure appeared with p values between 3 and 2.65 on Ag(111) in HI at potentials negative with respect to the peaks 2, 2' shown in Figure 4a.

The "rosette" LEED patterns were successfully observed on Ag(111) at potentials between the peaks 2-2' and 3-3'. The observation of the "rosette" patterns in LEED suggests that the iodine adlayer is more strongly attached on Ag(111) than on Au(111). It is now clear that the peaks 2-2' in Figure 4a correspond to the transition between the

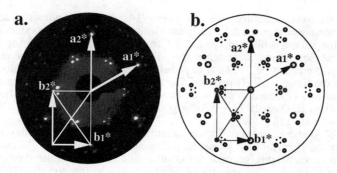

Figure 3. (a) LEED pattern shown in Figure 2b with reciprocal unit vectors a_1^*, a_2^* of Au(111) and b_1^*, b_2^* of the iodine adlattice. (b) Simulated LEED pattern of the c(2.5x√3R-30°) iodine adlattice. (Reproduced with permission from ref. 11. Copyright 1995 American Chem. Soc.)

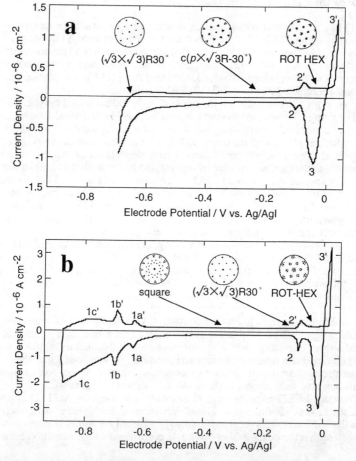

Figure 4. Cyclic voltammograms for Ag(111) in 0.1 mM HI (a) and in 0.1 mM KI buffered with 10 mM KF and 0.1 mM KOH at pH 10 (b).

c($px\sqrt{3}$R-30°) phase and the rot-hex phase. It is important to note that these two phases include the LEED patterns obtained by Salaita et al. in the same 0.1 mM HI solution (39). Their structure model for the former was an array of ($\sqrt{3}x\sqrt{3}$)R30° hexagon domains arranged with a periodicity of 17 Ag-Ag distances. Similar models had been given by Wieckowski et al. for the underpotential deposition (UPD) of Ag on an iodine-modified Pt(111) electrode (41). The triangular spritting in LEED patterns observed for the Ag UPD layer has also recently been explained by a new model proposed by Hubbard et al. (42). However, in our STM images in the c($px\sqrt{3}$R-30°) phase, we found no signs of domain boundaries. No boundary was found either during the UPD of Ag on I-Pt(111) as described in our previous paper (6). Note that Ocko et al. very recently examined iodine adlattices on Ag(111) in pure 0.1 M NaI by means of SXS (43). They also found a series of structures continuously varying from the ($\sqrt{3}x\sqrt{3}$)R30° phase, via c($px\sqrt{3}$R-30°) to the rot-hex phase during a positive potential sweep (43), which is consistent with our results obtained in the dilute HI solution.

A remarkable pH dependence was found for the adlayer structures on Ag(111) (13). Figure 4b shows well resolved four pairs of reversible peaks 1a, 1a'; 1b, 1b'; 2,2'; and 3, 3' in a 0.1 mM KI solution buffered with 10 mM KF + 0.1 mM KOH at pH 10. The peaks are labeled according to the CV reported by Salaita et al.(39). The cathodic peaks 1a, 1b, and 1c appearing in the negative potential range are due to the desorption of adsorbed iodine (I), the background current being due to the hydrogen evolution reaction. The peaks (2-2') seem to be an indication of the phase transition similar to that observed in HI. The anodic current (3'), which steeply increased at ca. 0 V, is due to the bulk formation of AgI, and the corresponding cathodic peak (3) is the reduction of AgI to form metallic Ag. The potential and current of each of these peaks are in good agreement with those reported in the solution with identical composition by Salaita et al. (39). Neverthless, it was surprising to find that the adsorbed iodine in the alkali solution formed an almost square adlattice even on the 3-fold symmetry of Ag(111), which was confirmed by both in situ STM and ex situ LEED (13). The ($\sqrt{3}x\sqrt{3}$)R30° structure seemed to be the most compressed adlayer at potentials negative of the peaks 2 and 2'. At more positive potentials than the peaks 2 and 2', the rot-hex phase was clearly discerned by STM and LEED, indicating that the peaks 2 and 2' are due to the phase transition.

Figure 5 summarizes the structures found on Ag(111) in acidic and alkaline solutions. It is noteworthy that all iodine atoms can be accomodated in the ($\sqrt{3}x\sqrt{3}$)R30°, c($px\sqrt{3}$R-30°), and also square ($\sqrt{3}x\sqrt{3}$R-30°) adlattices regardless of the value of p on the unit cell bisectors, which was defined by the traces across 2-fold bridge sites as shown in Figure 5. The in situ STM images of these structures look always very flat, indicating that all iodine atoms are located on physically equivalent positions. The model structures (a, b, and d) shown in Figure 5 can explain the flat adlayers of iodine observed by STM (12,13).

Adsorption of Organic Molecules

TMPyP on I-Au(111). Formation and characterization of ordered adlayers of organic molecules at electrode-electrolyte interfaces are important from the fundamental and technological points of view (14). The recently developed in situ STM and related techniques revealed exciting possibilities of the direct observation and characterization of electrode surfaces in solutions.

It was surprisingly found in our previous study that well-ordered TMPyP arrays were formed on the I-Au(111) surface in HClO₄ solution (22). The experimental procedure was rather simple. A well-defined Au(111) surface prepared by the flame annealing-quenching method was immersed into 1mM KI solution for several min and then thoroughly rinsed with 0.1M HClO₄ solution. The iodine modified electrode was

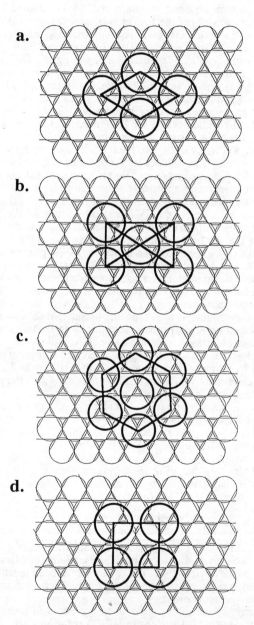

Figure 5. Model structures of the iodine adlattices observed on Ag(111). (a) (√3x√3)R30°, (b) c(px√3R-30°), (c) rot-hex, (d) (√3x√3R-30°).

installed in an electrochemical cell containing a pure HClO₄ solution for in stu STM measurements. Under potential control, the iodine adlayer structures can be determined by in situ STM (*12,31*). After achieving an atomic resolution, a dilute solution of TMPyP was injected into the HClO₄ solution. After the addition of TMPyP, STM images for the iodine adlayer usually became unclear within the first 5-10 min because of the adsorption of TMPyP, and then ordered adlayers became visible, extending over atomically flat terraces (*22*).

Molecular orientations, packing arrangements and even internal molecular structures of the adsorbed TMPyP molecules can be seen in high resolution STM images. Two typical STM images are presented in Figure 6a and 6b. These high resolution images directly demonstrate that the flat-lying TMPyP molecule can be recognized in the images as a square with four additional bright spots. The shape of the observed features in the image clearly corresponds to the chemical structure of TMPyP molecule. The characteristic four bright spots located at the four corners of a square correspond to the pyridinium units of TMPyP. The center to center distance between the bright spots was found to be 1.3 nm measured diagonally, which is nearly equal to the distance between two diagonally located pyridinium units. In addition to the internal structure, the STM image shown in Figure 6a reveals details of the symmetry and the packing arrangement. It can be seen that there are three different molecular rows marked by arrows **I, II, and III**. In the row marked **I**, all TMPyP molecules show identical orientation with an intermolecular distance of ca. 1.8 nm. An alternated orientation can be seen along the row **II** in which every second molecule shows the same orientation. On the other hand, the rotation angle of ca. 45° can be recognized between two neighboring molecules in row **III**.

The well-ordered TMPyP arrays were found to extend over atomically flat terraces. Figure 6b shows the STM image obtained in areas involving monoatomic step edges. It is clearly seen that the same molecular feature as that seen in Figure 6a extends over the lower and upper terraces. The ordered structure can be seen even on the relatively narrow terrace in the upper part of the image. It is also surprising to find that individual TMPyP molecules exist very near the monoatomic step. The result shown in Figure 6 indicates that the entire surface of the I-Au(111) is almost completely covered by ordered TMPyP molecules even near the end of the terraces. The same structure was consistently observed in a potential range between 0.6 and 1.0 V vs. a reversible hydrogen electrode in 0.1 M HClO₄.

Figure 7a and 7b illustrate a structural model showing a top view and a side view of the ordered TMPyP adlayer on the iodine adalayer of Au(111), respectively. We have also investigated the adsorption of TMPyP on a well-defined Au(111) in the absence of iodine adlayer (*44*). After achieving an atomic resolution of Au(111)-(1x1) structure in 0.1 M HClO₄, a dilute solution of TMPyP was injected into the solution in a manner similar to that described above. Although TMPyP molecules adsorbed directly on Au(111) could be seen by STM, the adsorbed molecules did not form ordered adlayers. Disordered adlayers formed on bare Au(111) suggest that strong interactions including chemical bonds between the Au substrate and the organic molecules prevented self-ordering processes from occurring, which must involve surface diffusion of the adsorbed molecules. The surface diffusion of the molecules adsorbed on bare Au was found to be very slow. Relatively weak van der Waals type interactions between the hydrophobic iodine adlayer and the organic molecules could be the key factor promoting self-ordering processes on the I-Au(111) substrate.

TMPyP on I-Ag(111). The I-Ag(111) electrode should be one of the most interesting iodine-modified electrodes, because the structures of iodine adlayers on Ag(111) have been well characterized as described above, and also because bare Ag electrodes have frequently been used as the substrate for the investigation of the

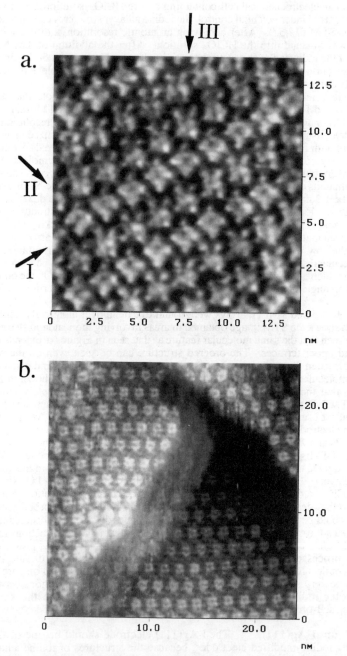

Figure 6. In situ STM images of TMPyP array on I-Au(111) in 0.1 M HClO4.

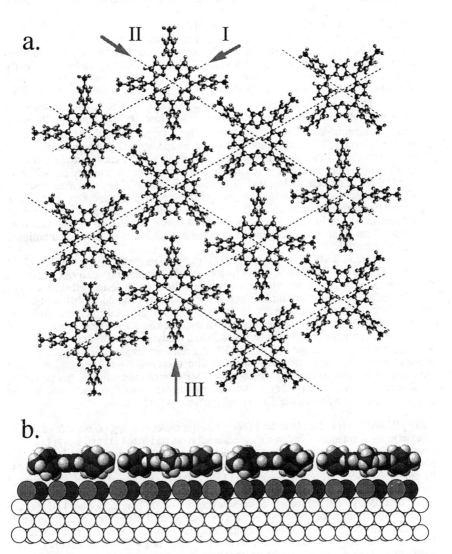

Figure 7. Structural model of TMPyP array. (a) Top view and (b) side view.

adsorption of organic molecules in solution (14). It was reported in our recent paper (24) that TMPyP molecules were irreversibly adsorbed and formed highly ordered molecular adlayers on the surface of the I-Ag(111) electrode within the same potential range and in the same alkaline KI solution at pH 10 as those used for the voltammetric study shown in Figure 4b.

Figure 8a shows a high-resolution STM image acquired in an area of 24x24 nm², revealing clear internal molecular structures and molecular orientations in the ordered adlayer. The STM image includes the TMPyP arrays formed on two adjacent terraces separated by a monoatomic step. An individual TMPyP molecule can be clearly recognized as a square with four additional bright spots at the corners. As described above, these bright spots can be attributed to the pyridinium units in the TMPyP molecule. The molecules along the [11$\bar{2}$] direction can be seen to align with the same orientation, resulting in the straight molecular rows. However, domains with the same molecular orientation are rather narrow, forming long stripes along the [11$\bar{2}$] direction. On the lower terrace, at the righthand side of the image, striped domains consisting of three molecular rows can be seen. Although all molecules in each domain show the same orientation, the molecules in the adjacent domains were rotated by ca. 45° with respect to those in the central domain. The structure of the TMPyP adlayer observed on the lower terrace is illustrated by the model shown in Figure 8b. The model represents three adjacent domains (A,B,A) in the TMPyP adlayer separated by domain boundaries indicated by the solid lines. In the domains denoted by A, the adsorbed molecules form an almost square lattice. On the other hand, a slightly tilted lattice is formed in domain B.

The structure of TMPyP on I-Ag(111) is clearly different from that on I-Au(111) shown in Figure 6a. In general, many factors should be taken into account to explain the adlayer structures and the difference between those on different substrates. The interactions between adsorbed molecules and iodine adlayers should be different on Ag and Au. Note that the architecture of TMPyP seemed to be primarily determined by the structure of the underlying iodine adlayer. In the alkaline KI solution at pH=10, we observed the continuously varying series of iodine adlayers from the square ($\sqrt{3}$x$\sqrt{3}$R-30°) structure to the triangular ($\sqrt{3}$x$\sqrt{3}$)R30° structure as described for Figure 4b. It was also found that one of three equivalent $\sqrt{3}$ directions was always unchanged during the electrocompression (13). Presumably, the TMPyP molecules align in this particular $\sqrt{3}$ direction on I-Ag(111) (24). An order-disorder transition was successfully followed by in situ STM for the TMPyP adlayer on I-Ag(111) (24).

TMPyP on I-Pt(111). In order to complete our understanding on the role of iodine adlayer in the formation of ordered molecular adlayers on I-Au(111) and I-Ag(111), we have extended our investigation to other iodine-modified electrodes such as Pt(111) and Rh(111) (25). The commensurate (3x3) and ($\sqrt{7}$x$\sqrt{7}$)R19.1° structures on Pt(111) (4-6) and the ($\sqrt{3}$x$\sqrt{3}$)R30° structure on Rh(111) (7,8) were previously well-characterized by ex situ LEED and in situ STM. In contrast to I-Au(111) and I-Ag(111), it was found that TMPyP did not form ordered adlayers on I-Pt(111) nor on Rh(111). Figure 9 shows an example of the STM images of adsorbed TMPyP molecules on an I-Pt(111) electrode with the (3x3) iodine structure in 0.1 M HClO$_4$ (25). The surface of I-Pt(111) was almost completely covered by flat-lying TMPyP molecules. The individual TMPyP molecules can be recognized as the characteristic square shape. However, it is clear that the adsorbed TMPyP did not form a highly ordered array on the I-Pt(111) surface, although several domains with a short-range ordering can be seen in the STM image. A prolonged imaging in the same area showed that the position of each TMPyP molecule was not altered, resulting in the same disordered adlayer. This observation indicates that the surface diffusion of the adsorbed TMPyP molecule on I-Pt(111) is rather slow. TMPyP molecules seem to be more strongly adsorbed on I-

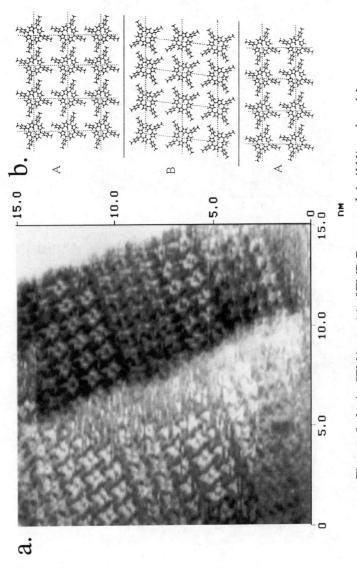

Figure 8. In situ STM image (a) of TMPyP array on I-Ag(111), and model structure (b). (Reproduced with permission from ref. 24. Copyright 1996 American Chem. Soc.)

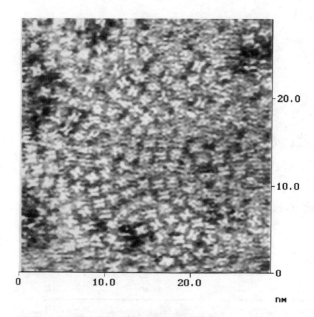

Figure 9. In situ STM image of TMPyP on I-Pt(111) in 0.1 M HClO₄.

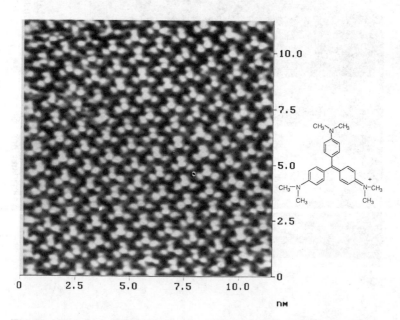

Figure 10. In situ STM image of crystal violet (CV) array on I-Au(111) in 0.1 M HClO₄.

Pt(111) than on the other iodine-modified electrodes. Similar disordered adlayers have also been found on the I-Rh(111) surface (25).

Other Molecules on I-Au(111). Here, we briefly describe further evidence that the I-Au(111) electrode can be employed as an ideal substrate for in situ STM imaging of various adsorbed organic molecules in solution. Organic substances investigated were water soluble cationic molecules purposely selected based on their characteristic shapes: triangular and linear. Hexamethylpararosaniline (crystal violet; CV) and 4,4'-bis(N-methyl-pyridinium)-p-phenylenedivinylene (PPV) were also found to form highly-ordered molecular arrays on top of the iodine monolayer adsorbed on Au(111) (23). In situ STM with near-atomic resolution revealed their orientation, packing arrangement and internal structure of each molecule.

A typical high-resolution STM image of the molecular arrays of CV is shown in Figure 10. It is also surprising to see that the STM image shows a distinctly characteristic, propeller-shaped feature for each molecule with highly-ordered arrays. Each CV molecule has three benzene rings located at the apexes of a triangle with an equal distance from the central carbon atom. Three bright spots seem to correspond to the benzene rings. The center of the spot is located at ca. 0.35 nm from the center of the triangle. An additional spot can also be seen at the position of the central carbon atom of CV. According to the STM image, it is clear that all molecules are oriented in the same direction. The unit cell shown in Figure 10 can be characterized by the lattice parameters: a=0.9 nm, b=1.1 nm and the angle of ca. 75°, indicating that the CV adlayer was slighly deformed from three-fold symmetry.

The third compound investigated is the highly symmetric cationic PPV molecule with two terminal pyridinium rings connected with the straight phenylene-divinylene core. The image shown in Figure 11 is a typical STM image for the ordered PPV adlayers formed on the I-Au(111) surface. The individual flat-lying PPV molecule can be seen as a linearly aligned feature consisting of three bright spots that can be attributed to the three aromatic rings in PPV. Three bright spots aligned in a straight line suggest that the PPV molecules adsorb on the I-Au(111) surface with a straight configuration. PPV molecules are expected to form straight or bent configurations, depending on the relative orientation of the two trans CH=CH double bonds. The STM image shown in Figure 11 indicates that the two trans CH=CH double bonds are located on the opposite side, forming the straight configuration in the adlayer shown in Figure 11. It is also clearly seen that the tightly packed arrangement forms long striped domains. In each domain, all molecules show the same orientation as indicated by the model in Figure 11. The width of each domain along the molecular axis was found to be ca. 2.1 nm from the STM image, which corresponds to the total molecular length of PPV. It is also interesting to note that a zig-zag arrangement appears alternately in these striped domains.

Summary

It is demonstrated in this chapter that complimentary use of in situ STM and ex situ LEED is a powerful technique for characterizing the atomic structure of iodine adlayers on Au(111) and Ag(111) surfaces. Two distinguished incommensurate phases of c($px\sqrt{3}$R-30°) and rot-hex are fully understood on both electrodes. The c($px\sqrt{3}$R-30°) and rot-hex adlattices are more compressed with increasing electrode potential. Uniaxial compression of the c($px\sqrt{3}$R-30°) phase was successfully followed in detail by LEED. In alkaline KI solution, a continuously varying series of flat adlattices from square ($\sqrt{3}x\sqrt{3}$R-30°) to ($\sqrt{3}x\sqrt{3}$)R30° was found in the double layer potential range. The remarkable pH dependence was found for the iodine adlayer on Ag(111).

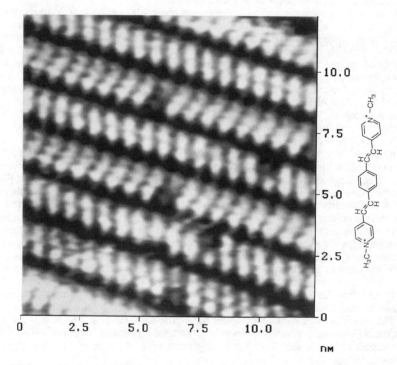

Figure 11. In situ STM image of PPV array on I-Au(111) in 0.1 M HClO4.

This chaper also described an extremely important new aspect of the iodine-modified electrodes. The organic molecules formed surprisingly highly-ordered arrays on the iodine-modified electrodes, which were successfully visualized with an extraordinarily high resolution by in situ STM, allowing the observation of the internal structure of the molecules as well as the packing arrangement in the ordered arrays. Although several factors were involved in the present achievement of high resolution, the most important factor must have been that the iodine adlayer on the metal surfaces allowed the adsorbed molecules to diffuse on the surface, resulting in the formation of highly-ordered arrays. Relatively weak interactions on the iodine adlayer seems to be a key factor in the formation of the ordered molecular arrays. The novel approach, using the iodine monolayer as an intermediate layer for the adsorption and formation of molecular arrays, has a great potential for applications in investigations of many organic molecules including native biological materials.

Acknowledgments

This work was carried out under the Exploratory Research for Advanced Technolgy (ERATO): Itaya-Electrochemiscopy program organized by the Research Development Corporation of Japan (JRDC).

Literature Cited

1. Hubbard, A. T. *Chem. Rev.* **1988**, *88*, 633.
2. Soriaga, M. P. *Progr. Surf. Sci.* **1992**, *39*, 325.
3. Lu, F; Salaita, G. N.; Baltruschat, H.; Hubbard, A. T. *J. Electroanal. Chem.* **1987**, *222*, 305.
4. Schardt, B. C.; Yau, S. -L.; Rinaldi, F. *Science*, **1989**, *243*, 981.
5. Vogel, R.; Kamphausen, I.; Baltruschat, H. *Ber. Bunsenges. Phys. Chem.* **1992**, *96*, 525.
6. Shinotsuka, N.; Sashikata, K.; Itaya, K. *Surf. Sci.* **1995**, *335*, 75.
7. Hourani, M.; Wasberg, M; Rhee, C.; Wieckowski, A. *Croat. Chem. Acta*, **1990**, *63*, 373.
8. Wan, L. -J.; Yau, S. -L.; Swain, G. M.; Itaya, K. *J. Electroanal. Chem.* **1995**, *381*, 105.
9. Soriaga, M. P.; Schimpf, J. A.; Carrasquillo, A.; Abreu, J. B.; Temesghen, W.; Barriga, R. J.; Jeng, J. -J.; Sashikata, K.; Itaya, K. *Surf. Sci.* **1995**, *335*, 273.
10. Ocko, B. M.; Watson, G. M.; Wang, J. *J. Phys. Chem.* **1994**, *98*, 897.
11. Yamada, T.; Batina, N.; Itaya, K. *J. Phys. Chem.* **1995**, *99*, 8817.
12. Batina, N.; Yamada, T.; Itaya, K. *Langmuir*, **1995**, *11*, 4568.
13. Yamada, T.; Ogaki, K.; Okubo, S.; Itaya, K. *Surf. Sci.* **1996** in press.
14. *Adsorption of Molecules at Metal Electrodes*, Lipkowski, J.; Ross, P. N., Eds.; VCH Publishers, 1992.
15. Chiang, S. In *Scanning Tunneling Microscopy I*, Guntherodt, H. -J.; Wiesendanger, R., Eds.; Springer-Verlag, Berlin, 1992; pp 181-205.
16. Tao, N. J.; DeRose, J. A.; Lindsay, S. M. *J. Phys. Chem.* **1993**, *97*, 910.
17. Tao, N. J.; Shi, Z. *J. Phys. Chem.* **1994**, *98*, 1464.
18. Tao, N. J. Phys. Rev. Lett. **1996** in press.
19. Cunha, F.; Tao, N. J. *Phys. Rev. Lett.* **1995**, *75*, 2376.
20. *Introduction to Surface Chemistry and Catalysis I*, Somorjai, A. T. Ed.; John Wiley & Sons Inc., New York, 1994.
21. Yau, S. -L.; Kim, Y. -G.; Itaya, K. *J. Am. Chem. Soc.* **1996** in press.
22. Kunitake, M.; Batina, N.; Itaya, K. *Langmuir*, **1995**, *11*, 2337.
23. Batina, N.; Kunitake, M.; Itaya, K. *J. Electroanal. Chem.* **1996**, *405*, 245.

24. Ogaki, K.; Batina, N.; Kunitake, M.; Itaya, K. *J. Phys. Chem.* **1996**, *100*, 7185.
25. Batina, N.; Kunitake, M.; Ogaki, K.; Kim, Y. -G.; Wan, L. -J.; Itaya, K. in preparation.
26. Bravo, B. G.; Michelhaugh, S. L.; Soriaga, M. P.; Villegas, I.; Suggs, D. W.; Stickney, J. L. *J. Phys. Chem.* **1991**, *95*, 5245.
27. McCarley, R. L.; Bard, A. J. *J. Phys. Chem.* **1991**, *95*, 9618.
28. Haiss, W.; Sass, J. K.; Gao, X.; Weaver, M. J.*Surf. Sci. Lett.* **1992**, *274*, L593.
29. Gao, X.; Weaver, M. J. *J. Am. Chem. Soc.* **1992**, *114*, 8544.
30. Tao, N. J.; Lindsay, S. M. *J. Phys. Chem.* **1992**, *96*, 5213.
31. Sugita, S.; Abe, T.; Itaya, K. J. Phys. Chem. **1993**, *97*, 8780.
32. Cochran, S. A.; Farrell, H. H. *Surf Sci.* **1980**, *95*, 359.
33. Toney, M. F.; Gordon, J. G.; Samant, M. G.; Borges, G. L.; Melroy, O. R.; Kau, L.-S.; Wiesler, D. G.; Yee, D.; Sorensen, L. B. *Phys.Rev. B* **1990**, *42*, 5594.
34. Toney, M. F.; Gordon, J. G.; Samant, M. G.; Borges, G. L.; Melroy, O. R.; Yee, D.; Sorensen, L. B. *Phys.Rev. B* **1992**, *45*, 9362.
35. Hossick Schott, J.; White, H. S. *J. Phys. Chem.* **1994**, *98*, 291.
36. Hossick Schott, J.; White, H. S. *J. Phys. Chem.* **1994**, *98*, 297.
37. Hossick Schott, J.; White, H. S. *Langmuir,* **1994**, *10*, 486.
38. Kawasaki, M.; Ishii, H. *Langmuir,* **1995**, *11*, 832.
39. Salaita, G. N.; Lu, F.; Laguren-Davidson, L.; Hubbard, A. T. *J. Electroanal. Chem.* **1987**, *229*, 1.
40. Sickney, J. L.; Rosasco, S. D.; Song, D.; Soriaga, M. P.; Hubbard, A. T. *Surf. Sci.* **1983**, *130*, 326.
41. Wieckowski, A.; Schardt, B. C.; Rosasco, S. D.; Stickney, J. L.; Hubbard, A. T. *Surf. Sci.* **1984**, *146*, 115.
42. Frank, D. G.; Chyan, O. M. R.; Golden, T.; Hubbard, A. T. J. Phys. Chem. **1994**, *98*, 1895.
43. Ocko, B. M.; Magnussen, O. M.; Wang, J. X.; Adzic, R. R.; Wandlowski, T. *Physica B,* **1996** in press.
44. Kunitake, M.; Akiba, U.; Batina, N.; Itaya, K. in preparation.

Chapter 14

Structure of the GaAs(100) Surface During Electrochemical Reactions Determined by Electrochemical Atomic Force Microscopy

Kohei Uosaki, Michio Koinuma, Namiki Sekine, and Shen Ye

Physical Chemistry Laboratory, Division of Chemistry, Graduate School of Science, Hokkaido University, Sapporo 060, Japan

Electrochemical atomic force microscope was employed to observe and modify the structure of the GaAs(100) surface and to investigate the electrodeposition of Cu on the p-GaAs(100) face in various electrolyte solutions under potential control. Atomic arrangement of the (100)-(1x1) structure was observed at both p-type and n-type GaAs electrodes. Sub-μm pattern was created by the scanning of an AFM tip. This is due to the tip-induced anodic dissolution reaction. How the Cu deposition proceeded strongly depended on the surface structure of the GaAs, the applied potential, and the concentration of Cu^{2+} ion in solution. Atomic arrangement corresponding to the Cu(111)-(1x1) was observed on top of the Cu deposits.

Control and observation of electrochemical and photoelectrochemical reactions at semiconductor electrodes are very important in establishing the electrochemical/ photoelectrochemical etching processes and stable photoelectrochemical cells (1,2). To understand the mechanism of the electrochemical and photoelectrochemical reactions, in situ information of morphological and electronic structures of semiconductor electrode surfaces with atomic resolution is essential. Although techniques such as electron microscopy and optical microscopy have been applied to examine the morphology of the surface of solid substrates, the former can be used only for ex situ examination and the latter has poor resolution (3,4).

Scanning tunneling microscope (STM) has become a popular tool for imaging electrode surface in situ with atomic resolution and has been proved to be a very useful aid to understand the fundamentals of electrode processes (5). This technique has, however, some limitation. Because STM uses tunneling current as a probe, only the surface of conducting materials can be imaged. Thus, STM measurement of semiconductor electrode is not possible under reverse bias because of the existence of space charge layer *(6, 7)*. On the other hand, atomic force microscope (AFM) can be used to observe the structure of solid surfaces in wide potential region as it is not affected by the sample conductivity. Furthermore, electrochemical processes on the STM tip which may cause serious problems in the STM measurements in electrolyte solutions can be avoided in the AFM measurements.

In this article, electrochemical and tunneling behaviors of p-GaAs/electrolyte interfaces were described and electrochemical AFM was employed to study the atomic structure of and the Cu electrodepostion on the GaAs(100) electrodes in various electrolyte solutions. Sub-μm pattern formation by scanning the AFM tip was also attempted.

Experimental

A NanoScope I (Digital Instruments, CA) was modified to carry out the electrochemical tunneling spectroscopy (ETS) measurement. ETS measurement was carried out by monitoring the tunneling current under a constant tip-sample distance. Details of the present ETS system were described elsewhere *(7)*. The potential of the tip and the sample were controlled independently by a home-built bi-potentiostat *(8)* with respect to a common Pd-H reference electrode which was charged sufficiently at a current density of 10 μA cm^{-2} before each measurement. Potentials were presented with respect an Ag/AgCl electrode in this article. The counter electrode was a Pt wire. The tip was a simply cut Pt wire (0.3 mm) insulated with apiezon wax.

AFM measurement were carried out using a NanoScope II or E (Digital Instruments, CA) operating in the constant (repulsive) force mode. Microfabricated Si$_3$N$_4$ cantilevers of 100 μm long with a spring constant of 0.58 N/m were used throughout. Typical force during imaging was less than 1 nN. Calibrations of the piezo scanner for the x and y planes were carried out by imaging a mica and a single crystalline Au(111) face in Milli-Q water. A glass fluid cell of 0.2 ml volume (Digital Instruments) was used for the electrochemical AFM measurements. The counter electrode and reference electrode were a Pt wire and a Ag/AgCl electrode, respectively. The area of exposed electrode in contact with the solution was about 0.32 cm^2.

GaAs(100) single crystals were donated by Mitsubishi Chemical Corporation. GaAs substrates were cleaned in hot acetone and in ethanol and rinsed with Milli-Q water (Millipore) before each experiment. Ohmic contacts were made by using an In-Zn alloy for p-type and an In metal for n-type substrates.

The electrolyte solutions were prepared by using reagent grade chemicals (Wako Pure Chemicals) and Milli-Q water. The solution was deaerated by passing purified N$_2$ gas for at least 20 min prior to the experiment.

Results and Discussion

Electrochemical and tunneling characteristics of the p-GaAs(100) electrode.
Figure 1 shows a current-potential relation of the p-GaAs electrode in 10 mM HCl in the dark. As expected for a p-type semiconductor electrode, a large anodic current was observed but only a small cathodic current flowed. Anodic current corresponding to the GaAs dissolution was observed at potentials more positive than +0.15 V vs. Ag/AgCl.

Figure 2 shows a typical tunneling current-potential relation at the p-GaAs(100) electrode in 2 mM HClO$_4$ solution *(7)*. Current due to the electron tunneling from the sample to the tip is denoted as a cathodic current and from the tip to the sample as an anodic current. The sample potential was first set in the forward bias region (+0.25 V) and the tip potential was set in the double layer region of Pt at +0.15 V. Pre-set tunneling current was 8 nA. The tip was approached to the sample surface and a tunneling state was achieved under ordinary STM feedback control. The STM was then switched into the external control mode and the potential of the GaAs electrode was swept while the tip potential and the tip-sample distance were kept constant. The tunneling current decreased quickly as forward bias was decreased and only very small tunneling current flowed at potentials more negative than +0.1 V. This behavior can be explained as follows. When the potential of the p-GaAs electrode became more positive than the flat-band potential, the p-GaAs electrode is under "accumulation" condition and there are enough carriers (holes) at the semiconductor surface. If the tip is brought close to the p-GaAs electrode surface under this condition with feedback circuit active, electron can tunnel from the tip to the p-GaAs electrode and the pre-set tunneling current flows. As the potential is swept negatively in external control mode, the Fermi level of the p-GaAs electrode moves upward and the hole concentration at the p-GaAs surface is decreased, leading to a rapid decrease of the anodic tunneling current corresponding to the electron tunneling from the tip to the p-GaAs electrode. The tunneling current decreases to zero near the flat-band potential and no or very small current is expected to flow at the potentials more negative than the flat-band potential as observed experimentally, since a space charge layer is formed within the semiconductor when the potential becomes more negative than the flat-band potential ("depletion' region). The space charge layer can be considered as an extra barrier for the electron tunneling from the tip to the p-GaAs electrode.

Atomically resolved structure of the GaAs(100) surface in electrolyte solutions obtained by electrochemical AFM. The results in the previous section clearly show that the STM cannot be used to image the surface morphology of the p-GaAs electrode at potentials more negative than the flat-band potential because of the formation of the space charge layer, i.e., an extra barrier. Thus, AFM was employed to observe the surface structure of the GaAs(100) surface in wide potential region as it uses the force between the tip and the sample as a probe and can image solid surfaces regardless of their conductivity. Although atomic arrangement was not observed in air as the GaAs surface was covered with native oxide, an atomic arrangement was observed when the electrode was immersed in electrolyte solutions and cathodic potential was applied, showing the oxide layer was reduced electrochemically.

Figures 3(a) and 3(b) show typical atomically resolved AFM images of a p- and an n-type GaAs(100) electrodes, respectively, obtained in HCl solution *(9)*. AFM images were captured near the open circuit potential, i.e., at 0 V for the p-type and at -0.5 V for the n-type GaAs(100) electrodes, after the p- and the n-type GaAs electrodes were kept at -0.8 V and -0.6 V, respectively, for about 10 min. The both images show that the atomic structure is of nearly four-fold symmetry with a nearest neighbor distance of about 0.40 nm, indicating that the surface has the GaAs(100)-(1x1) structure in HCl solution. This is in contrast to the fact that the GaAs(100) surface forms reconstructed structures such as (2 x 4) and c(4 x 4) in UHV condition because of the existence of the dangling bonds in the (1x1) structure *(10)*. The atomic images were clearer near the

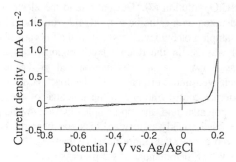

Figure 1. Current-potential relation of the p-GaAs(100) electrode in 10 mM HCl solution.

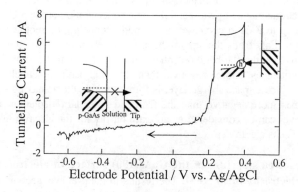

Figure 2. Potential dependence of the tunneling current at the p-GaAs(100) electrode in 2 mM HClO$_4$ solution with a fixed tip-sample distance. Tip potential was set at +0.15 V.

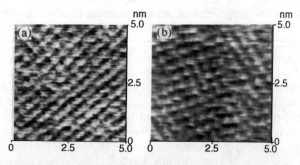

Figure 3. Atomically resolved AFM images of the (a) p- and the (b) n-type GaAs(100) electrodes in 10 mM HCl obtained at 0 V and -0.5 V, respectively.

open circuit potential than in more cathodic potential region. The XP spectrum shows that the Ga/As ratio was higher and the Cl peak was stronger at samples which were removed from the cell after a positive potential was applied than at samples without anodic treatment Thus, we speculate that the topmost atoms were Ga and the (1x1) structure was stabilized by the termination of the dangling bond with Cl (Cl$^-$).

Atomic arrangement showing the GaAs(100)-(1x1) structure was also observed in H_2SO_4 solution. Figure 4(a) shows a current-potential relation of the p-GaAs(100) in 10 mM H_2SO_4 solution. Atomic images were obtained more clearly at more cathodic potential region which was close to the potential of H_2 evolution than at the potentials near the open circuit potential as shown in Figure 4 (b)*(11)*. Atomic arrangement was very clear at -0.75 V (Figure 4(b-i)) but became less clear at -0.4 V (Figure 4(b-ii)). It was hardly observed at +0.1 V which was very close to the open circuit potential (Figure 4(b-iii)). These results suggest that SO_4^{2-} ion cannot stabilize the (100)-(1x1) structure by terminating the dangling bonds of the GaAs(100) surface but the GaAs(100)-(1x1) structure is stabilized in cathodic potential region as a result of the termination of the dangling bond with H (H$^+$). Itaya et al. also observed the (1x1) structure at n-GaAs(100) in H_2SO_4 solution in cathodic potential region by using electrochemical STM and proposed that the termination of the dangling bond by hydrogen is responsible for (1x1) structure *(12)*.

To clarify the role of the adsorbed ions/atoms on the GaAs surface, the surface was treated with S, which is known to decrease surface state density *(7, 13)*. Atomic images of the p-GaAs(100) surface were observed in Na_2S solution in very wide potential region where almost no current flowed, i.e., between the onset potential of hydrogen evolution (-0.7 V) and anodic dissolution (+0.05 V). Figure 5 shows (a) a current-potential relation and (b) atomically resolved AFM images at various potentials of the p-GaAs(100) electrode in 2 mM Na_2S solution. All the AFM images obtained in Na_2S solution showed identical structure regardless of the electrode potential. The structure showed four-fold symmetry with nearest neighbor distance of about 0.40 nm. The atomic distance and the directions of the atom rows were the same as those obtained in 10 mM H_2SO_4 solution at -0.8 V before the solution was replaced by 2 mM Na_2S solution. Thus, it is clear that the p-GaAs(100) surface has the (1x1) structure also in Na_2S solution. This is again in contract to the result obtained in the UHV condition where sulfur atoms form a (2x1) adlayer when it was exposed to H_2S gas *(14)*. Once the GaAs(100) surface was treated in Na_2S solution, the atomic arrangement of the GaAs(100)-(1x1) structure was observed easily and in wide potential region even after Na_2S solution was replaces by H_2SO_4 solution. These results should be related to the fact that the GaAs electrode shows ideal characteristics of semiconductor electrode if the surface is treated with sulfur *(6)*.

A question arises what actually we imaged by AFM. Are they the topmost layer of the adsorbed species, i.e., Cl (Cl$^-$), H (H$^+$) and S (S^{2-}), or the underlying GaAs(100) surface? If the atomically resolved AFM images are of the adsorbed species we should see not the (1x1) structure but the (1x2) structure because the GaAs(100) surface has two dangling bonds and Cl (Cl$^-$) and H (H$^+$) can make a bond with each dangling bond except S (S^{2-}) which can make bonds with two dangling bonds. Thus, we think the AFM images are of the underlying GaAs(100) except the S terminated surface where the AFM image may be of the adsorbed S layer. The atomically resolved stable image of the S terminated surface in wide portential region even after Na_2S solution was replaced by H_2SO_4 shows the very strong interaction between S and the GaAs surface.

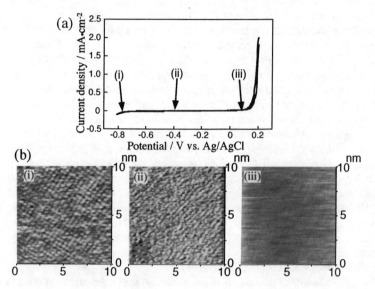

Figure 4. (a) Current-potential relation of the p-GaAs(100) electrode in 10 mM H$_2$SO$_4$ solution. (b) AFM images of the p-GaAs(100) surface at (i) -0.75 V, (ii) -0.4 V, and (iii) +0.1 V.

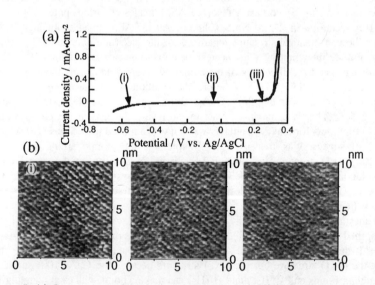

Figure 5. (a) Current-potential relation of the p-GaAs(100) in 2 mM Na$_2$S solution. (b) AFM images of the p-GaAs(100) surface at (i) -0.55 V, (ii) -0.05 V, and (iii) +0.2 V.

Electrodeposition of Cu on the GaAs(100) surface. Effects of surface defects on the electrochemical characteristics of semiconductor electrode have been known for long time but microscopic investigation on the effects has not been carried out. In this section, we investigated the electrochemical deposition of Cu on the p-GaAs(100) surfaces of various surface structures, i.e., with various defect densities, by using electrochemical AFM to visually demonstrate the effects of surface defects on the reactivity. Effects of applied potential and the concentration of Cu^{2+} ion in solution on the way how Cu deposition took place were also studied.

Figure 6 shows a current-potential relation of the p-GaAs(100) electrode in a solution containing 9 mM HCl + 1 mM $CuCl_2$. Cu deposition took place at potentials more negative than -0.1 V and anodic stripping peaks were observed at +0.01 V and +0.09 V. How the electrodeposition of Cu proceeded was strongly dependent on the structure of the substrate *(15)*. Typical examples are shown in Figures 7 and 8. Figure 7 shows a series of AFM images taken before (a) and during (b) - (f) the bulk deposition of Cu on a relatively flat surface of the p-GaAs(100) in 9 mM HCl + 1 mM $CuCl_2$ solution. Potential was stepped from +0.1 V to -0.15 V at the time indicated by a thick arrow in Figure 7(b) . Figure 7(b) clearly shows that immediately after potential was stepped to -0.15 V, a large number of small grains were generated with spacings of several tens of nm. These initial deposits of Cu on the surface seemed to act as effective nucleation centers and the initial growth of these grains seemed to be three-dimensional. AFM images were continuously captured at -0.15 V and are shown in Figures 7(c) - (e). As time progressed the grains of Cu overlapped with each other and finally truncated pyramidal structures of relatively uniform size were formed (Figure 7(e)). Upon stepping the potential back to +0.1 V as indicated by a thick arrow in Figure 7(f), Cu deposits were removed immediately within the time domain of AFM measurement and the surface was returned to a state similar to that before the Cu deposition (Figure 7(f)).

When potentials more positive than 0 V were applied to the GaAs electrode in HCl solution, truncated pyramidal structures were formed on the surface as a result of the anodic dissolution of the GaAs electrode. Actually, it was hard to obtain an atomically flat surface over a wide range if the electrode was kept in the anodic potential region where no Cu deposition took place. The surface shown in Figure 7(a) was one of the flattest surfaces we have observed. To examine the effect of the surface structure on Cu deposition, we also monitored the deposition process of Cu on the surface with the pre-formed truncated pyramids of relatively uniform size as shown in Figure 8(a). Totally different time sequences were observed for the electrodeposition of Cu on this surface as shown in Figures 8(b) - (e). As soon as stepping the potential to the bulk deposition region (-0.15 V) as indicated by a thick arrow in Figure 8(b), the electrochemical deposition of Cu started to occur along the pre-formed structure (Figure 8(b)). Grains observed in Figure 7 were not seen in this image. The truncated pyramids grew with the progress of deposition mainly in vertical direction as shown in Figures 8(c)-(e). Thus, the Cu deposition process seemed to occur predominantly at the defects. Again Cu deposits were removed immediately within the time domain of AFM measurement and the surface was returned to a state similar to that before the Cu deposition (Figure 8(f)).

The effect of the applied cathodic potential on the way how Cu electrodeposition proceeded was investigated on the flat p-GaAs(100) surface in a solution containing 10

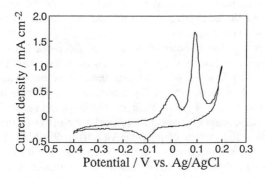

Figure 6. Current-potential relation of p-GaAs(100) in 9 mM HCl + 1 mM CuCl$_2$ solution.

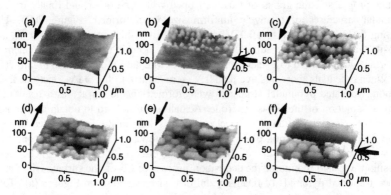

Figure 7. Sequentially obtained AFM images of the p-GaAs electrode in the relatively flat region in 9 mM HCl + 1 mM CuCl$_2$ solution (a) at +0.1 V vs. Ag/AgCl, (b) while the potential was pulsed to -0.15 V, (c)-(e) at -0.15 V, and (f) while the potential was pulsed back to the initial potential (+0.1 V). Time after the application of -0.15 V at the beginning of imaging was (c) 4 s, (d) 12 s, (e) 20 s, and (f) 74 s. Thick arrows indicate the onset of deposition (b) and stripping (f) of Cu. Arrows beside the figure indicate the scan direction of the tip.

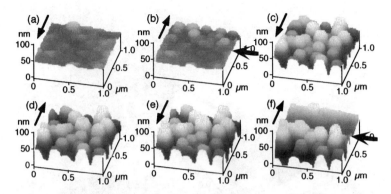

Figure 8. Sequentially obtained AFM images of the p-GaAs surface with pre-formed truncated pyramidal structures in 9 mM HCl + 1 mM CuCl$_2$ solution (a) at +0.1 V vs. Ag/AgCl, (b) while the potential was pulsed to -0.15 V, (c)-(e) at -0.15 V, and (f) while the potential was pulsed back to the initial potential (+0.1 V). Time after the application of -0.15 V at the beginning of imaging was (c) 36 s, (d) 116 s, (e) 196 s, and (f) 204 s. Thick arrows indicate the onset of deposition (b) and stripping (f) of Cu. Arrows beside the figure indicate the scan direction of the tip.

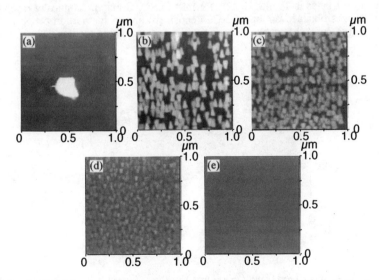

Figure 9. AFM images of the p-GaAs(100) electrode obtained 30 s after potential was pulsed from +0.05 V to (a) -0.05 V, (b) -0.1 V, (c) -0.2 V, (d) -0.3 V, and (e) -0.4 V for 30 sec in 10 mM H$_2$SO$_4$ + 1 mM CuSO$_4$ solution.

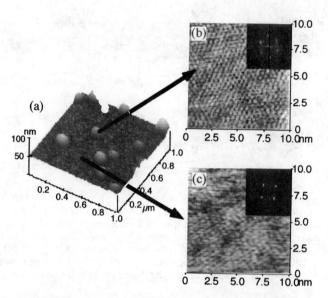

Figure 10. (a) AFM images (1 μm x 1 μm) of the p-GaAs(100) surface with several Cu islands in 10 mM H_2SO_4 + 1 mM $CuSO_4$ solution. Atomically resolved AFM images obtained on top of a Cu deposit (b) and of the portion between Cu deposits (c). Inset shows two-dimensional Fourier spectra of the images.

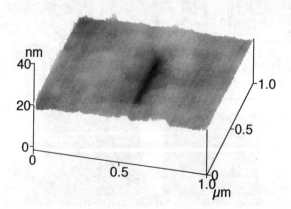

Figure 11. AFM image (1000 x 1000 nm^2) of the p-GaAs(100) surface in 10 mM H_2SO_4 at the open circuit potential after only x-direction was scanned for 30 min with stronger force (10 nN).

mM H_2SO_4 + 1 mM $CuSO_4$. Figures 9(a) - (e) are the AFM images obtained 30 s after the potential was pulsed from +0.05 V to various potentials. When the electrode potential was pulsed to -0.05 V which is almost equal to the onset potential of the bulk Cu deposition (Figure 9(a)), small number of large clusters of Cu were formed on the surface. As the applied potential became more negative, the number of Cu clusters increased but the size of each cluster decreased. Cu clusters were not observed on the GaAs(100) surface when -0.4 V was applied as shown in Figure 9(e). In this case, Cu deposition seemed to occur almost uniformly on the whole GaAs(100) surface. These results show that we were able to control the size and the numbers of the Cu clusters formed on the GaAs(100) surface by adjusting the applied overpotentials. The Cu clusters grew in size with time but the total number of Cu clusters was constant from the initial stage of the deposition. These results confirm that the initial deposits of Cu on the surface acted as effective nucleation centers.

The size and numbers of Cu clusters were also influenced by the concentration of Cu^{2+} ion.

Figure 10(a) shows an AFM image of the p-GaAs(100) surface with several Cu deposits in 10 mM H_2SO_4 + 1 mM $CuSO_4$ solution at -0.2 V. Cu islands of relatively uniform size were found on the GaAs surface. High resolution AFM images of the surface of the top of a Cu island and the portion between Cu islands are shown in Figures 10(b) and (c), respectively. Figure 10(b) shows that topmost atoms have a hexagonal structure with a nearest-neighbor distance of 0.26 ± 0.04 nm, which is almost equal to the known lattice constant (0.256 nm) of bulk Cu in the (111) basal plane. Thus, Cu deposits seemed to have a closed packed structure. On the other hand, the atomic arrangement of square lattice with atomic distance of about 0.4 nm corresponding to the underlying structure of the GaAs (100)-(1x1) structure was observed at the region between Cu deposits. The direction of atom rows of the Cu(111) structure on top of each Cu cluster was different from clusters to clusters. This result suggests that the bulk deposition of Cu on the GaAs(100) surface proceeded without strong influence of the orientation of the underlying GaAs(100) surface.

AFM tip induced modification of GaAs(100) surface structure. Nano-fabrication of semiconductor surface by using scanning probe microscope has been widely studied, although most of the studies have been carried out in UHV condition *(16-18)*. Here, we were successful to modify the GaAs(100) surface by an AFM tip in 10 mM H_2SO_4 solution at various electrode potentials *(19)*. Figure 11 shows a typical AFM image of the p-GaAs(100) surface at an open circuit potential (1 μm x 1 μm) after the tip was scanned only x-direction of 500 nm with a scan rate of 25 lines/s at an open circuit potential for 30 min. While the typical force for the imaging was less than 1 nN, 10 nN was applied for the surface modification. A wedge is observed at the central part of the image where the tip was initially scanned with the strong force. The depth of the wedge was proportional to the number of scans. Figure 12 shows a relation between the electrode potential and the depth of the fabricated wedge which was created by 45000 scans. If the potential was kept in the cathodic region, surface was not modified at all. The more positive the potential, the larger the anodic current and the deeper the wedge. However, when the electrode potential became more positive than +0.075 V, the wedge structure was not clearly observed and large hill-and-valley structures were formed everywhere on the surface because the dissolution process on the p-GaAs surface

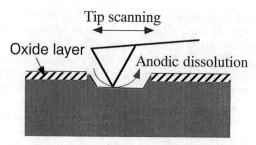

Figure 12. The mean depth of the wedge and the steady current as a function of the electrode potential for the p-GaAs(100) electrode in 10 mM H_2SO_4 solution.

proceeded violently. These results suggest that the formation of wedge structure on the surface requires both the scanning of the tip and the flow of anodic current, i.e., the modification is the result of the tip-induced electrochemical etching. The wedge structure was formed even at the open circuit potential (Figure 11), maybe because small anodic current corresponding to the GaAs dissolution flowed to compensate the cathodic photocurrent which was generated by laser light used for sensing the deflection of the cantilever. Figure 13 shows a proposed model how the tip-induced electrochemical etching proceeds. In anodic potential region, the GaAs surface is covered with oxide/hydroxide which passivates the electrochemical dissolution. As these layers are mechanically weak, they were easily perturbed or removed by the scanning of the AFM tip. Dissolution of the GaAs proceeds selectively where the oxide layer is removed.

When Cu was electrochemically deposited onto the GaAs surface, part of which was modified by the AFM tip, Cu was deposited selectively at the scratched portion (12). This result may lead to a novel technique for the sub-micron fabrication of semiconductor.

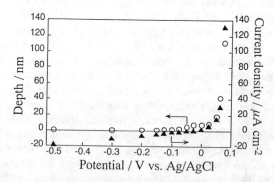

Figure 13. A proposed model for the AFM tip-induced electrochemical etching process.

Acknowledgment

We are grateful to Mr. H. Fujita of Mitsubishi Chemical Corporation for providing with GaAs single crystals. This work was partially supported by Grant-in-Aids for Scientific Research (04555191) and for Priority Area Research (05245202, 06236201, 07225202, 06239204, 07241203) from the Ministry of Education, Science, Sports and Culture, Japan and by the Kato Science Foundation.

References

(1) *Electrochemistry at Semiconductor and Oxidized Electrode;* Morrison, S. R., Ed; Plenum Press: New York, 1980.

(2) Gerischer, H. *J. Electroanal. Chem.* **1975**, *58*, 263.

(3) Damjanovoic, A.; Setty, T. H. V.; Bockris, J. O' M. *J. Electrochem. Soc.* **1966**, *113*, 429.

(4) Hottenhuis, M. H. J.; van den Berg, A. L. M.; van den Eerden, J. P. *Electrochim. Acta* **1988**, *33*, 1519.

(5) For example, *Nanoscale Probes of the Solid/Liquid Interface*; Gewirth, A. A.; Siegenthaler, H., Eds; Kluwer Acd. Publ., Dordrecht, 1995.

(6) Uosaki, K.; Koinuma, M. *Faraday Discuss.* **1992**, *94*, 361.

(7) Uosaki, K.; Ye, S.; Sekine, N. *Bull. Chem. Soc. Jpn.* **1996**, *69*, 275.

(8) Uosaki, K.; Kita, H. *J. Electroanal. Chem.* **1989**, *259*, 301.

(9) Koinuma, M.; Uosaki, K. *Surf. Sci. Lett.* **1994**, *311*, L737.

(10) Biegelesen, D. K.; Bringans, R. D.; Northrup, J. E.; Swartz, L. E. *Phys. Rev. B* **1990**, *41*, 5701.

(11) Koinuma, M.; Uosaki, K. *J. Electroanal. Chem.* **1996**, in press.

(12) Yao, H.; Yau, S. L.; Itaya, K.*Surf. Sci.* **1995**, *335*, 166.

(13) Fan, J. F.; Oigawa, H.; Nannichi, Y. *Jpn. J. Appl. Phys.* **1988**, *27*, L1331.

(14) Mönch W.; Gant, H.*J. Vac. Sci. Technol.* **1980**, *17*, 1094.

(15) Koinuma, M.; Uosaki, K.*Electrochem. Acta* **1995**, *40*, 1345.

(16) Dagata, J. A.; Schneir, J.; Harary, H. H.; Evans, C. J.; Postek, M. T.; Bennett, J. *Appl. Phys. Lett.* **1990**, *56*, 2001.

(17) Snow, E. S.; Campbell, P. M.; McMarr, P. J. *Appl. Phys. Lett.* **1993**, *63*, 749.

(18) Whidden, T. K.; Allgair, J.; Jenkins-Gray, A.; Kozicki, M. N. *J. Vac. Sci. Technol. B* **1995**, *13*, 1337.

(19) Koinuma, M.; Uosaki, K.*Surf. Sci.* **1996**, in press.

Chapter 15

Microfabrication and Characterization of Solid Surfaces Patterned with Enzymes or Antigen–Antibodies by Scanning Electrochemical Microscopy

H. Shiku, Y. Hara, T. Takeda, T. Matsue[1], and I. Uchida

Department of Applied Chemistry, Graduate School of Engineering, Tohoku University, Sendai 980–77, Japan

The microfabrication and characterization of glass surfaces patterned with enzymes (diaphorase, horseradish peroxidase(HRP)) or antigen-antibodies (carcinoembryonic antigen (CEA), human chorionic gonadotropin (HCG) and human placental lactogen (HPL)) were studied using scanning electrochemical microscopy (SECM). Localized enzymes and antigen-antibody complexes with labeled enzymes were characterized on the basis of detection of catalytic current for ferrocenylmethanol by SECM. The SECM technique was extended to the enzyme-linked immunosorbent assay (ELISA). This method detects as low as ~ 10^4 CEA molecules in a single microspot. We also demonstrated a novel dual assay using microfabricated glass substrates with anti-HCG and anti-HPL microspots.

Scanning electrochemical microscopy (SECM), a member of the scanning probe microscopies (SPM), uses an ultramicroelectrode as the probe to characterize the localized electrochemical properties of the surfaces (1-3). Although the lateral resolution of SECM is inferior to that of STM or AFM because it is difficult to fabricate a small probe microelectrode with atom-size radius, SECM has capabilities of detecting chemical reactions. In addition, it can induce chemical reactions in an extremely small volume. Using the unique properties of SECM, a single molecule (4) or a radical with a short lifetime (5) was recently detected.

We report here the SECM characterization of glass surfaces micropatterned with enzymes or antigen-antibodies. Patterned enzymes or enzyme-labeled complexes of antigen and antibody were viewed by SECM based on monitoring catalytic reactions of the patterned enzymes. We have drawn lines and spots of diaphorase, horseradish peroxidase (HRP), HRP-labeled antigen-antibody complexes with carcinoembryonic antigen (CEA), human chorionic gonadotropin (HCG) and human placental lactogen (HPL) on glass substrates by a motor-driven glass capillary pen. The dimension of the patterns can be controlled by changing the outer tip radius of the pen. It is also possible to create a pattern with more than two proteins in a very small area with μm dimensions. We also demonstrate here a dual assay of HCG and HPL at a photofabricated glass substrate.

---

[1]Corresponding author

Experimental

Materials. NADH (> 98%, Sigma Chemical Co.), HRP, (Funakoshi), bovine serum albumin (BSA, Wako Pure Chemical Industries), glutaraldehyde (GA, 25%, Wako Pure Chemical Industries), (3-aminopropyl)triethoxysilane (GR grade, Tokyo Chemical Industry), n-octadecyltrichlorosilane (GR grade, Tokyo Chemical Industry), and Tween 20 (Kanto Chemical Co.) were used as received. Ferrocenylmethanol (FMA) was synthesized by reduction of ferrocenecarboxyaldehyde (98%, Aldrich Chemical Co.). Completion of the reaction was confirmed with thin-layer chromatography. Diaphorase purified from *Bacillus stearothermophilus* (EC 1. 6. 99. -) was donated by Unitika. CEA, HCG, HPL, anti-CEA, anti-HCG, anti-HPL, HRP-labeled anti-CEA, HRP-labeled anti-HCG, and HRP-labeled anti-HPL were donated by Mochida Pharmaceutical Co. CEA is a glycoprotein and its molar mass is ca. 180,000 to 200,000 Da. Both HCG and HPL are polypeptide hormones, and their molar masses are 38,000 and 22,000 Da, respectively.

Micropatterning with Biomaterials on Glass Substrates. A Pyrex glass tube (Narishige, G-1) were pulled with a capillary maker (Narishige, PD-5). The tip of the capillaries were cut flat to the desired size. Typical tip radii of the capillaries were $3 \sim 30$ μm. This capillary was used as a pen for drawing micropatterns. The capillary pen filled with a biomaterial buffer solution as ink was attached to the motor-driven XYZ stage (Chuo Precision Industrial Co., M-9103). The capillary was in contact with glass substrate vertically and then moved along the substrate under observation with an optical microscopy. The line width and spot size depend on the tip diameter of the capillary pen.

Preparation of Diaphorase-Lined Substrate. A glass slide was successively dipped into a 10 mM (3-aminopropyl)triethoxysilane/benzene and a 1 % (v/v) GA/water solutions for 30 min. Between these steps, the glass substrate was washed thoroughly with water under ultrasonication. A 0.1 mM diaphorase/phosphate buffer solution was used as ink for the glass capillary pen (tip radius, 10 μm). The capillary was scanned over the substrate repeatedly with 50 μm distance. The substrate was then washed thoroughly with water under ultrasonication.

Preparation of HRP Immobilized Substrate. A glass slide was dipped into a 10 mM (3-aminopropyl)triethoxysilane/benzene solution for 8 h. A phosphate buffer solution containing a 530 units/ml HRP, a 1.7 mg/ml BSA, and a 0.8 % (v/v) GA was spotted on the silanized glass slide at intervals of 100 μm by a glass capillary pen (tip radius, 20 μm). The substrate was then washed thoroughly with buffer solution.

Preparation of the Microspotted CEA Substrate. The antigen-antibody was immobilized by a sandwich method. Antigen molecules have an affinity for the hydrophobic surface and adsorb physically on the glass slide treated with long-chain alkylsilane. A glass slide was successively dipped into a 10 mM n-octadecyltrichlorosilane/benzene solution for 8 h and a 500 μg/ml anti-CEA/phosphate buffer solution for 2 h. A CEA/phosphate buffer solution was then spotted on the anti-CEA adsorbed glass surfaces at intervals of 100 μm by the glass capillary pen. The optical microscopic observation demonstrated that the sizes of the spots were almost uniform and that the average radius of the spots was ca. 20 μm. If we assume that the spot is hemispherical, the volume of the spot is 17 picoliters. After washing with water, the substrate was soaked into a 15 μg/ml HRP-labeled anti-CEA/0.1 % Tween 20/phosphate buffer solution for 20 min, followed by a thorough washing with water.

Preparation of Lined Anti-CEA-Adsorbed-Polystyrene Beads Substrate.
A glass slide was successively dipped into a 10 mM n-octadecyltrichlorosilane/
benzene solution for 8 h and a 500 μg/ml anti-CEA/phosphate buffer solution for 2 h.
Onto the substrate we drew lines at intervals of 30 μm with a glass capillary (tip
radius, 5 μm) filled with a 2.0 μg/ml CEA/phosphate buffer solution. The substrate
was then washed thoroughly with water. Polystyrene beads (Kodak, diameter 10
μm) adsorbing anti-CEA were prepared by soaking the bear beads into a 500 μg/ml
anti-CEA /phosphate buffer solution for 2 h, followed by washing three times in a
0.1 M phosphate buffer solution. The anti-CEA adsorbed polystyrene
beads/phosphate buffer solution was dispersed on the CEA patterned substrate. The
bead-adsorbed substrate was gently washed with water.

Preparation of Dual Assay Substrates. A positive photoresist (Tokyo Oka,
OFPR-5000) was spin-coated on a glass slide, followed by prebaking at 80 ℃ for
30 min. The photoresist-coated substrate was irradiated with a xenon lamp through a
photomask with square patterns. The image was obtained in a developer (Tokyo Oka,
NMD-W). The resist-patterned substrate was then etched chemically in a 10 % HF
solution for 5 min to form uplifted flat squares. The average height of the steps was
ca. 2 μm by atomic force microscopy (Digital Instruments, Dimension 3000). The
substrate was then dipped into a 10 mM n-octadecyltrichlorosilane/ benzene solution
for 8 h. Onto the left and right plateaus were, respectively, spotted a 760 μg/ml anti-
HCG/phosphate buffer and a 540 μg/ml anti-HPL/phosphate buffer solutions by
glass capillary pens. The substrate was soaked into a 10 mg/ml BSA/water solution
for 2 h. A sample solution containing the corresponding antigen was dropped onto
the substrate which subsequently stood for 1 h. Typically, the sample solution was a
20 IU/ml HCG or a 1.0 μg/ml HPL phosphate buffer solution and the sample
volume was 10 μl. The substrate was then dipped into a phosphate buffer solution
containing a 20 μg/ml HRP-labeled anti-HCG, a 7.0 μg/ml HRP-labeled anti-HPL,
and a 0.1 % Tween 20 for 20 min.

Measurement System. The microelectrode for SECM was fabricated as follows:
A Pt wire (radius, 7.5 μm) was etched electrochemically in saturated $NaNO_3$ to
sharpen the wire, and inserted into a softglass capillary. The tip was fused in a small
furnace at 320 °C in vacuo to coat the Pt filament with the soft glass. The tip was
then polished with a diamond grinder (# 5000) on a turntable (Narishige, Model EG-
6). The Pt disk radius including the insulating part was ca. 30 μm. The Pt disk
radius was determined from the steady-state voltammogram and found to be 2.4 μm.
The measurements were carried out in a two-electrode configuration. The current was
amplified with a Keithley model 428 amplifier. Movement of the microelectrode tip
was performed by means of a motor-driven XYZ stage (Chuo Precision Industrial
Co., M-9103). The resolution of the XYZ stage was 0.1 μm. The measurement
solution was 0.1 M KCl/0.1 M phosphate buffer containing the corresponding
enzyme substrate (5 mM NADH or 0.5 mM H_2O_2) and FMA as the electron mediator.
The potential of the tip electrode was held at 0.4 or 0.05 V vs. Ag/AgCl. The tip was
scanned over the substrate at 9.8 μm/s. The distance between the electrode and the
substrate surfaces was estimated using a working curve which represents a
theoretical oxidation current vs. distance profile (6).

Results and Discussion

Figure 1 shows a two dimensional SECM image of the diaphorase-immobilized line-and-space and a cross-section of the SECM image in 0.5 mM FMA/5 mM NADH/0.1 M KCl/0.1 M phosphate buffer (pH 7.5). In the SECM measurements, the distance between the tip and the substrate was maintained at 8 μm. The tip potential was set at 0.4 V vs. Ag/AgCl to monitor the oxidation current of FMA. The bright part indicates the area where the enzyme activity is high. One can see the enzyme line clearly at the glass substrate at intervals of 50 μm. The oxidation current of FMA increases when the microelectrode is scanned above the diaphorase-line. At the microelectrode tip, FMA is oxidized to FMA$^+$ which diffuses into the substrate and is reduced back to FMA by the diaphorase catalyzed reaction. Therefore, the enzyme reaction occurs only when the tip runs above the enzyme line (feedback mode (7)). In the feedback mode, it is possible to calculate the local diaphorase activity from the catalytic oxidation current (8, 9). In previous paper, we calculated the steady-state current at various tip-substrate distances by digital simulation (9). Since the steady-state current vs. distance profile depends on the surface concentration of diaphorase, the apparent surface concentration is estimated from the oxidation current. The surface concentrations of diaphorase at the center of the line and the gap were estimated to be $\sim$ 2 x 10^{-12} and $\sim$ 6 x 10^{-13} mol cm^{-2}, respectively.

Figure 2 shows an SECM image (tip-substrate distance, 8 μm) and its cross-section of the substrate spotted with HRP in 1.0 mM FMA/0.5 mM H$_2$O$_2$/0.1 M KCl/0.1 M phosphate buffer (pH 7.0). HRP catalyzes the oxidation of FMA by H$_2$O$_2$ to generate FMA$^+$. In this case, the tip potential was set at 0.05 V vs. Ag/AgCl to detect the reduction current of FMA$^+$ which was generated continuously from the active HRP area and diffused into the solution (generation mode). When the FMA$^+$ is used as a mediator, it is also possible to visualize the substrate in the feedback mode. In that case, the redox cycling between the tip and the substrate enhances the reduction current observed at the tip. In a previous paper, we reported the difference in the SECM images between the two detection modes (10). Although the generation mode gives lower background currents and larger responses, it loses the positional resolution and makes the quantitative analysis complicated. Nevertheless, the generation mode is suitable for characterizing surfaces low enzyme activities. The concentration of H$_2$O$_2$ had almost no influence on the reduction current profile when the concentration is above 0.1 mM.

Figure 3 shows an SECM image of the substrate with CEA/HRP-labeled anti-CEA spots in 1.0 mM FMA/0.5 mM H$_2$O$_2$/0.1 M KCl/0.1 M phosphate buffer (pH 7.0). The potential of the tip was 0.05 V vs. Ag/AgCl to monitor the reduction current of FMA$^+$. The appearance of the spots with increased reduction current is caused by the HRP-catalyzed reaction, which in turn indicates the localized presence of CEA. The reduction currents depend on the CEA concentration of the solution making a CEA spot. In this figure, the concentration of CEA was 0.2 μg/ml. Since the volume to make a spot was estimated as 17 picoliters, the number of CEA molecules in a single spot was as small as $\sim 10^4$. When CEA was spotted on untreated glass or silanized glass surfaces, no clear image was observed. The SECM system combined with the enzyme-linked immunosorbent assay (ELISA) can be integrated to a novel system for trace analysis of biomolecules.

Figure 4 shows an optical microscopic photograph showing alignment of anti-CEA-adsorbed polystyrene beads at the CEA immobilized lines at the glass substrate. The lines of the beads appeared at intervals of 30 μm. Although the pattern was wiped out by washing vigorously in water, the beads pattern was reproduced by dispersing the beads onto the substrate. It will be possible to pattern with living cells on the basis of the interaction between functional groups at the glass surface and receptors at the cell membranes.

The micropatterned substrates with several biomaterials can be used in an

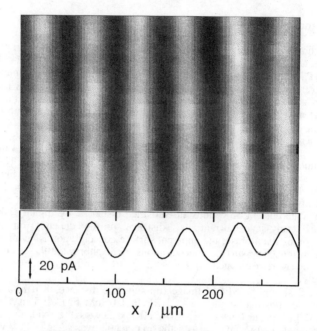

Figure 1.   An SECM image of the substrate with diaphorase immobilized (upper) and its cross-section (lower).

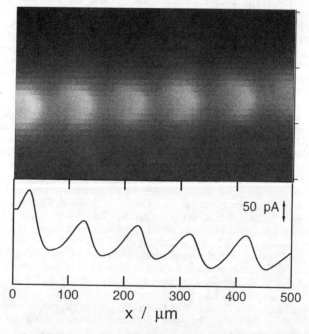

Figure 2.   An SECM image of the substrate with HRP immobilized spots (upper) and its cross-section (lower).

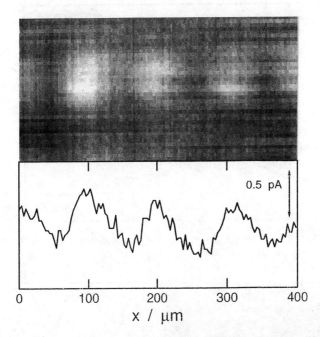

Figure 3.   An SECM image of CEA microsppotted substrate (upper) and its cross-section (lower). The CEA concentration of the solution for making spot is 0.2 μg/ml.

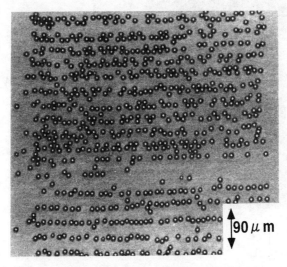

Figure 4.   A microscopic photograph showing alignment of anti-CEA-adsorbed polystyrene beads at a glass substrate with CEA immobilized lines.

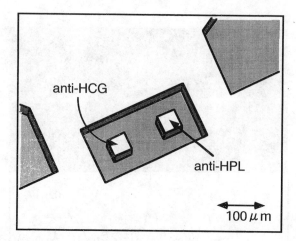

Figure 5.   The schematic drawing of the photofabricated glass substrate used in dual assay for HCG and HPL.

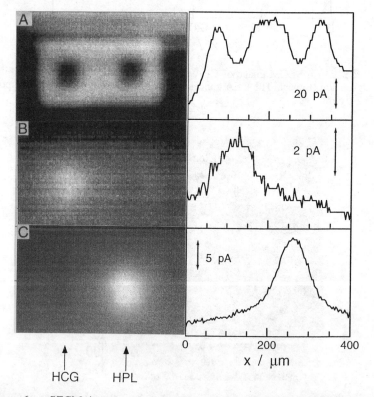

Figure 6.   SECM images of the substrates in a 1.0 mM FMA/0.5 mM H$_2$O$_2$/0.1 M KCl/0.1 M phosphate buffer solution (pH 7.0) and their cross-sections. A; no H$_2$O$_2$, potential, 0.4 V vs. Ag/AgCl. B; treated with 20 IU/ml HCG, potential, 0.05 V vs. Ag/AgCl. C; treated with 1.0 μg/ml HPL, potential, 0.05 V vs. Ag/AgCl.

integrated multianalyte system. We have prepared glass substrates with two microfabricated antibody spots for a novel combination SECM-ELISA system. Figure 5 shows the schematic drawing of the substrate used in the present study; two 50 x 50 μm^2 plateaus were fabricated in the 300 x 150 μm^2 valley. We microspotted an anti-HCG/buffer solution at the left plateau and an anti-HPL solution right plateau. A sample solution containing HCG or HPL was dropped on the substrate, which was subsequently treated with the solution containing HRP-labeled anti-HCG and anti-HPL. Figure 6 shows SECM images and their cross-sections of the substrate in 1.0 mM FMA/0.1 M KCl/0.1 M phosphate buffer (pH 7.0). In figure 6A, the tip potential was set at 0.4 V vs. Ag/AgCl. At this potential, we monitor the oxidation current of FMA which depends on the tip-substrate distance. Therefore, the resulting image reflected the topography of the substrate. Two plateaus are visible as the lower current area. After adding 0.5 mM H_2O_2, we switched the tip potential to 0.05 V vs. Ag/AgCl to monitor the local FMA^+. Since FMA^+ is generated by the catalytic reaction of the labeled HRP, the SECM measurement based on the reduction of FMA^+ gives the image of localized HCG and HPL at the substrate. The SECM image of the substrate treated with a 20 IU/ml HCG or a 1.0 μg/ml HPL solution clearly shows the bright spot at the corresponding plateau (Fig. 6B, C). If we microfabricate many plateaus immobilized with different antibodies, the present SECM-ELISA system can be integrated into a rapid and compact multianalysis system for biosamples with small volumes.

References

1. Arca, M.; Bard, A. J.; Horrocks, B. R.; Richards, T. C.; Treichel, D. A. *Analyst,* **1994**, 119, 719.
2. Bard, A. J.; Fan, F. -R. F. *Faraday Discuss.,* **1991**, 94, 1.
3. Engstrom, R. C.; Pharr, C. M. *Anal. Chem.,* **1989**, 61, 1099A.
4. Bard, A. J.; Fan, F. -R. F. *Science,* **1995**, 267, 871.
5. Zhou, F.; Bard, A. J. *J. Am. Chem. Soc.,* **1994**, 116, 393.
6. Yamada, H.; Shiku, H.; Matsue, T.; Uchida, I. *Bioelectrochem. Bioenerg.,* **1994**, 33, 91.
7. Pierce, D. T.; Bard, A. J. *Anal. Chem.,* **1993**, 65, 3598.
8. Pierce, D. T.; Unwin, P. R.; Bard, A. J. *Anal. Chem.,* **1992**, 64, 1795.
9. Shiku, H.; Takeda, T.; H. Yamada, H.; Matsue, T.; Uchida, I. *Anal. Chem.,* **1995**, 67, 312.
10. Shiku, H.; Matsue, T.; Uchida, I. *Anal. Chem.,* **1996**, 68, 1276.

Chapter 16

Electroactive Polymers: An Electrochemical and In Situ Scanning Probe Microscopy Study

P. Forrer, G. Repphun[1], E. Schmidt, and H. Siegenthaler

Departement für Chemie und Biochemie, Universität Bern, Freiestrasse 3, CH 3012 Bern, Switzerland

The electrochemical and morphological properties of Polythiophene (pT) and Polyhydroxyphenazine (pOPh) films, electropolymerized in varying electrolyte solutions at different electrochemical polarization conditions, are investigated by voltammetry and in-situ SPM. Depending on the electropolymerization solution, pT films with compact or heterogeneous morphology are formed. Cyclic and linear sweep electropolymerization of pOPh yields films with globular morphology with different thickness and diameter of the globules. Both films can be imaged by SFM, whereby enhanced resolution is achieved by changing from contact to tapping mode. STM imaging of pOPh is only possible at reduced films. The charge exchanged during the redox process amounts to only 1-5% of the charge consumed during electropolymerization in the case of pT films, and to 100% in the case of pOPh films.

During recent years the research on electrochemically active polymers has widely increased, due to their various applications in electrochemical, optical and electrical systems and devices, summarized and discussed in several review articles (*1-6*). According to Inzelt (*6*) these polymers can be classified in two classes: A first category includes polymers with covalently bound redox sites, that are either an intrinsic part of the polymer chain or are attached as pendant groups. In several systems (e.g. polyaniline, polypyrrole and polythiophene) electronic conductivity prevails along the polymer chains by conjugated π-systems with an associated conduction band. In the other polymers, without conjugated π-systems (e.g. polyhydroxyphenazine, tetrathiofulvalene and tetracyanoquinodimethane), electron transfer is presumably achieved by electron hopping between localized redox sites in sufficient proximity. A similar mechanism may hold for electron transfer between polymer chains. In a second category, without covalently bound redox sites, the polymer matrix contains ionogenic groups (e.g. sulphonate groups), to which redox groups are attached electrostatically. Examples include protonated polyvinylpyridine

[1]Corresponding author

with ionogenic bound $Fe(CN)_6^{3-/4-}$ or Nafion with $Ru(bpy)_3^{3+/2+}$. In the electrochemically active polymers the maintenance of electroneutrality usually involves additional ion exchange processes between the polymer network and electrolyte.

A large number of investigations reported in the literature concerns electrochemical studies such as voltammetry, chronoamperometry and frequency response analysis (*6*). Another important aspect is the micro- and nano-morphology of the polymers in relation to their electrochemical properties. With the advances of in-situ scanning probe techniques (SFM, STM) the interest in this aspect has grown. First morphological studies of polymers with electrochemical STM under potential control were performed by *Fan et al.* (*7*) at polypyrrole films on Pt. Recently, STM has been used by *Fan et al.* (*8*) to determine the thickness of a polymer film on an electrode by monitoring the tunneling current during the approach of the tip from the electrolyte through the film to the substrate. *Nyffenegger et al.* investigated polyaniline-coated carbon electrodes by in situ SFM and STM (*9, 10*). Thereby, the growth of the polymer coating and its swelling during reduction and oxidation could be monitored directly.

The present paper reports voltammetric and in-situ scanning probe microscopy (SFM, STM) results of polythiophene and polyhydroxyphenazine films. In both systems the redox sites are covalently bound, but they differ in their conduction mechanism:

Polythiophene (pT) represents an example of a polymer with a conjugated π-system along the chains forming a conduction band. PT can be grown electrochemically by anodic polymerization from organic (*11-18*) and aqueous (*19, 20*) solutions on noble metal or carbonaceous substrates. The following reaction scheme for the electropolymerization, involving oxidized oligomers as intermediates, has been suggested by *Roncali* (*21*):

$$n \; \underset{S}{\diagup\!\!\!\diagdown} \longrightarrow \left[\underset{S}{\diagup\!\!\!\diagdown} \right]_n + 2\,n\,H^+ + 2\,n\,e^-$$

(1)

Polythiophene is redox-active, but different values (ranging from 1 electron per 5 to 1 electron per 2 thiophene units) have been proposed in the literature for the charge stoichiometry (*22*).

Polyhydroxyphenazine (pOPh) is investigated as an example of a polymer without conjugated π-system. It is formed by anodic electropolymerization of hydroxyphenazine on noble metal or carbonaceous substrates (*23-26*) in a polymerization reaction

(2)

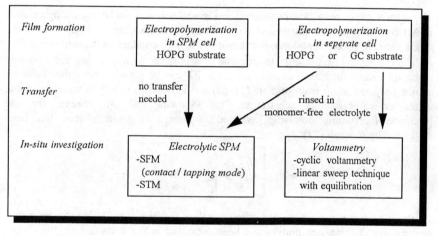

Figure 1: General experimental procedures for the measurements.

where the monomers are assumed to be linked via oxygen bindings. The polymers can be oxidized / reduced with a charge stoichiometry of ca. 2 electrons per unit (*24*), and it is believed, that the electron transport occurs in an electron hopping process between the redox centers.

The morphology of pOPh has not yet been investigated, whereas some morphological experiments are reported for pT: *Caple et. al.* (*15*) *Yang et. al.* (*27*) and *Dong et. al.* (*28*) have imaged the surface of thin polythiophene, poly(3-methylthiophene) and poly(3-bromothiophene) films by STM in the molecular range (1-50nm). They observed helical and coil-like structures in molecular and submolecular resolution in air and solution, but without potential control. Differences between preoxidized and prereduced films have been shown (*28*), and the authors have observed that oxidized films give "clearer" images. Also atomic resolution of the thiophene ring is reported. *Porter et. al.* (*29*) have performed true non-contact SFM experiments on thick (> 1μm) pT films and have found string-like structures, that cannot be observed by contact mode SFM. They explain this fact with the high forces applied to the surface in contact mode SFM.

Experimental Procedures

Electrodes and Electrolyte Solutions. All solutions were prepared from ultrapure chemicals and Milli-Q water. The polymerization solutions for different types of polythiophene (pT) and polyhydroxyphenazine (pOPh) films (specified below) were:
– 85% H_3PO_4 emulsified with thiophene. Films prepared from this solution are named in this report pT1 films .
– 1M H_3PO_4 / acetonitrile with the ratio 2:1 and 3.8mM thiophene for pT films, named pT2.
– 1M H_2SO_4 with 0.5mM hydroxyphenazine (HOPhH) for the pOPh films.
For the voltammetric analysis of the polymer films the following solutions were used:
– 85% H_3PO_4 for pT films of type pT1.
– 1M H_3PO_4 for pT films of type pT2.
– 1M H_2SO_4 for the pOPh films.
Both mechanically polished glassy carbon (GC) and freshly cleaved highly oriented pyrolytic graphite (HOPG) were used as film substrates. HOPG substrates were chosen for most of the SPM measurements in order to enable a better discrimination between the bare substrate and the polymer film.

Instrumentation. The electrochemical measurements were performed by means of a PC-controlled potentiostat and signal generator (Metrohm E611 and E612). Morphological studies were done in-situ by STM, contact and tapping mode SFM. A Nanoscope III instrument was used for the tapping mode SFM experiments, and the STM and most of the contact mode SFM measurements were carried out on a Park Universal instrument.

General Procedure for the Measurements. The general experimental procedure applied for the measurements is shown schematically in Figure 1. For the SPM measurements, the film was polymerized either directly in the SPM cell and imaged in

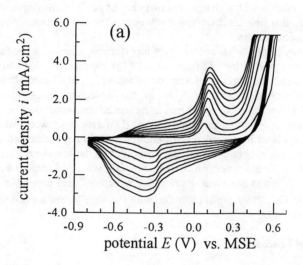

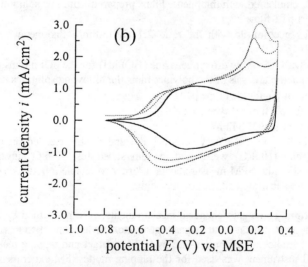

Figure 2: Cyclic voltammograms of pT type pT1 films during electropolymerization in 85% H_3PO_4 emulsified with thiophene (a), and after electropolymerization in monomer-free 1M H_3PO_4 (b). The dashed lines and the solid line in Figure (b) correspond to the first 3 cycles and the 1000th cycle, respectively. Scan rate 20mV/s.

presence of the monomer, or was prepared in a separate cell and imaged after transfer into a monomer-free electrolyte. In addition, purely voltammetric measurement routines were performed with films electropolymerized in a separate cell.

Electropolymerization Routines and Cyclic Voltammetry

Polythiophene. PT was polymerized by cyclic sweep polarization between -0.8V vs. MSE and different polymerization potentials specified below:

- For *polymer films of type pT1*, a polymerization potential of ca. 0.6V vs. MSE was applied. During potential cycling between -0.8V and 0.6V vs. MSE the current density i in the voltammogram increases with the number of cycles (Figure 2a). After applying 10 polymerization cycles and transfer of the electrode into the monomer-free electrolyte, the subsequently recorded cyclic voltammograms (Figure 2b). show an irreversible oxidation peak, that decreases with each cycle. The reduction peak also diminishes and shifts towards more positive potentials. After ca. 1000 cycles the voltammogram shows a reversible and stable redox behaviour with a peak at -0.25V vs. MSE, and the amount of charge exchanged in the redox reaction of the stabilized pT1 film is approximately 5% of the charge consumption during polymerization.

- For preparing *pT2 type films*, a polymerization potential > 1.47V vs. MSE was chosen. During cyclic polymerization between -0.78V and 1.9V vs. MSE (Figure 3a), the voltammogram does not show any increase of i in the potential region of the redox reaction (-0.4V < E < 0.2V vs. MSE), and only 1 oxidation peak is observed at ca. 1V vs. MSE, that changes with the number of cycles. After 10 polymerization cycles with a total anodic charge of ca. 0.6Ascm^{-2} and transfer of the film into monomer-free 1M H$_3$PO$_4$, the cyclic voltammograms recorded between -0.8V and 0.2V vs. MSE change only slightly within 1000 polarization cycles. A typical example is shown in Figure 3b with a reduction and oxidation peak at -0.2V and -0.05V vs. MSE, respectively. By cycling in this limited range, the maximum electrochemical charge exchanged in the redox reaction is 2 - 6 mAs/cm^2, which represents about 1% of the charge consumption during polymerization.

Polyhydroxyphenazine. POPh was polymerized in HOPhH solutions by two different polarization routines:

1. LSP type films: Linear potential sweep from 0.15V to 0.65V vs. MSE with equilibration at 0.65V vs. MSE.
2. CSP type films: Repetitive cyclic potential sweeps between -0.85V and 0.65V vs. MSE.

The formation of the two different film types is associated with the following voltammetric behaviour:

- During the linear sweep polymerization of a *LSP type film* (Figure 4a), the current density i increases strongly at potentials E > 0.35V vs. MSE. In the subsequent equilibration stage i decreases proportionally to $t^{-1/2}$. After transfer into monomer-free electrolyte the cyclic voltammogram of the film shows reversible redox peaks (Figure 4b) associated with a charge transfer of 2 electrons per monomer unit following the equation:

$$-OPhH^+- + 2e^- + H^+ \rightarrow -OPhH_2- \qquad (3)$$

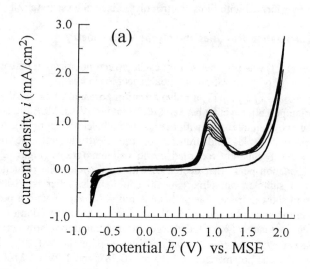

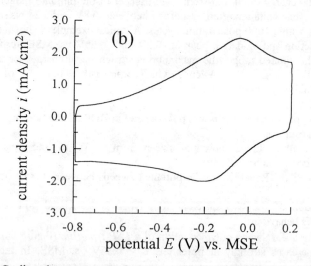

Figure 3: Cyclic voltammograms of pT type pT2 films in 1M H_3PO_4 / acetonitrile (2:1) + 3.8mM thiophene (a), and after electropolymerization in monomer-free 1M H_3PO_4 (b). Scan rate 20mV/s.

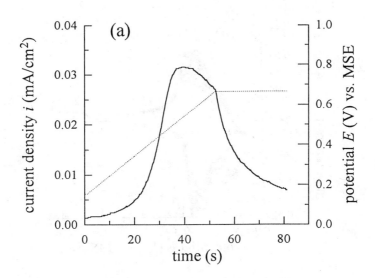

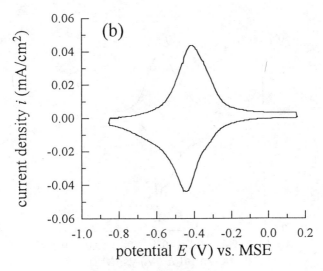

Figure 4: (a): Potential-time plot (right axis, dashed line) and current-time plot (left axis, solid line) of the electropolymerization of pOPh type LSP in 1M H$_2$SO$_4$ + 0.5mM HOPhH. (b): Cyclic voltammogram of pOPh type LSP in monomer-free 1M H$_2$SO$_4$. Scan rate 10mV/s.

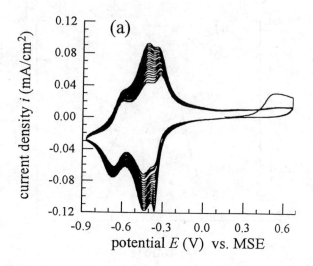

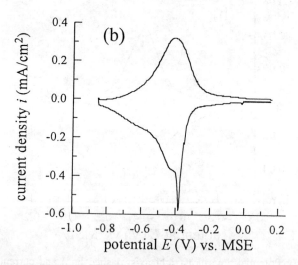

Figure 5: Cyclic voltammograms of pOPh type CSP films during the first 30 electropolymerization cycles (a), and after 1000 polymerization cycles in monomer-free 1M H_2SO_4 (b). Scan rate 10mV/s.

The resulting LSP type films grow only to a limited thickness reflected in the redox charge of the film of 2mAscm^{-2}. However, the charge consumed during electropolymerization equals the redox charge of the obtained film indicating that all built-in monomer groups are involved in the 2-electron redox reaction of the polymer.
• To prepare thicker polymer layers than LSP type films, the potential has to be cycled between -0.85V and 0.65V vs. MSE, leading to films of the *CSP type*. The cathodic potential sweep in each polymerization cycle obviously enables a further film growth in the subsequent anodic sweep of the next polymerization cycle. This leads to films growing in each polymerization cycle, as shown for the first 30 cycles in Figure 5a. The increase of redox charge is seen in the increase of the current density i in the cyclic voltammogram. As a remarkable feature, after applying ca. 20 cycles, a sharp cathodic peak appears at $E \approx 0.38$V vs. MSE. The redox behaviour of a CSP type film after 1000 polymerization cycles and transfer into monomer-free electrolyte is shown in Figure 5b: Whereas the oxidation peak resembles the shape observed at LSP type films, the reduction peak is much sharper. As in the case of LSP films, the charge exchanged in the redox reaction is close to the charge consumption during polymerization. This suggests, that even for the thicker CSP type films practically 100% of the built-in monomer groups participate in the redox process.

Equilibrium Charges and Voltammetric Capacities

To measure the equilibrium charges per unit area Δq, the polymer films were polarized as follows:
Starting at an initial potential E_S beyond the range of the redox peaks, the potential was first shifted in a linear sweep to a selected potential value E within the redox interval, followed by equilibration. In a subsequent linear sweep in the opposite direction the potential was then changed to the initial value E_S followed again by equilibration. For each such loop polarization routine (involving a cathodic and an anodic sweep) 2 values of the current density time integral

$$\Delta q(E) = \int_0^{\Delta t} idt, \qquad (4)$$

were determined, corresponding to the cathodic and anodic polarization. Δt designs the entire time interval of the potential sweep + subsequent equilibration time. The routine was repeated for various values of E throughout the entire redox range. Due to the differences in the voltammetric characteristics of the polymers the potentials were chosen as follows:
• In the case of the pT films E_S was selected at the cathodic potential limit of the redox interval: $E_S = -0.5$V and $E_S = -0.8$V vs. MSE for pT1 and pT2 type films, respectively. E was varied between E_S and 0.3V vs. MSE.
• In the case of the pOPh films the anodic potential limit E_S of the redox interval was set to $E_S = 0.15$V vs. MSE. E was varied between -0.85V vs. MSE and E_S.
From the experimentally determined values $\Delta q(E)$ the voltammetric capacity $C_{volt}(E)$ was calculated by differentiating Δq numerically with respect to E:

$$C_{volt}(E) = \partial \Delta q(E) / \partial E \qquad (5)$$

From these voltammetric capacity data an "equilibrium voltammogram"

$$i_{eq} = C_{volt}(E) \cdot \partial E / \partial t \qquad (6)$$

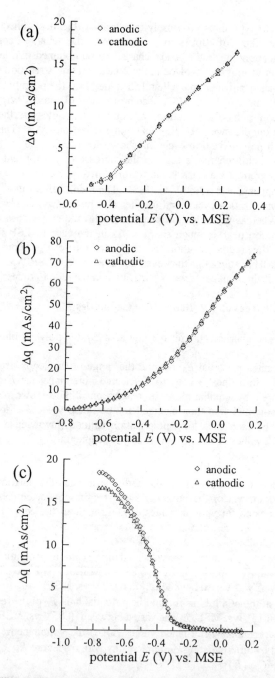

Figure 6: Equilibrium charges $\Delta q(E)$ of pT and pOPh films. (a): PT films type pT1 in 85% H_3PO_4. (b): PT films type pT2 in 1M H_3PO_4. (c): POPh films type CSP in 1M H_2SO_4.

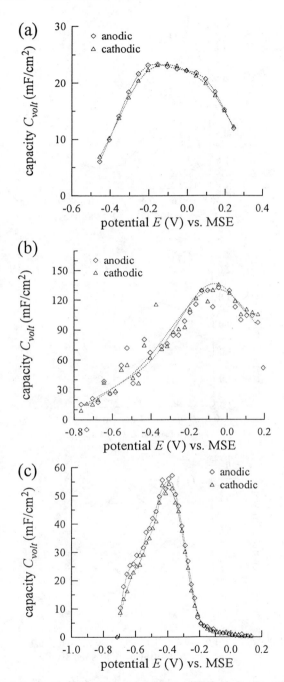

Figure 7: Voltammetric capacities $C_{volt}(E)$ of pT and pOPh films. (a): PT films type pT1 in 85% H_3PO_4. (b): PT films type pT2 in 1M H_3PO_4. (c): POPh films type CSP in 1M H_2SO_4.

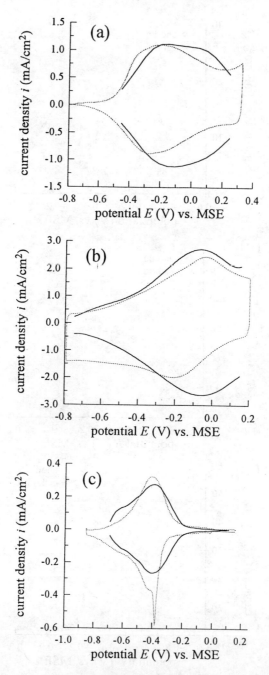

Figure 8: Comparison of measured (dashed line) and "equilibrium" (solid line) voltammograms of pT and pOPh films. (a): PT films type pT1 in 85% H₃PO₄. (b): PT films type pT2 in 1M H₃PO₄. (c): POPh films type CSP in 1M H₂SO₄.

was derived, that represents the non-faradayic current density $i_{eq}(E)$ of the redox process at any given potential scan rate $\partial E / \partial t$. These "equilibrium voltammograms" were then compared with the experimentally determined voltammograms measured with the same potential scan rates.

For the *pT films* a comparison of $\Delta q(E)$ and $C_{volt}(E)$ between the cathodic and anodic potential sweeps is shown in Figures 6a,b and 7a,b: Whereas the pT1 type films show a broad maximum of C_{volt} between -0.3V and 0.1V vs. MSE, a more narrow peak between -0.2V and 0V vs. MSE is found for the pT2 type films. For both film types, there is good agreement between the cathodic and anodic values of Δq and C_{volt}.

A comparison between the measured and "equilibrium" voltammograms (Figures 8a,b) shows a disagreement in the cathodic sweep with a marked shift of the experimental redox peaks to more negative potentials.

Figure 6c and 7c show the Δq and C_{volt} values for *CSP type* pOPh films. In difference to the pT films, the equilibrium charges for the anodic and cathodic sweeps of CSP type films agree only in the potential range $E > -0.6V$ vs. MSE. (Figure 6c). At more negative potentials the equilibrium charges of the anodic sweep exceed the corresponding cathodic values markedly. This may be due to irreversible processes or hydrogen overvoltage and has apperently no influence upon the voltammetric capacity, where no differences between the cathodic and anodic values are found at negative potentials. Similarly to the pT2 film, C_{volt} exhibits a peak at -0.4V vs. MSE.

As observed in the case of pT films, the measured and "equilibrium" voltammograms (Figure 8c) deviate also in the cathodic sweep with a marked shift of the experimental redox peaks to more negative potentials. In addition the sharp reduction peak observed in the experimental voltammogram is not seen in the equilibrium case. These differences may be due to non-equilibrium effects.

Morphological Investigations

Polythiophene. Preliminary morphological investigations of pT films were carried out ex-situ by optical microscopy and SEM. Electrodes covered by a pT1 film show large circular areas without polymer coverage. It is assumed that this is due to the inhomogeneity of the polythiophene emulsion. The pT2 type films cover the electrode completely.

More detailed investigations have been performed at pT2 type films by in-situ SPM (contact mode SFM and STM) on GC and HOPG substrates in 1M H_3PO_4 electrolyte solutions. Figure 9 shows the typical film morphology of a pT2 film on GC, as imaged by SFM: The film consists of globules with diameters of ca. 1μm and heights of ca. 150nm. As a remarkable feature, interspersed bubble-like domains of ca. 10μm diameter are found, that are mostly arranged along lines. On these bubbles, the globular film morphology is maintained. Occasionally, larger bubbles with diameters up to 75μm and the same film morphology are observed, that are compressed by the scanning SFM tip. Similar bubbles are also seen in the optical microscope. In-situ SFM images of pT2 films on HOPG show similar features in the dimensional range 1-5μm as the films grown on GC. However, in difference to the films on GC no bubble-like features have been observed up to now.

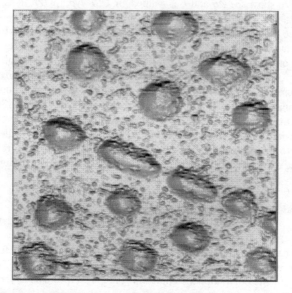

Figure 9: Contact mode SFM image of a pT2 type film on GC in 1M H_3PO_4. Scan size 60μm, greytone scale 1.59μm.

The applied contact mode SFM technique has not enabled the resolution of morphological features below ca. $0.2\mu m$, and no differences have been observed in the morphology of oxidized and reduced pT2 films.

Figure 10 shows an in-situ STM image of a reduced and an oxidized pT2 type film on HOPG. In difference to contact mode SFM imaging, the resolution of the STM images are markedly enhanced to a few tens of nm. In both examples small string-like features are observed. The reduced film (Figure 10a) appears smoother than the oxidized film (Figure 10b). From the profiles shown in Figure 10 it is evident, that the overall morphology is not affected by the oxidation state. However, in the case of the oxidized film the surface shows a remarkable enhancement of small scale features in the dimensional scale of a few nm.

Polyhydroxyphenazine. In a first set of experiments, the progress of electropolymerization on HOPG has been monitored by in-situ contact mode SFM directly in the polymerization solution containing $1M\ H_2SO_4 + 0.5mM\ HOPhH$.

An example of the SFM study of the *formation of a LSP film* is shown in Figure 11: Starting the linear sweep polarization at $E = 0.15V$ vs. MSE (where the atomic surface structure of HOPG can be imaged), the first stages of film formation (at ca. $E = 0.35V$ vs. MSE, see Figure 4a) are only indirectly observed by the loss of resolution in the SFM imaging. Such imaging conditions are presumably due to the formation of an initial polymer layer, that does not adhere well to the substrate surface. At $E = 0.55V$ vs. MSE (potential of the peak in the voltammogram of Figure 4a) the polymer apperently adheres better to the substrate, but is still mechanically affected in an irreversible way by the SFM probe. This is shown in the SFM image of Figure 11a, where the film coverage is damaged repeatedly (scan lines marked with an arrow) during the recording of the image. Finally, at the anodic equilibration potential ($E = 0.65V$ vs. MSE in Figure 4a) the surface is completely covered with a stable polymer film, that is no longer damaged by the SFM tip. As shown in Figure 11b, it consists of globules about 50-100nm diameter. From additional SFM measurements including uncovered parts of graphite, an approximate thickness of 10nm has been determined for these LSP type polymer films.

Contact mode SFM has also been used to follow the progress of the *formation of CSP type films* during electropolymerization. In these experiments, the morphology of the formed CSP films has been imaged in dependence of the number of polymerization cycles in a large interval from 10 to 1000 cycles. In general the films exhibit a globular morphology, that is also observed when imaged in a monomer-free electrolyte and will be shown in more detail in Figure 13. The influence of the number of polymerization cycles upon the morphological features of the films consists mostly in an alteration of the diameter of the observed globules. This is demonstrated in Figure 12, where the diameter of the globules as measured by in-situ SFM, and the charge exchanged in the redox process are presented in dependence on the number of polymerization cycles. It is clearly seen, that the diameter of the globules increases with increasing number of polymerization cycles, together with the increase of film thickness shown indirectly in the charge Δq. The film coverage is principally unlimited, but the amount of polymer formed during one cycle apparently decreases with each cycle, as indicated in the plot of Δq.

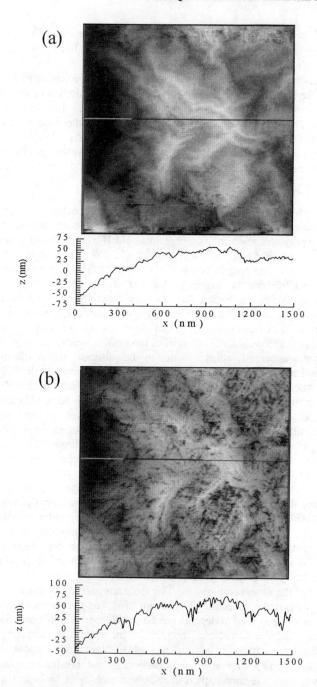

Figure 10: STM images and profiles of pT2 type films on HOPG in 1M H_3PO_4. Scan size 1.5μm. (a): Reduced film (E = -0.7V vs. MSE), greytone scale 137nm. (b): Oxidized film (E = 0.2V vs. MSE), greytone scale 175nm.

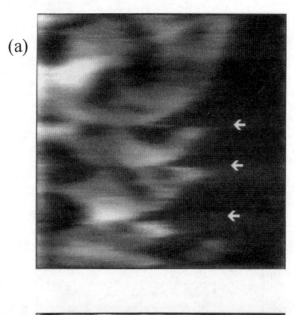

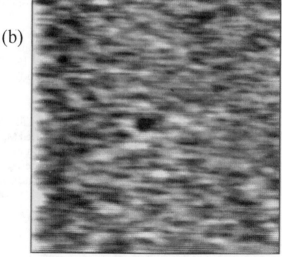

Figure 11: Contact mode SFM imaging of different stages of electropolymerization of a LSP film on HOPG in 1M H_2SO_4 + 0.5mM HOPhH. Scan size 0.5μm, greytone scale 10nm. (a): Film imaged at $E = 0.55V$ vs. MSE. The scan lines marked by arrows design the stages in the SFM image, where the film is damaged by the SFM tip. (b): Film imaged at $E = 0.15V$ vs. MSE after equilibration at $E = 0.65V$ vs. MSE.

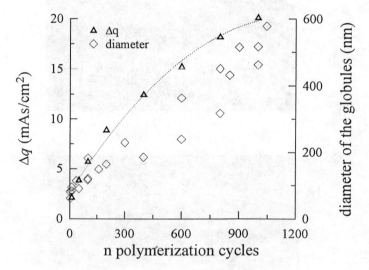

Figure 12: Characterization of the electropolymerization of CSP type films on HOPG in 1M H_2SO_4 + 0.5mM HOPhH by contact mode SFM, and charge measurements in dependence on the number of polymerization cycles. Diamond symbols and right-hand axis: Diameter of the globules observed at the film surface. Triangular symbols and left-hand axis: Charge exchanged in the redox process.

In a second set of experiments, previously electropolymerized films have been transferred into a monomer-free 1M H_2SO_4 electrolyte for SPM imaging.

LSP films show the same morphology as observed by contact mode SFM in the electropolymerization solution and are presented in Figure 11b. No changes of morphology or thickness as function of the potential have been observed.

SFM images of *CSP films* have been performed in both contact mode and tapping mode. An example of a film transferred after 800 polymerization cycles is shown in Figure 13: The contact mode image (Figure 13a) shows the typical globular morphology with globules of ca. 300nm diameter and a thickness of ca. 200nm. As in the case of the LSP films no potential dependent changes in morphology and thickness are found. If the imaging mode is changed from contact mode to tapping mode (Figure 13b) the film morphology changes in a similar way as observed on pT2 type films after changing from contact mode SFM to STM: The resolution is enhanced to about 20nm, resulting in the appearrance of smaller features at the film surface.

In-situ STM experiments on pOPh have shown, that an appropriate tunneling mode can be established only at reduced films ($E < -0.4V$ vs. MSE), with maximum tunneling currents I_T of ca. 100pA. It appears, that the tip moves into the polymer at higher I_T or more positive E and destroys the film. This is seen in Figure 14a, where I_T is plotted against the vertical tip dislocation for a pOPh film of CSP type: Upon approaching the tip to an electrode with an oxidized pOPh film, the tip penetrates into the polymer, as indicated by the large noise in region a. As the tip reaches the electrode substrate in region b I_T increases strongly. In the reduced state of the film ($E < -0.4V$ vs. MSE), a weak increase in I_T is observed (region a). This is assigned to a tunneling mode between tip and polymer, presumably with an electrolytic tunneling barrier. Further approach of the tip to the substrate leads to an increase in I_T reaching ca. 1nA close to the electrode. This increase is assigned to the penetration of the tip into the polymer, whereby the contact area tip-polymer increases and the polymer thickness decreases with increasing penetration depth. Finally, in region c , a strong increase of I_T is observed and associated with contact of the tip with the substrate. If STM experiments are performed at a reduced film ($E < -0.4V$ vs. MSE) and at sufficiently small tunneling currents ($I_T < 100pA$), it is possible to establish the tunneling conditions corresponding to region a of the reduced film. In this case reproducible STM imaging can be achieved, and the resulting topography of the film surface (Figure 14b) resembles closely the images obtained by tapping mode SFM, with small features on top of the globules.

Discussion

Voltammetric Behaviour. From a comparison of the charge consumed during electropolymerization with the charge exchanged in the redox transfer, the percentage of redox active units in the polymer films can be estimated: In the case of pT film oxidation / reduction, different values have been claimed for the charge stoichiometry (*22*) ranging between 1 electron per 5 thiophene units and 1 electron per 2 thiophene units. The experimentally determined charges exchanged in the redox process amount to ca. 5% of the charge consumed in the electropolymerization for pT1 type films, and to ca. 1% for pT2 type films. If it is assumed, that 1 electron is exchanged in the redox reaction per 5 thiophene units (acting as 1 redox center), this would mean, that every

(a)

(b)

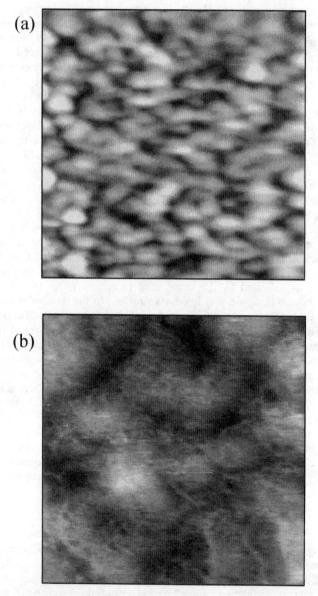

Figure 13: SFM images of a CSP film on HOPG in 1M H_2SO_4. $E = 0.15V$ vs. MSE, greytone scale 70nm. (a): Contact mode SFM, scan size 2.5μm. (b): Tapping mode SFM, scan size 1μm.

(a)

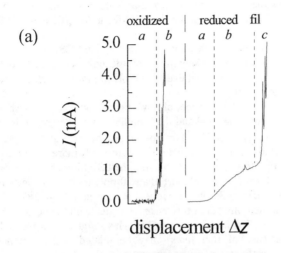

displacement Δz

(b)

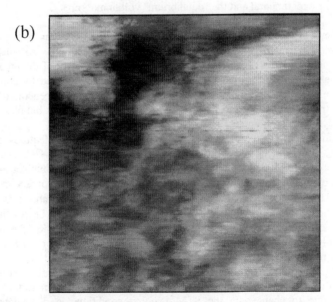

Figure 14: In-situ STM of a CSP type film on HOPG in 1M H₂SO₄. (a): Current-distance plot at an oxidized (E = 0.15V vs. MSE) and reduced (E = -0.55V vs. MSE) film. Further explanations in text. (b): STM image, scan size 2μm, greytone scale 70nm, I_T = 100pA, E = -0.45V vs MSE.

2nd redox center is active in the redox transfer of pT1 type films, and only every 10th redox center in the case of pT2 type films. However, if the maximum literature value of the redox charge stoichiometry (1 electron per 2 thiophene units) is chosen, only every 5th and every 25th redox center would be active in the pT1 and pT2 type films, respectively.

In contrast to this behaviour, pOPh films of both LSP and CSP types exhibit a 1:1 correlation between the charge stoichiometry of the polymerization and the 2-step redox reaction (24). This suggests that every HOPhH monomer built into the polymer participates in the redox reaction.

Except for the pOPh films of LSP type, all investigated polymer films show marked differences between the experimental and the "equilibrium" voltammograms calculated from the equilibrium charges $\Delta q(E)$: For both types of pT films the experimental voltammograms show a shift of the reduction peak towards more negative potentials. In the case of the pOPh films, it is remarkable that the LSP films show reversible redox peaks, while the CSP films formed by more than 20 polymerization cycles exhibit the previously mentioned sharp spike in the reduction peak at E = -0.38V vs. MSE. It should be noticed that the potential of this spike coincides with the positive limit of the potential interval where STM imaging is possible. This suggests that the spike (that is not observed in the case of thin films) may be related with a change in film conductivity. Further experiments are necessary to explain the observed differences between the experimental and the "equilibrium" voltammograms.

Morphological Properties. With the exception of the pT films of type pT1, all investigated polymer films cover the electrode substrate completely. The inhomogeneity of the pT1 films is assigned to the heterogeneity of the emulsion used for their electropolymerization. The bubble-like features observed by contact mode SFM on pT2 type films on glassy carbon are assumed to be generated along the polishing traces of the GC substrate (e.g. by the formation of gas bubbles during the electropolymerization), and may be filled with electrolyte or gas.

This assumption is also confirmed by the absence of similar features on HOPG substrates, and from ex-situ SEM images of the dried films on GC, where special features are noticed that can be interpreted as relicts of the bubbles presumably damaged by the evaporation process. While the resolution of contact mode SFM-images of the pT films on GC and HOPG is limited to features above ca. 0.2μm, a significant enhancement of resolution is obtained also by in-situ STM of the films on HOPG in aqueous 1M H_3PO_4, although a similar resolution as achieved ex-situ and in organic solvents at pT films on metallic substrates in previous work (15, 27, 28) has not been possible yet. As shown in Figures 10a and b, oxidation of the film is accompanied by an apparent enhancement of nm scale features in the STM imaging. Whether this is due to real morphological modifications of the film surface in the nm scale, or due to changes in the STM imaging conditions in dependence on the oxidation state (e.g. changes of film conductivity, variations of the barrier heights, possible mechanical interactions between tip and polymer film) cannot be clarified from the observed results.

In the case of pOPh films, a similar resolution is achieved in contact mode SFM as at pT films. As a remarkable feature of these films, the diameter of the characteristic globules increases with increasing film thickness, ranging from < 100nm in the case of

LSP films to ca. 600nm in the case of thicker CSP films. In difference to previous experiments by *Nyffenegger et al.* (*10*) on polyaniline, no changes in film thickness and morphology upon film oxidation / reduction can be observed by contact mode SFM. In contrast to these observations, SFM imaging of pOPh films in the tapping mode technique results in a remarkable enhancement of resolution into the range of ca. 20nm, showing the previously mentioned string-like features on top of the globules. A similar difference in resolution has been observed between contact and true non-contact mode SFM by *Porter et.al* (*29*) at pT films in air. Evidently, in-situ STM imaging of pOPh is possible, but only at reduced films and low tunneling currents, resulting in image patterns that resemble very closely the features recorded by tapping mode SFM.

The remarkable enhancement in the resolution obtained by the application of tapping mode SFM may be explained by smaller vertical and lateral force interactions between the SFM probe and the substrate under the conditions prevailing in the larger distance range of the tapping mode measurements. In the case of "soft" samples, it is expected that the strong force interactions applied in the contact mode will lower the resolution of small features due to mechanical deformation of the sample surface. An additional important aspect is the control of the force interactions by both, the surface charge of sample and SFM tip, and the electrolyte concentration, as investigated theoretically by *Hartmann* (*30*). This requires further experimental clarification.

Conclusions

In the present work it has been shown, that the surface morphology and growth behaviour of electropolymerized polythiophene (pT) and polyhydroxyphenazine (pOPh) films can be imaged in the nm-scale in the electrolytic environment and under potential control by appropriately chosen in-situ SPM methods: In the case of the pT films, both oxidized and reduced films can be investigated by STM, although observed differences in the resolution of nm-scale features suggest a considerable influence of the potential-dependent doping state of the film upon the prevailing tunneling conditions, that is not elucidated yet. In the case of the pOPh films the influence of the oxidation state upon STM imaging is much more pronounced, enabling stable tunneling conditions only at reduced films in a narrow potential range. This observation is interpreted tentatively with the assumption that the film conductivity is diminished markedly in the oxidized state. In-situ SFM is therefore considered a more appropriate technique to monitor the morphology of pOPh films. A comparison of the applied contact and tapping modes gives clear evidence that the surface morphology is affected by the strong mechanical interactions prevailing under the conditions of contact mode imaging, whereas tapping mode images are thought to provide a more accurate information about film morphology. In both systems, the film growth and morphology is affected by the electropolymerization conditions, such as by the polarization dynamics and the chemical system chosen for electropolymerization. Further studies in progress, involving also the nm-scale investigation of the initial film formation and the growth of ultra thin films, are aimed at the elucidation of the growth mechanism.

The observed remarkable differences between the experimental voltammograms and the "equilibrium" voltammograms calculated from the equilibrium charges cannot be explained yet and are the subject of further work in progress.

Acknowledgments

The authors thank F. Niederhauser for technical assistance, and L. Schlapbach and D. Chartouni? (Université de Fribourg) for access and support with the tapping mode SFM. Financial support by the Swiss Federal Office of Energie and the Swiss National Foundation is gratefully acknowledged.

Literature Cited

1. Murray, R. W. In *Electroanalytical Chemistry*; Bard, A. J., Ed.; Marcel Dekker: New York, 1984 ,Vol. 13; pp. 191-368.
2. Fujihira, M., In *Topics in Organic Electrochemistry*; Fry, A. J. and Britton, W. E., Eds.; Plenum Press: New York, 1986 ,Vol. 255.
3. Hillman, A. R., In *Electrochemical Science and Technology of Polymers*; Linford, L. G., Ed.; Elsevier Applied Science: London, 1987; pp. 103-241.
4. Evans, G. P., In *Advances in Electrochemical Science and Engineering*; Gerischer, H. and Tobias, C. W., Eds.; VCH Press: Weinheim, 1990 ,Vol. 1; p. 1.
5. Murray, R. W. *Molecular Design of Electrode Surfaces*; J. Wiley: New York, 1992.
6. Inzelt, G., In *Electroanalytical Chemistry*; Bard, A. J., Ed.; Marcel Dekker: New York, 1994 ,Vol. 18; pp. 90-241.
7. Fan, F.-R. F.; Bard, A. J. *J. Electrochem. Soc.* **1989**, *136*, p. 3216.
8. Fan, F.-R. F.; Mirkin, M. V., and Bard, A. J. *J. Phys. Chem.* **1994**, *98*, p. 1475.
9. Chomakova-Haefke, M.; Nyffenegger, R.; Schmidt, E. *Appl. Phys. A* **1994**, *59*, p. 151.
10. Nyffenegger, R.; Ammann, E.; Siegenthaler, H.; Kötz, R.; Haas, O. *Electrochim. Acta* **1995**, *40*, p. 1411.
11. Tourillon, G.; Garnier, F. *J. Electroanal. Chem.* **1982**, *135*, p. 173.
12. Otero, T. F.; Rodriguez, J. *J. Electroanal. Chem.* **1991**, *310*, p. 219.
13. Wang, S.; Tanaka, K.; Yamabe, T. *Synth. Met.* **1989**, *32*, p. 141.
14. Johnson, B. W.; Read, D. C.; Christensen, P. Hammett, A.; Armstrong, R. P. *J. Electroanal. Chem.* **1994**, *364*, p. 103.
15. Caple, G.; Wheeler, B. L.; Swift, R.; Porter, T. L.; Jeffers, S. *J. Phys. Chem.* **1990**, *94*, p. 5639.
16. Sunde, S.; Hagen, G.; Odegard, R. *Synth. Met.* **1991**, *41-43*, p. 2983.
17. Downard, A. J.; Pletcher, D. *J. Electroanal. Chem.* **1986**, *206*, p. 147.
18. Diaz, A. F.; Crowley, J.; Bargon, J.; Gardini, G. P.; Torrance, J. B. *J. Electroanal. Chem.* **1981**, *121*, p. 355.
19. Dong, S.; Zhang, W. *Synth. Met.* **1989**, *30*, p. 359.
20. Bazzaoui, E. A.; Aeiyach, S.; Lacaze, P. C. *J. Electroanal. Chem.* **1994**, *364*, p. 63.

21. Roncali, J. *Chem. Rev.* **1992**, *92*, p. 711.
22. Servagent, S.; Vieil, E. *J. Electroanal. Chem.* **1990**, *280*, p. 227, and literature cited therein.
23. Haas, O.; Zumbrunnen, H.-R. *Helv. Chim. Acta* **1981**, *64*, p. 854.
24. Haas, O.; Rudnicki, J.; McLarnon, F. R.; Cairns, E. J. *Faraday Trans.* **1991**, *87*, p. 939.
25. Miras, M. C.; Barbero, C.; Kötz, R.; Haas, O. *J. Electrochem. Soc.* **1992**, *338*, p. 279.
26. Semling, T., *Diploma Thesis*, Universität Bern, 1993.
27. Yang, R.; Evans, D. F.; Christensen, L.; Hendrickson, W. A. *J. Am. Chem. Soc.* **1990**, *94*, p. 6117.
28. Dong, S. Z.; Cai, Q.; Liu, P.; Zhu, A. R. *Appl. Surf. Sci.* **1992**, *60/61*, p. 342.
29. Porter, T. L.; Minor, D.; Zhang, D. *J. Phys. Chem.* **1995**, *99*, p. 13213.
30. Hartmann, U. *Ultramicroscopy* **1992**, *42-44*, p. 59.

Chapter 17

X-ray Photoelectron Spectroscopy and Scanning Tunneling Microscopy Studies of Thin Anodic Oxide Overlayers on Metal and Alloy Single-Crystal Surfaces

P. Marcus and V. Maurice

Laboratoire de Physico-Chimie des Surfaces, Unité de Recherche Associée 425, Centre National de la Recherche Scientifique, Université Pierre et Marie Curie, Ecole Nationale Supérieure de Chimie de Paris, 11 rue Pierre et Marie Curie, 75231 Paris Cedex 05, France

Recent advances in the chemistry and structure of thin oxide overlayers (passive films) on metal and alloy surfaces have been achieved by using XPS and STM. XPS data have shown that the passive film formed on Ni in acid solution consists of a thin oxide layer (NiO) covered by hydroxyls. The total thickness of the passive film is about 1 nm. On Ni(111), *ex situ* STM imaging after passivation revealed the presence of steps and terraces on the oxide layer surface. Atomic resolution imaging provided evidence that the oxide film is crystalline and epitaxially grown on the substrate. On Fe-22Cr(110) surfaces, the passive film formed in acid solution is between 1 and 2 nm thick and markedly enriched in Cr^{3+}. This enrichment increases with polarization time over periods of several hours. STM imaging revealed that the passive film is non-crystalline after 2 hours of polarization and becomes more crystalline after longer periods of polarization. The atomically resolved images of the surface give evidence of the presence of crystalline areas of α-Cr_2O_3 epitaxially grown ($Cr_2O_3(0001) \parallel$ Fe-22Cr(110)). The role of chromium hydroxide in cementing the chromium oxide grains is discussed.

Thin anodic oxide overlayers (passive films) can be highly resistant against corrosion. They also play a major role in adhesion, tribology, catalysis and micro-electronics. A better understanding of the relationship of the chemistry and structure to the properties of metal surfaces covered with thin oxide overlayers is needed.

The following points are key factors in the growth, stability and breakdown of thin anodic oxide layers : (i) chemical composition, (ii) chemical states, (iii) thickness, (iv) stratification of phases (layered structure), (v) cristallinity, (vi) atomic structure, (vii) epitaxy and (viii) defects. Many data on points (i) through (iv) above have been provided by angle-resolved X-ray Photoelectron Spectroscopy (AR-XPS), using the binding energies and the intensities of core levels (see (*1*) and references therein). New possibilities in obtaining structural data (points (v) through (viii) above) have been opened up by the advent of Scanning Tunneling Microscopy(STM) (see (*2*) and references therein).

The combined use of XPS and STM for the study of passive films is illustrated by the examples of a pure metal, nickel, and an alloy, Fe-22Cr. This combined approach applied to well-defined single crystal surfaces is expected to provide key information to the understanding and prediction of the stability and breakdown of thin anodic oxide layers.

XPS Studies of Passive Films on Metals and Alloys

Figure 1 shows schematically the angle-resolved XPS measurements on an oxide layer having a thickness of the same order of magnitude as the attenuation length of the analysed photoelectrons, a situation often found for natural oxides and passive films on metals and alloys such as Ni, Cr, Fe-Cr alloys (ferritic stainless steels) and FeCrNi alloys (austenitic stainless steels). Photoelectrons emitted by the core levels of both the thin oxide layer and the near interface region of the metal or alloy under the oxide layer are detected and their binding energies (E_b) as well as their intensities (I) are measured. The core level binding energy shift between the metal and the cation in the film ($\Delta E_b(M^0\text{-}M^{n+})$) enables the indentification of the chemical state of M in the surface film. The intensities of the signals emitted by the metal (I_M) and the cation ($I_{M^{n+}}$) can be analysed in a quantitative manner, using the equations indicated in Fig. 1 where k is a spectrometer constant, D is the atomic fraction, Y is the photoelectron sensitivity factor (or yield), λ is the attenuation length of the photoelectrons, θ is the angle of the surface with the direction in which the electrons are analysed and d is the thickness of the oxide. The equations are used to determine the composition and the thickness of the oxide layer.

Figure 2 shows the polarisation curve of Ni in 0.05M H_2SO_4 and the model proposed in an early XPS study of passivated nickel (*3*). The model is supported by the presence of Ni2$p_{3/2}$ signals at 854.4eV, assigned to Ni^{2+} in NiO, and at 856.6eV, assigned to Ni^{2+} in Ni(OH)$_2$. The O1s spectra, showing peaks at 529.9eV (O^{2-}) and at 531.7eV (OH$^-$), confirm this finding. The Ni2$p_{3/2}$ signal from the metal is detected at 852.8eV, which indicates that the passive film is very thin. The oxide layer thickness was calculated, using the equations presented in Fig. 1 and was found to be 1 nm. This was the first investigation establishing the bilayer structure of the passive film, which is schematically represented in Fig. 3 and appeared later to be quite general.

The chemical composition and the chemical states of passive films formed in 0.5M H_2SO_4 on the (110) face of a single crystal of Fe-22Cr alloy have been investigated by AR-XPS as a function of polarization potential and time (*4*). Figure 4 shows the XPS spectra of Fe2$p_{3/2}$, Cr2$p_{3/2}$ and O1s recorded after different times of polarisation at 500mV/SHE (20 minutes, 2 hours and 22 hours). The spectra measured at two different take-off angles (90° and 40°) are shown for 22 hours of polarisation.

The spectra reveal the presence of Cr^{3+} in an oxide (Cr_2O_3-type) and an hydroxide (Cr(OH)$_3$) with an increasing contribution of the oxide when the time of polarisation increases. This is an essential finding for the understanding of the passivation of Cr-containing alloys : aging of the passive film in the electrolyte causes the enrichment of Cr^{3+} in the oxide layer. A similar conclusion can be obtained in examining the O1s spectra also shown in Fig. 4. The angle resolved spectra show that the relative contribution of Cr^{3+}(ox) with respect to Cr^{3+}(hy) is higher at a take off angle of 90°, which reveals that the chromium oxide is located in the inner part of the film on the alloy surface, whereas the chromium hydroxide is located on the electrolyte side (outer part of the film). This finding is in agreement with the view that passive films are generally stratified, i.e. they consist of a bilayer such as the one shown above in Fig. 3.

The quantitative analysis of the core level intensities allowed us to calculate the concentrations of Cr^{3+} and the thickness of the passive layer. After passivation for 22 hours at 500mV/SHE, the Cr^{3+} concentration in the oxide layer was found to be 95 % and the thickness of this oxide layer was 12Å (the total thickness of the passive film, including the surface layer of hydroxide, was 14Å).

These data are important to the development of a full understanding of the properties of the passive layers. The resistance of the iron-chromium alloys to passivity breakdown and localized corrosion can be improved by prolonged polarisation, which is directly related to the Cr^{3+} enrichment in the oxide layer (*5*).

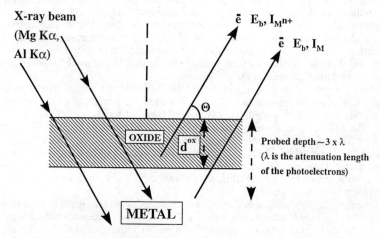

.Core level binding energy shift:
$$\Delta E_b(M - M^{n+})$$

.Quantitative analysis of the intensities:

$$I_M = k\, D_M\, Y_M\, \lambda_M^M\; \sin\Theta\; \exp\left[-d^{ox}/(\lambda_M^{ox}\; \sin\Theta)\right]$$

$$I_{M^{n+}} = k\, D_{M^{n+}}\, Y_M\, \lambda_{M^{n+}}^{ox}\; \sin\Theta\; \left\{1- \exp\left[-d^{ox}/(\lambda_{M^{n+}}^{ox}\; \sin\Theta)\right]\right\}$$

Figure 1. Principle of AR-XPS studies of thin oxide overlayers.

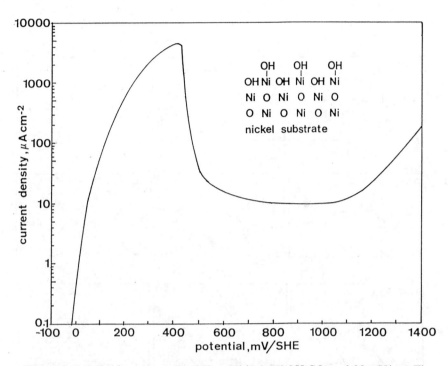

Figure 2. Polarisation curve recorded on Ni in 0.05M H$_2$SO$_4$ at 0.28 mV/sec. The insert shows the model of the passive film deduced from XPS measurements. Reproduced with permission from reference (3).

Copyright © 1979 Societe Francaise des Microscopies

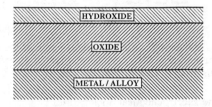

Figure 3. Bilayer model for passivated metals and alloys.

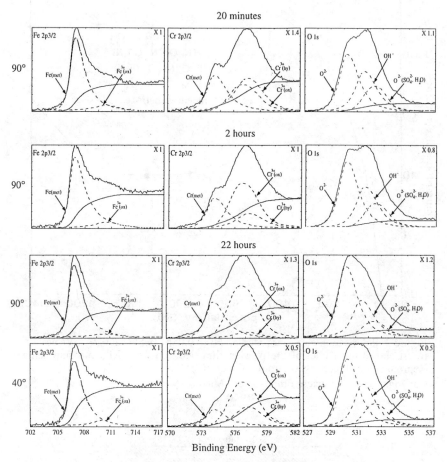

Figure 4. XPS spectra of the Fe2p$_{3/2}$, Cr2p$_{3/2}$ and O1s regions recorded at 90° and 40° take-off angles on the Fe-22Cr(1$\bar{1}$0) surface passivated at +500 mV/SHE for 20 minutes, 2 and 22 hours in 0.5 M H$_2$SO$_4$. Reproduced with permission from reference (*4*).

Copyright © 1996 The Electrochemical Society, Inc.

STM Studies of the Structure of Passivated Metal and Alloy Surfaces

Ex situ STM imaging has been applied to the investigation of the passive films formed on Ni (111)-oriented single crystal surfaces in 0.05 M H_2SO_4 at three different values of the passivation potential (+550, +650 and +750 mV/SHE) (4,5). STM analyses were performed in air, in which conditions the passive films were stable. Modifications of the passivated Ni surfaces with respect to the non-passivated one were recorded on two different lateral scales.

On a mesoscopic scale corresponding to scans of hundred nanometers, islands were observed that completely covered the surface. Their size was found to decrease and their density was found to increase with increasing passivation potential (from 2 x 10^{10} cm^{-2} to 1 x 10^{11} cm^{-2}). Their shape varied from trigonal contours with ledges oriented along the main crystallographic directions of the substrate, to hexagonal contours still with ledges oriented along the main crystallographic directions, and finally to non-symmetrical contours with non-oriented ledges after passivation at +550, +650 and +750 mV/SHE, respectively. These variations have been assigned to competition, during the passivation treament, between metal dissolution and formation of the passive film. The roughening effect due to dissolution increases with potential and produces a higher density of islands with less oriented ledges.

On these submicroscopic islands, a stepped crystalline lattice was imaged on the atomic scale. A typical image is shown in the left part of Fig. 5. The lattice parameters measured on the terraces (0.32 ± 0.2 nm) of the stepped lattice are in good agrrement with the lattice parameters of the (111) orientation of NiO (0.295 nm), the inner component of the passive film. These lattice parameters and the step density were found to be independent of the passivation potential. The density and height of the steps correspond to an average tilt of 8(±5)° between the surface of the film and the (111) orientation of the terraces. The resulting epitaxy relationships with the substrate are: NiO(433)//Ni(111) with NiO[0-11]/ /Ni[0-11] and NiO(765)//Ni(111) with NiO[1-21]//Ni[1-21]. The tilt is thought to result from a relaxation of the strained epitaxy due to a mismatch of 16% between the lattice parameters of the oxide film and those of the metal substrate and/or from a relaxation of the polar NiO(111) terraces. A possible atomic model of the interface between a 1 nm thick NiO(111) layer tilted by 8° and the Ni(111) substrate is shown in Fig. 6. It is constructed from bulk parameters in the case of the NiO(433) // Ni(111) with NiO[0-11] // Ni[0-11] epitaxy. This model illustrates different chemical terminations of the oxide film and possible atomic displacements (marked by arrows) at the interface towards the nearest threefold hollow sites of the substrate. The presence of terraces and steps of similar width at the surface of the passive film for the different potentials tested suggests a preferential dissolution at steps in the passive state. In addition to steps and kinks at the surface of the passive film, other crystalline defects such as point defects possibly related to vacancies have been imaged. One example is also shown by the image in the right hand part of Fig. 5. The presence of crystalline defects at the surface of the passive film may play a key role in the resistance to breakdown. The bottom of step and kinks correspond to sites of reduced thickness of the passive film (as illustrated in Fig. 6) where the barrier property of the film is expected to be diminished. These defects may also constitute sites of preferential adsorption of aggressive ions such as chloride ions, when present in solution.

The thickness of the hydroxide layer present in the outer part of the film was measured to be about 0.6 nm (3), and corresponds to the thickness of about one monolayer of β-Ni(OH)$_2$ (0001) (0.46 nm as calculated from lattice parameters). This hydroxide layer could not be observed in a distinct manner in the STM images. This results from the good agreement between the measured lattice parameters with both the lattice parameters of β-Ni(OH)$_2$ (0001) and NiO(111), 0.32 nm and 0.295 nm respectively. A likely situation is that the crystalline passive film is NiO (111) oriented and terminated by a β-Ni(OH)$_2$ (0001) hydroxide layer in (1x1) epitaxy on the oxide layer. These results on the cristallinity and orientation of the passive film have been confirmed by *in situ* STM for a different crystallographic orientation of the electrode and in a different electrolyte: Ni(100) in 1M NaOH (8). This shows that independently of the substrate orientation (Ni(100) or Ni(111))

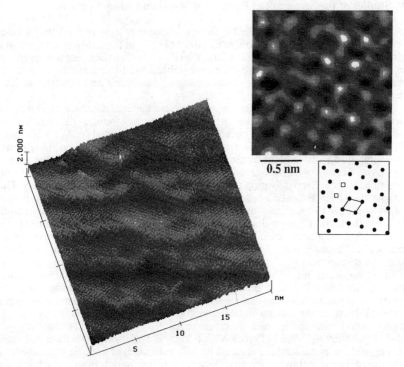

Figure 5. STM images of the passive film formed on Ni(111) in 0.05M H₂SO₄ at +750 mV/SHE. Reproduced with permission from reference (7).

Copyright © 1994 Elsevier Science—NL

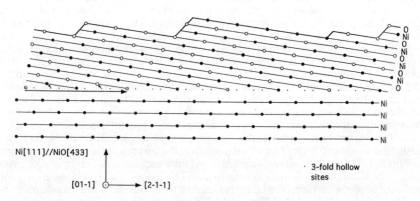

Figure 6. Section profile along the [2-1-1] direction of the interface between the thin oxide film and the metal substrate. The atomic planes and nods are indicated. Reproduced with permission from reference (7).

Copyright © 1994 Elsevier Science—NL

and of the conditions of investigation, the passive film is crystalline (NiO (111) oriented) and its orientation measured *in situ* or *ex situ* is similar.

The structure of the passive film on a Fe-22Cr alloy (*4*) was investigated with (110) oriented single crystal surfaces. The passivation was performed in 0.5 M H_2SO_4 at potential values of +300, +500 and +700 mV/SHE. Different time periods of polarization were investigated (20 minutes, 2, 22 and 63 hours). STM measurements in air showed that the terrace topography of the substrate surface is maintained after passivation.

On the same mesoscopic scale corresponding to scans of hundred nanometers where islands were formed on passivated Ni, the topography of the passivated Fe-22Cr alloy was characterized by the presence of small protrusions (up to 10 nm accross in lateral size). As a function of aging under polarisation, these protrusions were observed to follow a fusing process. This process is attributed to the coalescence of islands of the passive film occuring during film growth and aging. It shows that the formation of the passive film likely obeys a nucleation and growth mechanism where each protrusion of nanometer dimension could correspond to a nucleus of the passive film. Vertical variations of 0.4 to 0.8 nm amplitude result from the presence of the protrusions. In terms of resistance to breakdown, these topography variations also induce sites of reduced thickness of the passive film, the average thickness of which is about 1.5 nm. At these sites the barrier property of the film is expected to be diminished.

On the atomic scale, the passive film has been found to be non-crystalline after polarization for 2 hours at +500 mV/SHE. Aging for longer periods under polarization favors a crystallization process evidenced by the presence of epitaxial crystalline areas consistent with the structure of α-Cr_2O_3. This is illustrated in fig. 7 where the left image shows a crystalline area with possibly the emergence of a screw dislocation and where the right image shows the quasi-hexagonal lattice recorded. This lattice is in good agreement with the basal plane (0001) of Cr_2O_3 which indicates that the polar orientation of the epitaxial areas is: Cr_2O_3(0001) || Fe-22Cr(110). On this passivated alloy, three different azimuthal orientations of the oxide islands have been found: Cr_2O_3[01-10] || Fe-22Cr[1-1-1], Cr_2O_3[21-30] || Fe-22Cr[-11-1] and Cr_2O_3[-1100] || Fe-22Cr[1-10]. The quite thin hydroxide layer detected by XPS (about 0.2 nm of equivalent thickness) likely corresponds to the non-crystalline areas detected as in the case of the passive film formed on pure Cr(110) (*9*). On the crystalline areas, deviations from a quasi-hexagonal arrangement of the corrugations are measured as illustrated in the right hand image of Fig. 7. This is possibly due to the presence of a monolayer of OH groups.

These data show that the crystallization is not complete in these conditions and the topography of the passive film is intermediate between that recorded on passivated Ni(111) (complete crystallization with large crystals) and that recorded on passivated Cr(110) (nanocrystals cemented by non-crystalline areas) (*9*). It shows the presence of both crystalline defects in crystalline areas and non-crystalline areas. It is therefore possible that the amorphous structure of the thin hydroxide does not completely cover the crystalline areas of the oxide. The defects in these crystalline areas covered by hydroxide may be cemented by the thin hydroxide layer and offer higher resistance to film breakdown. In addition, the amorphous structure of the hydroxide is expected to minimize the variations of coordination of the surface atoms at crystalline defects and therefore to induce a higher chemical passivity at these sites. Hence, the role of cement played by the chromium hydroxide would be a key factor in the protective character of the passive films formed on Cr-containing alloys.

Conclusion

AR-XPS has been extensively used to investigate the chemical composition, the chemical states and the thickness of thin anodic oxide overlayers (passive films) formed on well-defined metal and alloy single crystal surfaces. More recently direct imaging of the surface structure by STM with atomic resolution provided new data on the crystallinity, the epitaxy and the nature of the structural defects existing in the thin oxide layers. Such data are useful

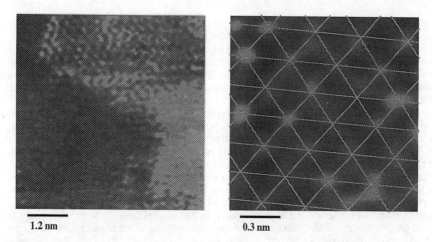

1.2 nm   0.3 nm

Figure 7. STM images of the passive film formed on Fe-22Cr(110) in 0.5M H_2SO_4 at +500 mV/SHE and aged 63 hours under polarization. Reproduced with permission from reference (4).
Copyright © 1996 The Electrochemical Society, Inc.

for a better understanding on the atomic or molecular level of the key factors ruling the reactivity of metal surfaces covered by thin anodic oxide overlayers.

Literature cited

(1) *Corrosion Mechanisms in Theory and Practice*, P. Marcus and J. Oudar, Editors, Marcel Decker Inc., New York (1995).
(2) *Scanning Tunneling Microscopy and Spectroscopy*, Ed. D. A. Bonnell, VCH Publishers, Inc., New York (1993).
(3) Marcus, P.; Oudar, J.; Olefjord, I. *J. Microsc. Spectrosc. Electron.* **1979,** *4,* 63.
(4) Maurice, V.; Yang, W.; Marcus, P. *J. Electrochem. Soc.* **1996,** *143,* 1182.
(5) Yang, W.; Costa, D.; Marcus, P. *J. Electrochem. Soc.* **1994,** *141,* 111.
(6) Maurice, V.; Talah, H.; Marcus, P. *Surf. Sci.* **1993,** *284,* L431.
(7) Maurice, V.; Talah, H.; Marcus, P. *Surf. Sci.* **1994,** *304,* 98.
(8) Yau, S.-L.; Fan, F.-R.F.; Moffat, T.P.; Bard, A.J. *J. Phys. Chem.* **1994,** *98,* 5493.
(9) Maurice, V.; Yang, W.; Marcus, P. *J. Electrochem. Soc.* **1994,** *141,* 3016.

Chapter 18

Structure and Catalysis of Rh–Pt(100), Rh–Pt(110), Pt–Rh(100), and Pt–Rh(110) Surfaces Prepared by Electrochemical Metal Deposition

K. Tanaka, Y. Okawa, A. Sasahara, and Y. Matsumoto

Institute for Solid State Physics, University of Tokyo,
7–22–1 Roppongi, Minato-ku, Tokyo 106, Japan

Structure and catalysis of Rh/Pt(100), Rh/Pt(110), Pt/Rh(100) and Pt/Rh(110) surfaces prepared by electrochemical deposition were studied. Rh/Pt(100) heated in O_2 at T > 340 K gives a clear $p(3\times1)$ LEED pattern. Similar $p(3\times1)$ structure is established on Pt/Rh(100) by heating in O_2 at ca. 600 K. Catalytic activity for NO + H_2 is markedly enhanced by the formation of $p(3\times1)$ structure. Rh/Pt(110) and Pt/Rh(100) surfaces are also activated by the chemical reconstruction caused by O_2 or NO. STM image proved that the $p(3\times1)$ $Pt_{0.25}Rh_{0.75}(100)$ surface is composed of combined array of one Rh-O and two Pt rows along the <011> directions, which is responsible for catalysis of NO + H_2 reaction.

So far, it has been accepted that molecules or atoms are adsorbed on sites being in registry with the substrate lattice, and the catalysis is explained by their reaction over the surface. By the adsorption of molecules or atoms, surface metal atom undergoes so often short distance shift from their equilibrium position, which is the adsorption induced reconstruction. It should be pointed out that such short distance shift of the surface atoms from their original position is principally reversible, that is, the (1×1) surface will be recovered by desorption of molecules.

One important development of surface chemistry and/or physics of metal given by the scanning tunneling microscope (STM) is that surface metal atoms undergo chemical reaction sometimes with adsorbed molecules. For example, when Cu(110) (1), Ni(110) (2), and Ag(110) (3-5) are exposed to O_2, metal atom at the step edges reacts with oxygen at room temperature, and one dimensional metal-oxygen strings are grown on the surface in the [001] direction. As increasing the strings on the surface, the (n×1) arrangement of the strings is established over the surface. Such surface prepared by self-assembly of the strings is undoubtedly different from the adsorption induced reconstruction of the substrate surface but is a chemical reconstruction. Adsorption of H_2 on Ni(110) surface is a good example. When

Ni(110) surface is exposed to H_2 at room temperature, one dimensional Ni-H strings grow in the [1$\bar{1}$0] direction and make an arrangement of the (1×2) structure (6-7), which is evidently different from the adsorption induced p(1×2) reconstruction of Ni(110) surface brought about at T < 220 K (8-10). Taking these facts into account, an idea of " **the formation of quasi-compounds and their self-ordering on the metal surfaces** " was proposed and the surface phenomena were rationalized by this new idea (11).

Chemical reconstruction of alloy and/or bimetallic surfaces will be more complex, and it is obvious that the catalysis of alloy and/or bimetallic surface can not be explained by the traditional idea of ligand effect and ensemble effect for the sites. That is, chemical reconstruction of alloy surface will occur by selective segregation or selective reaction of metal atoms. As a result, a new surface will be prepared sometimes , which is responsible to the prominent catalytic activity and/or selectivity of alloy and/or bimetallic surfaces. If this is the case, the catalysis of alloy surface is entirely different form the idiomatic idea of " ligand effect " and " ensemble effect ".

Pt-Rh alloy or Pt/Rh bimetal is well known catalyst by the name of three way catalyst for NO_x reduction. This is a typical example of the alloy or bimetallic catalyst, that is, a small amount of Rh is indispensable to get prominent catalytic activity for NO_x reduction. The role of Rh in activation of the surface, however, has not been settled. Pt and Rh make a randomly mixed bulk alloy and Pt-Rh(100) alloy surface gives a p(1×1) LEED pattern, which suggests that the Rh and Pt atoms are randomly distributed on the surface. Contrary to this, the depth profile of Pt and/or Rh depends markedly on the annealing temperature. For example, the topmost layer of a $Pt_{0.45}Rh_{0.55}$ alloy tip surface annealed at 973 K (700 °C) is enriched with Pt atoms and the 2nd layer is depleted with Pt atoms (Rh-enriched) (12-13). It is the same on $Pt_{0.25}Rh_{0.75}$(100) surface. Enrichment of Pt atom on the surface by annealing has been explained by the softening of Pt bond at the surface.

If this is the case, it is expected that the higher the annealing temperature, the higher the Pt-fraction in the topmost layer. As it was proved on $Pt_{0.25}Rh_{0.75}$(100) alloy surface, it is not the case but the Pt-fraction of the surface decreases monotonously upon increasing the annealing temperature and takes nearly the bulk composition at 1300 K (14). Therefore, it may be deduced that the lower the annealing temperature the higher the Pt fraction in the surface. In addition, it was shown that the surface composition is hardly equilibrated at the annealing temperature lower than 950 K. Taking these facts into account, the apparent discrepancy of the two experiments is responsible to the difficulty for equilibration of Pt fraction at lower than 900 K. As a result, an apparent maximum Pt fraction is attained at around 900 K - 1000 K (ca. 700 °C) as is the case of ref. (12).

As discussed above, the segregation as well as the equilibration of Pt or Rh fraction of the clean $Pt_{0.25}Rh_{0.75}$(100) surface are very slowly brought about at lower than 900 K. In contrast, when the surface is heated in O_2 or NO, segregation of Rh on the surface takes place easily even at 500 K and a clear p(3×1) LEED pattern appears (15-17). This phenomenon is caused by a chemical reconstruction of the alloy surface, that is, Rh atoms enriched in the 2nd layer by annealing are extracted by reacting with oxygen and they make a p(3×1) structure. Formation of same p(3×1) structure was proved on Rh deposited Pt(100) (Rh/Pt(100)) surface as well as on Pt deposited Rh(100) (Pt/Rh(100)) surface by heating in O_2 (18-20). Furthermore, it was shown that the catalytic activity of Pt/Rh(100) surface for the reaction of NO with H_2 is evolutionarily enhanced by the formation of the p(3×1) structure (21-22). This fact indicates that the catalytic activity of Rh/Pt(100), Rh/Pt(110), Pt/Rh(100), and Pt/Rh(110) surfaces is responsible for the p(3×1) surface prepared by the chemical reconstruction during the reaction of NO with H_2. As discussed in this paper, the bimetallic surfaces of Pt/Rh(100), Pt/Rh(110), Rh/Pt(100), and Rh/Pt(110) surfaces used in this paper were prepared by electrochemical deposition of metal ions and the structures and catalysis of these surfaces for the reaction of NO + H_2 were studied in UHV chamber.

Experimental.

In the experiment except the STM, the sample crystal disk was fixed by spot welding between the two Ta-wires and the sample was heated by direct current through the Ta-wire. The surface of Rh(100), Rh(110), Pt(100) and Pt(110) disks was made clean in the UHV chamber. In the case of Rh(100) and Rh(110), the surface was heated in O_2 (5×10⁻⁸ Torr) at 760 -780 K for 20 min., and then repeated the annealing at 1000 - 1200 K in UHV and Ar-ion sputtering. Clean Pt(110) and Pt(100) surfaces were obtained by Ar-ion sputtering and annealing in the UHV chamber at 1100 K. After the annealing, the Pt(110) surface gave a $p(1×2)$ LEED pattern and the Pt(100) gave a (5×20) LEED pattern, respectively. When the cleaning procedure was accomplished, the sample holder was pushed into a small volume cubic cell which made a perfect seal between the UHV chamber and the cubic cell. After then, the cubic cell was filled with one atmospheric pressure of highly pure Ar gas, and an electrochemical quartz cell containing electrolyte solution of 0.05 M H_2SO_4 + 5×10⁻⁵ M $PtCl_4$ (or $RhCl_3$) was lifted into the cubic cell to make contact with the one side of the crystal disk. After the electrochemical deposition of Pt ions or Rh ions, the crystal was washed with highly pure water and transferred back to the UHV chamber for measurement of LEED and XPS as described in the preceding paper (*18-19*). Temperature programmed reaction (TPR) was carried out in the cubic reactor by raising the temperature in a rate of 1 K/sec in a flow of a mixture of $5.8 × 10^{-9}$ Torr of NO and $1.6 × 10^{-8}$ Torr of H_2, where the reaction was monitored with a quadrupole mass spectrometer.

The voltammogram of the clean surface as well as of the surface after the adsorption or the reaction were obtained by contacting the surface to 5×10⁻² M H_2SO_4 solution in Ar atmosphere with respect to a saturated calomel electrode (SCE).

The STM used in this experiment was a Rasterscope-3000 (DME Co.). A mechanically polished $Pt_{0.25}Rh_{0.75}(100)$ disk was mounted on the STM holder and was made clean by repeating Ar^+ ion sputtering and heating in O_2 of a 10⁻⁸ Torr at 1000 K. The sample was heated by electron bombardment from the back side of the crystal and the temperature was monitored with a chromel-alumel thermocouple. All the STM measurement was performed at room temperature. The $p(3×1)$ surface was prepared by heating a clean $p(1×1)$ surface in O_2 (10⁻⁴ - 10⁻⁷Torr) at 600 K for 30 sec. to several minutes. In-situ STM measurements were performed in H_2 at room temperature.

Results and Discussion.

Preparation of Bimetallic Surfaces by Electrochemical Deposition. Figure 1a-d show cyclic voltammograms for the clean surfaces of (5×20) Pt(100), (1×2) Pt(110), (1×1) Rh(100), and (1×1) Rh(110) surfaces measured in a 0.05M H_2SO_4 solution, where the surfaces were cleaned in the UHV chamber. The voltammogram in Figure 1d is very similar to that reported by Gomez *et al.* (*23*). The peak at -0.21 V (SCE) in our experiment is less sharp compared to that (0.13V RHE) of their voltammogram. The difference may be due to imperfection of the surface. In fact, our Rh(110) surface prepared in the UHV chamber showed a clear (1×1) LEED pattern but the spots were little diffused, which may indicate imperfect flatness of the surface. The quronic charge for the desorption of H$^+$ ion in Figure 1d is about 161 µCcm^{-2} by subtracting the double-layer charge, which is almost equal to a value of 157 µCcm^{-2} calculated by assuming (1×1) flat surface and one hydrogen atom for each Rh atom.

The cyclic voltammogram for Pt(110) surface in Figure 1b is very similar to that of literature (24). Main desorption peak has about 200 µCcm^{-2} by correcting the double-layer charge, which is apparently larger than the calculated value of 146 µCcm^{-2} calculated by assuming (1×1) flat surface and one hydrogen atom for each Pt atom. In the literature, Armand et al. (24-25) reported 200 µCcm^{-2} for a Pt(110) surface but Yamamoto et al. (26) reported 150 µCcm^{-2} for the desorption peak of hydrogen on Pt(110) surface. We could say that when a heated p(1×2) Pt(110) surface is cooled in air or p(1×2) Pt(110) surface is immersed into the solution under the open circuit potential, the (1×2) reconstruction of the Pt(110) surface may be lifted. In our experiment, the Pt(110) sample was transferred to the cubic reactor in vacuum and then the cubic reactor was filled with Ar. After then, the surface was immersed at - 0.28 V of emersion potential. Michaelis et al. (27) reported that reconstructed (1×2) Pt(110) surface immersed at 0.2 V (Ag/AgCl, 0.18 vs SCE) was stable up to 0.6 V (Ag/AgCl, 0.58 V vs SCE) in 0.5 M H$_2$SO$_4$ solution. It is known that Au(110) surface undergoes potential induced reconstruction at rather negative potential and it is lifted by sweeping to positive potential, but no evidence has been shown for the potential induced reconstruction of Pt surfaces. Contrary to the result of Michaelis et al., we confirmed that the Pt(110) surface after emersion at - 0.28 V (SCE), the surface gave p(1×1) LEED pattern with high background intensity. Therefore, the difference between the calculated value of 146 µCcm^{-2} and the experimental value of 200 µCcm^{-2} may be due to the adsorption of bisulfate ions on the (1×1) Pt(110) surface.

So far a number of paper have been reported on the voltammogram of Pt(100) surface. Relative peak height reflecting the adsorption and desorption reaction of H$^+$ ions in Figure 1a appeared at - 0.05 and +0.06 V (SCE) on our (5×20) reconstructed Pt(100) surface is slightly different from that of the literature on Pt(100) surface (28-29), although the (5×20) Pt(100) surface used in this experiment had no detectable contaminants by the XPS analysis. Scortichini et al. (30) pointed out that the peak corresponding to that at 0.06 V in Figure 1a is due to strongly adsorbed hydrogen at the defects. On the other hand, Zei et al. (31) reported that the reconstruction of Pt(100) surface is lifted at the potential higher than 0.2 V (SCE) by adsorption of bisulfate anions. So in our experiment, the (5×20) Pt(100) surface may be converted into the (1×1) surface by sweeping the potential to 0.5 V (SCE). In fact, after the emersion at - 0.28 V, the Pt(100) surface gave a p(1×1) LEED pattern with high background intensity. The total charge for the two desorption peaks was about 195 µCcm^{-2} by correcting double-layer charge, which is close to a value of 208 µCcm^{-2} calculated by assuming one hydrogen atom for each surface Pt atom. This value is almost equal to the values reported by Scortichini et al. 210 ±10 (30) and Clavilier et

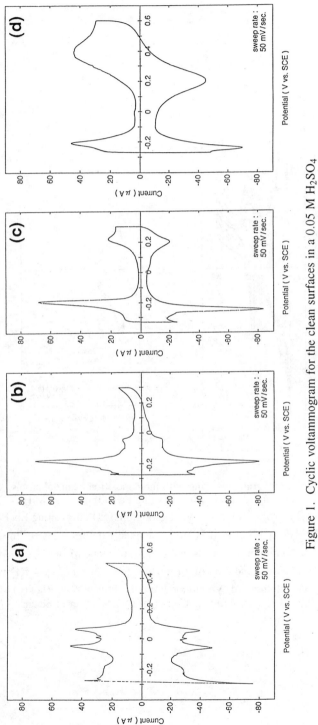

Figure 1. Cyclic voltammogram for the clean surfaces in a 0.05 M H_2SO_4 solution. (a) (5×20) Pt(100), (b) (1×2) Pt(110), (c) (1×1)Rh(100) (Adapted from ref. 22.), (d) Rh(110). Copyright © 1995 Elsevier Science—NL

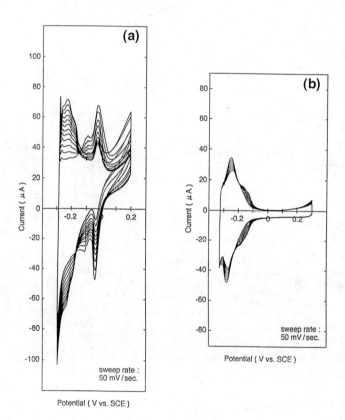

Figure 2. Voltammogram changing with the deposition of metal atoms. (a) In situ cyclic voltammogram for Rh depositing Pt(100) in a 0.05 M H_2SO_4 + 5×10^{-5}M $RhCl_3$. (b) In situ cyclic voltammogram for Pt depositing Rh(100) in a 0.05 M H_2SO_4 + 5×10^{-5}M $PtCl_3$. (Adapted from ref. 22.) (c) In situ cyclic voltammogram for Rh depositing Pt(110) in a 0.05 M H_2SO_4 + 5×10^{-5} M $RhCl_3$. (c') Voltammogram of a Rh deposited Pt(110) (θ_{Rh} = 1.1) in a 0.05 M H_2SO_4 solution. (d) Voltammogram of a Pt deposited Rh(110) (θ_{Pt} = 1.0) in a 0.05 M H_2SO_4 solution. Copyright © 1995 Elsevier Science—NL

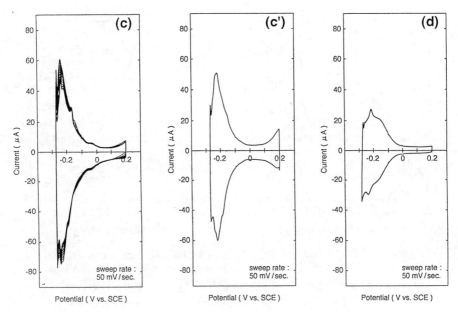

Figure 2. *Continued*

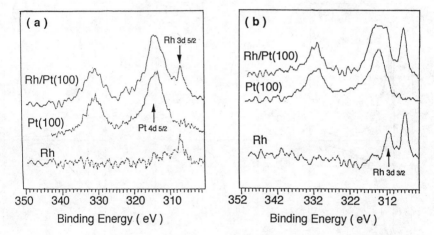

Figure 3. XPS spectra for Rh deposited Pt(100) surface, clean Pt(100) surface and their differential spectrum for separating $Rh3d_{5/2}$ (307.0eV) and $Rh3d_{3/2}$ (311.8eV) peaks from $Pt4d_{5/2}$ (316eV) peak. (a) $\theta_{Rh} = 0.4$ and (b) $\theta_{Rh} = 1.0$. (Adapted from ref. 18.) Copyright © 1993 Elsevier Science—NL

al. 205 ± 5 μCcm^{-2} (*32*), where they prepared the Pt(100) crystal surface by a flame annealed method.

In contrast to Pt(100) surface, only two papers have reported the voltammogram of Rh(100) surface, one is given by Wasberg *et al.* (*33*) in $HClO_4$ solution and the other is given by Rhee *et al.* (*34*). Rhee *et al.* scrutinized the effect of anions on the voltammogram of Rh(100) in perchlorate solution. The shape of the cyclic voltammogram in Figure 1c is different from that in $HClO_4$ solution, but is very similar to that in 0.1 M $HClO_4$ + 0.1 mM H_2SO_4. The quronic charge for the hydrogen desorption peak of the Rh(100) surface in Figure 1c is about 237 μCcm^{-2}, which is almost equal to a value of 222 μCcm^{-2} calculated by assuming a (1×1) flat surface and one hydrogen atom for each Rh atom.

Repeating the cyclic potential sweep in a solution of 0.05 M H_2SO_4 + 5×10^{-5} M $RhCl_3$ (or $PtCl_3$), the voltammogram gradually changes as depositing the metal ions. Figure 2a is in-situ voltammogram of Pt(100) surface as depositing Rh atoms (final θ_{Rh} =0.8) (*18-19*), and 2b shows that of Rh(100) surface changing with deposition of Pt atoms (final θ_{Pt} =1.1) (*21-22*). Figure 2c is in-situ voltammogram of Pt(110) surface with deposition of Rh atoms (final θ_{Rh} =1.1) and 2c' is the voltammogram of Rh deposited Pt(110) surface (θ_{Rh} =1.1) in the absence of Cl^- ions in 0.05 M H_2SO_4 solution (Sasahara, A.; Tamura, H.; Tanaka, K., submitted to publication.), and Figure 2d shows a voltammogram of Rh(110) surface with Pt atoms (θ_{Pt} =1.0) in 0.05 M H_2SO_4 solution (Sasahara, A.; Tanaka, K., submitted to publication.).

Amount of deposition of metal atoms was controlled by monitoring the in-situ voltammogram. After then the single crystal surface was washed with ultra-pure water replaced in the quartz cell, and then it was transferred into the main chamber for the LEED and XPS or AES analysis. The Rh/Pt(100) and Pt/Rh(100) surfaces gave (1×1) LEED pattern with high background intensity, but the Rh/Pt(110) and Pt/Rh(110) surfaces gave no LEED pattern with high background intensity. Figure 3 shows XPS analysis for Rh/Pt(100) surfaces with different amount of Rh deposition (θ_{Rh} =1.0 and 0.4). As it was mentioned above, when a clean $Pt_{0.25}Rh_{0.75}(100)$ alloy surface is annealed at 1000 K in vacuum, Pt atoms are enriched in the topmost layer. Taking this fact into account, thermal stability of the Pt/Rh(100) and Rh/Pt(100) surfaces in vacuum is quite interesting. Figure 4a shows the XPS analyses for a Pt/Rh(100) surface annealed at ca. 1000 K and 1050 K, which indicates that the Pt-peak changes little by annealing at 1000 K for 20 min. (spectrum (ii)) and at 1050 K for 10 min. (spectrum (iii)). Contrary to this, when a Rh/Pt(100) surface is heated in UHV at 1000 K for 15 min., Rh signal completely disappears in the XPS as shown in Figure 4b. Therefore, we can deduce that Rh atom on Pt(100) surface is not so stable at ca. 1000 K and is quickly diffused into the Pt bulk whilst Pt on Rh(100) is quite stable at this temperature.

When the Rh/Pt(100) surface is heated in O_2, however, Rh atoms are stabilized on the surface by making ordered arrangement of Rh-O compound. On the other hand, when the Pt/Rh(100) surface is heated in O_2, Rh atoms are extracted from the 2nd layer by reacting with oxygen and an ordered arrangement of Rh-O compound is completed on the surface. As it will be shown latter, a common surface is prepared on either Pt/Rh(100) or Rh/Pt(100) surface by reacting Rh with O_2, that is, a new surface composed of a combined arrangement of Rh-O and Pt rows in a (3×1) structure is prepared by the chemical reconstruction. The $p(3 \times 1)$ surface is also commonly formed on $Pt_{0.25}Rh_{0.75}(100)$ alloy surface by heating in O_2, and it is an active surface for NO + H_2 reaction.

Characterization of Bimetallic Surfaces.
Comparison of Rh/Pt(100) and Pt/Rh(100) Bimetallic Surfaces to Pt$_{0.25}$Rh$_{0.75}$(100) Alloy Surface. When Pt$_{0.25}$Rh$_{0.75}$(100) alloy surface is heated in O$_2$, segregation of Rh onto the surface occurs at temperature higher than ca. 450 K and a clear $p(3\times1)$ LEED pattern appears. It was confirmed that the $p(3\times1)$ surface prepared on Pt$_{0.25}$Rh$_{0.75}$(100) alloy surface is stable in vacuum at ca. 750 - 800 K, but it undergoes decomposition at ca. 900 K with the desorption of O$_2$, which results in decreasing Rh atoms from the surface (15-17).

In order to shed light on the mechanism for the formation of the $p(3\times1)$ structure, a model surface was prepared by depositing Rh ions on Pt(100) surface with electrochemical method and was subjected to annealing in a 10^{-7} Torr of O$_2$ or NO. As shown in Figure 5 the $p(3\times1)$ pattern appears at temperature as low as 340 K and gives very sharp spots at 400 K (18-20), which is undoubtedly the same $p(3\times1)$ structure observed on Pt$_{0.25}$Rh$_{0.75}$(100) surface. This result indicates that the formation of Rh-O compound on Pt(100) surface is responsible to the $p(3\times1)$ surface. Therefore, it could be deduced that when Pt$_{0.25}$Rh$_{0.75}$(100) alloy surface is heated in O$_2$, Rh atoms are extracted from the 2nd layer by reacting with oxygen, which make the $p(3\times1)$ surface on Pt enriched layer. It was confirmed that such reactive segregation of Rh with O$_2$ is brought about even on four Pt-layers deposited Rh(100) surface at 780 K (20).

The $p(3\times1)$ surface is established on Rh deposited Pt(100) surface at lower temperature than that appears on Pt$_{0.25}$Rh$_{0.75}$(100) alloy or Pt/Rh(100) surface because of easier chemical reconstruction. It should be pointed out that we can get a sharp $p(3\times1)$ pattern on Rh/Pt(100) surfaces with lower than one monolayer Rh atoms (θ_{Rh} = 0.4 - 1.0 monolayer). The $p(3\times1)$ surface formed on the Rh/Pt(100), Pt/Rh(100) and Pt$_{0.25}$Rh$_{0.75}$(100) alloy surfaces is commonly reactive with H$_2$, that is, the $p(3\times1)$ pattern is readily converted to the (1×1) surface at room temperature by exposing to a 10^{-7} Torr of H$_2$ and the (3×1) surface is recovered at room temperature by exposing to a 10^{-7} Torr of O$_2$ as shown schematically in Figure 6. This fact suggests that the (3×1) arrangement of Rh atoms seems to be retained on the (1×1) surface after the hydrogenation. In order to shed light on this interesting $p(3\times1)$ Pt$_{0.25}$Rh$_{0.75}$(100) surface, it was studied by the scanning tunneling microscopy (STM).

STM study of alloy surfaces has get behind that of single metal surfaces. Since Varga et al. (35) succeeded in distinguishing the elements on Pt$_3$Ni(111) alloy and other alloy surfaces by STM (36-38), the STM study of alloy and/or bimetallic surfaces becomes an interesting subject. Figure 7a shows an STM image for a clean (1×1) Pt$_{0.25}$Rh$_{0.75}$(100) surface, where the bright spots are not so uniformly distributed and the STM image shows not so perfect arrangement of atoms on the alloy surface (39). As discussed above, when the Pt$_{0.25}$Rh$_{0.75}$(100) surface is annealed in O$_2$, Rh atoms are segregated on the surface. Changing the surface composition by heating in O$_2$ indicates that the brighter spots in Figure 7a may reflect the Rh atoms (39). A $p(3\times1)$ surface was prepared on a Pt-enriched Pt$_{0.25}$Rh$_{0.75}$(100) surface by heating in 10^{-4} Torr of O$_2$ at ca. 600 K for 30 sec and was subjected to the STM study. Figure 7b shows a wide area STM image for the $p(3\times1)$ surface, where the protrusion of the (3×1) arrangement for the two domains

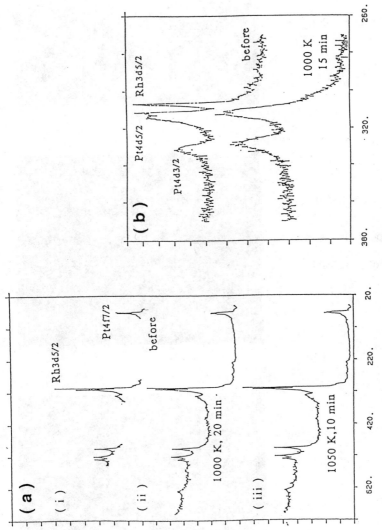

Figure 4. XPS of (a) Rh/Pt(100) ($\theta_{Rh} = 1.0$) surface annealed at 1000 K and 1050 K and (b) that of Pt/Rh(100) ($\theta_{Pt} = 1.0$) surfaces annealed at 1000 K. (Adapted from ref. 20.) Copyright © 1994 American Chemical Society

p(1×1)

Rh deposited Pt(100)

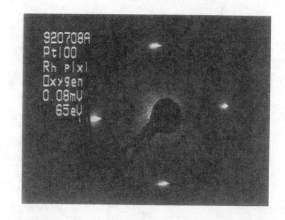

Streaky p(3×1)

10^{-7}Torr of O$_2$ at 340 K

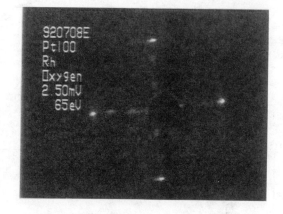

p(3×1)

10^{-7}Torr of O$_2$ at 400 K

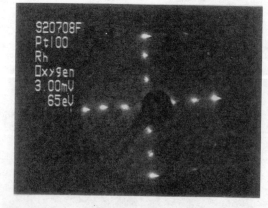

Figure 5. LEED pattern change of a Rh/Pt(100) surface heated in a 10^{-7} Torr of O$_2$.

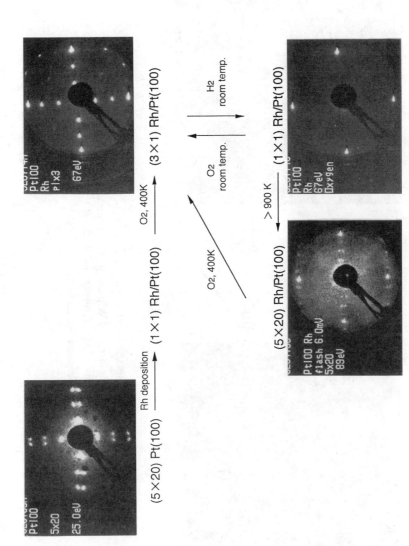

Figure 6. Structural change of Rh/Pt(100) surface by exposing to O_2 or H_2. (Adapted from ref. 20.) Copyright © 1994 American Chemical Society

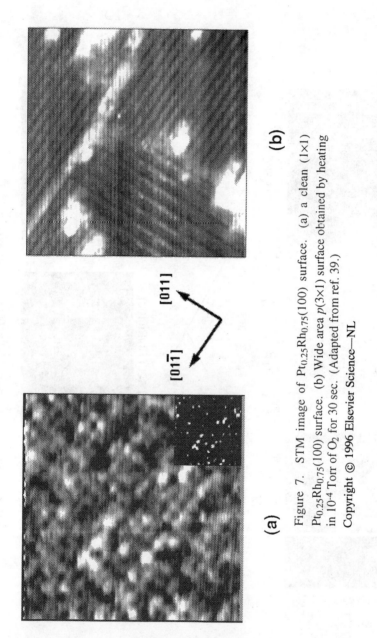

(a)

(b)

[011]

[01$\bar{1}$]

Figure 7. STM image of $Pt_{0.25}Rh_{0.75}(100)$ surface. (a) a clean (1×1) $Pt_{0.25}Rh_{0.75}(100)$ surface. (b) Wide area $p(3\times1)$ surface obtained by heating in 10^{-4} Torr of O_2 for 30 sec. (Adapted from ref. 39.)

Copyright © 1996 Elsevier Science—NL

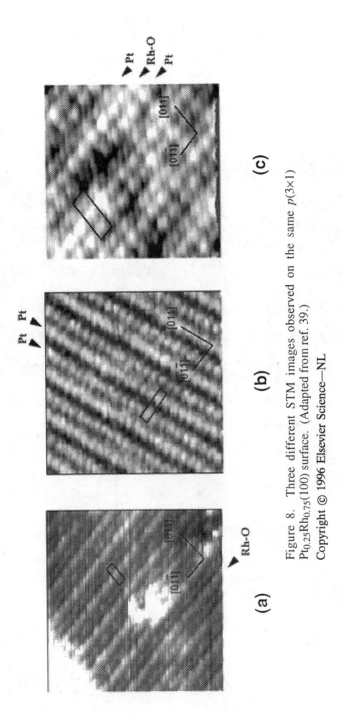

Figure 8. Three different STM images observed on the same $p(3 \times 1)$ $Pt_{0.25}Rh_{0.75}(100)$ surface. (Adapted from ref. 39.) Copyright © 1996 Elsevier Science—NL

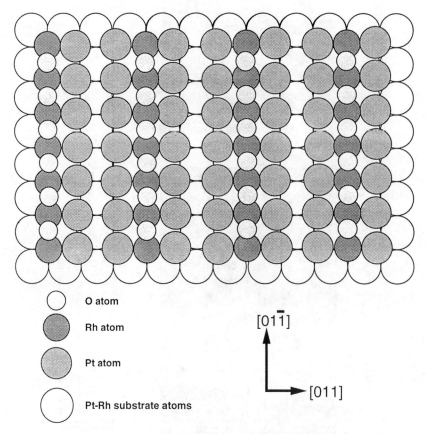

○ O atom

◐ Rh atom

◯ Pt atom

○ Pt-Rh substrate atoms

[01$\bar{1}$]

[011]

Figure 9. A model structure of $p(3\times1)$ surface commonly formed on Pt$_{0.25}$Rh$_{0.75}$(100), Pt/Rh(100) and Rh/Pt(100) surfaces by reacting with O$_2$. (Adapted from ref. 39.) Copyright © 1996 Elsevier Science—NL

runs along the [011] and [01$\bar{1}$] directions, respectively (*39*). We found a very interesting phenomenon during scanning the $p(3\times1)$ surface, that is, the three different STM images such as shown in Figure 8a, b, and c were accidentally obtained depending on tip condition. The STM image shown in 8c represents the all surface metal atoms, but only the middle line of the three rows can be seen on the image of 8a and the other two lines are seen in the STM image of 8b. These facts indicate that the $p(3\times1)$ surface is composed of the two kinds of lines, and a model for the $p(3\times1)$ surface shown in Figure 9 has been proposed (*39*). It is clear that the $p(3\times1)$ structure formed on the $Pt_{0.25}Rh_{0.75}(100)$ surface is different from missing row or added row structure which is popularly observed on Cu(100) (*40-42*), Cu(110) (*1, 43-45*), Ag(110) (*3-5, 46-48*), Ni(110) (*2, 49*), and Rh(110) (*50-51*) surfaces when they are exposed to O_2, but is a new combined surface prepared by the chemical reconstruction.

As described above, when this $p(3\times1)$ surface is exposed to H_2, the $p(3\times1)$ LEED pattern changes to $p(1\times1)$ pattern at room temperature and the $p(3\times1)$ surface is recovered by exposing to O_2 at room temperature. That is, a reaction of $p(3\times1) \rightarrow p(1\times1)$ by H_2 and $p(1\times1) \rightarrow p(3\times1)$ by O_2 takes place at room temperature (*18-20*). The STM image for a $p(1\times1)$ surface obtained by H_2 had more bright spots gathered along the <011> directions compared to that on the starting clean (1×1) surface annealed at 1000 K.

It is an interesting question that whether the cyclic voltammogram will change or not when the $p(3\times1)$ surface is converted to the $p(1\times1)$ surface by treating with H2. The voltammograms for a Rh deposited Pt(100) (θ_{Rh} = 0.8 by XPS) surface, both the $p(1\times1)$ and $p(3\times1)$ surfaces, measured in a 0.05 M H_2SO_4 solution are shown in Figure 10a and b. It is known that the voltammogram of the (1×1) surface is very similar to that of the (3×1) surface. The characteristic sharp peaks reflecting the reactions of $H(a) \rightarrow H^+ + e$ and $H^+ + e \rightarrow H(a)$ appear at -0.15 V and -0.17 V (SCE), which are different from the corresponding peaks on Rh(100) surface being at -0.19 V and -0.26 V shown in Figure 1c (*21-22*) and the peaks on Pt(100) surface shown in Figure 1a.

As it was mentioned above, as Rh deposited Pt(110) surface (θ_{Rh}= 1.5 by XPS) gave no LEED pattern, but when the Rh deposited Pt(110) surface was heated in 1×10^{-7} Torr of O_2 at 760 K, the surface gave a clear (1×2) LEED pattern. The cyclic voltammogram of as deposited Rh/Pt(110) surface was rather similar to that of Rh(110) surface shown in Figure 1d, that is, a peak appears at -0.22 V (SCE) in 0.05 M H_2SO_4 solution as shown in Figure 11a. However, the cyclic voltammogram of the (1×2)Rh/Pt(110) surface obtained by O_2 treatment at 760 K is apparently different from that of as deposited Pt(110) and Rh(110) surface. It gives the peak at -0.20 V (SCE) as shown in Figure 11b.

As it was shown in Figure 4, Rh on Pt(100) surface is not stable in vacuum at high temperature but Pt deposited on Rh(100) surface is stable at high temperature. When this stable Pt/Rh(100) surface is heated in O_2, however, Rh atoms are readily extracted by reacting with oxygen and a clear $p(3\times1)$ structure is established at 500 K - 600 K (*21-22*). Figure 12 shows the segregation of Rh atom on the Pt/Rh(100) surface by heating in O_2 and the formation of the $p(3\times1)$ structure in it. The cyclic voltammogram for the Pt deposited Rh(100) surface is apparently different from that

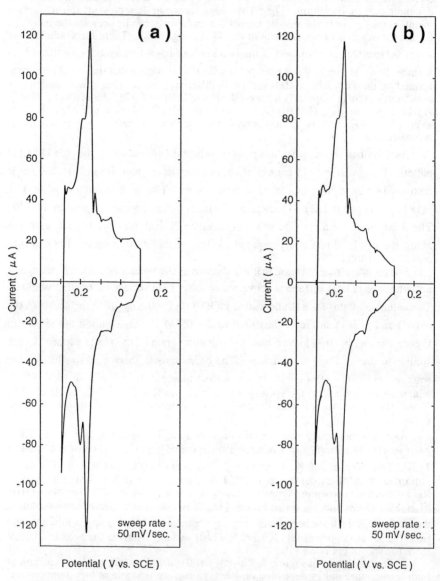

Figure 10. Cyclic voltammogram for (a) $p(3\times1)$ Rh/Pt(100) surface and (b) its $p(1\times1)$ surface in a 0.05 M H_2SO_4 solution ($\theta_{Rh} = 0.8$) .

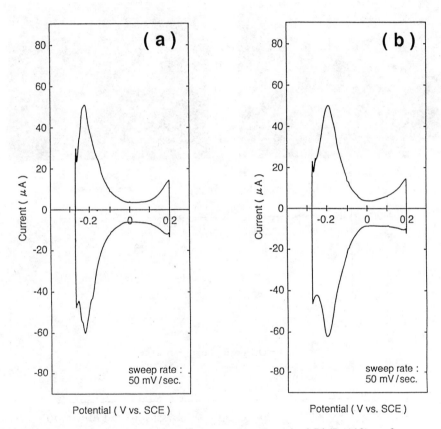

Figure 11. (a) Cyclic voltammogram of as deposited Rh/Pt(110) surface, and (b) (1×2) reconstructed Rh/Pt(110) surface obtained by heating in a 10^{-7} Torr of O_2 at 760 K. Voltammogram was measured in a 0.05 M H_2SO_4 solution.

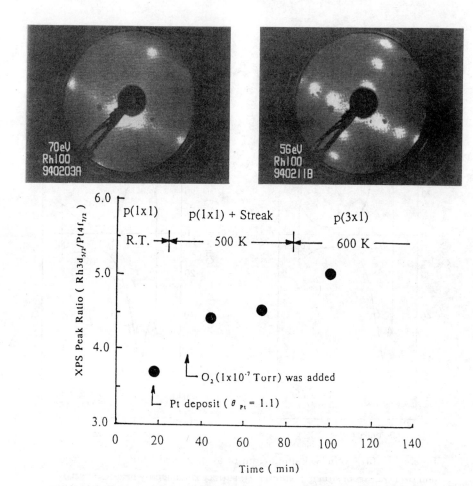

Figure 12. Reactive segregation of Rh atoms on Pt/Rh(100) surface (θ_{Pt} = 1.1) by heating in O_2 and formation of the $p(3\times1)$ structure on it. (Adapted from ref. 22.) Copyright © 1995 Elsevier Science—NL

of either Rh(100) or Pt(100) surface as shown in Figure 2b. It is note worthy fact that the $p(3\times1)$ Pt/Rh(100) surface obtained by annealing in O_2 at about 600 K and the $p(3\times1)$ Rh/Pt(100) surface give almost the same voltammogram as shown in Figure 10. Accordingly, we can conclude that a common $p(3\times1)$ structure composed of Rh-O and Pt-rows is established on $Pt_{0.25}Rh_{0.75}(100)$ alloy surface as well as on Rh/Pt(100) and Pt/Rh(100) surfaces when they are heated in O_2 or NO. As it will be discussed below, the $p(3\times1)$ surface is active surface for the catalytic reaction of NO $+ H_2$.

These results indicate that the minimum amount of Rh required for preparation of the $p(3\times1)$ structure is about 0.3 monolayer. In fact, 0.4 monolayer of Rh on Pt(100) surface gave the clear $p(3\times1)$ pattern, which suggests an important guide line for the practical catalyst.

Pt/Rh(100) and Pt/Rh(110) Surfaces. Cyclic voltammogram of Rh(110) surface measured in a 0.05 M H_2SO_4 solution has sharp peaks at -0.21 V and -0.24 V (SCE) as shown in Figure 1d, which are attributable to the reaction of $H^+ + e \rightarrow H(a)$ and $H(a) \rightarrow H^+ + e$. It should be pointed out that the peaks observed on Rh(110) are very similar to those on Rh(100) surface (*21-22*), which indicates that the hydrogen atoms adsorbed on Rh(100) and Rh(110) surfaces undergo the redox reactions of $H(a) \rightarrow H^+ + e$ and $H^+ + e \rightarrow H(a)$ at the same potentials. This fact is a remarkable contrast to the cyclic voltammogram of Pt(100) and Pt(110) surfaces, where the voltammograms are completely different as shown in Figure 1a and b.

As it was mentioned above, just after the deposition of Pt on Rh(110) surface, the surface gave no LEED pattern although the XPS spectrum of this surface gave a ratio of $Rh3d_{5/2}/Pt4d_{3/2} = 1.1$ with no contamination. The deposited amount of Pt atoms was estimated to be about 1.5 monolayer. The ratio of $Rh3d_{5/2}/Pt4d_{3/2}$ of the Pt/Rh(110) surface changed little by heating up to ca. 760 K in UHV but the ratio was increased by heating at 1000 K for 30 min. and the (1×1) LEED pattern appeared, which is a remarkable contrast to the Pt/Rh(100) surface where the Pt fraction on the Rh(100) surface does not change even at 1050 K as shown in Fig. 4.

The voltammogram for an as deposited Pt/Rh(110) surface ($\theta_{Pt} = 0.9$) was measured in a 0.05 M H_2SO_4 solution, where the effect of Cl ion on the voltammogram was avoided. As shown in Figure 13a, a characteristic peak at -0.21 V (SCE) for Rh(110) surface (Figure 1d) is suppressed by the deposition of Pt and a shoulder appears at ca. -0.15 V (SCE). As mentioned above, the ratio of $Rh3d_{5/2}/Pt4d_{3/2}$ changed little by heating at 760 K in UHV but the voltammogram was markedly changed by annealing the Pt/Rh(110) surface at 760 K, where a new peak appears at -0.15 V (SCE) as shown in Figure 13b. The peak appeared at -0.15 V in the voltammogram (b) may be responsible to a partly ordered Pt-layer on Rh(110) surface.

As mentioned above, when the Pt/Rh(110) surface was annealed at 1000 K for 30 min., the $p(1\times1)$ LEED pattern appeared and $Rh3d_{5/2}/Pt4d_{3/2}$ ratio was increased. When the (1×1) Pt/Rh(110) surface was heated in 1×10^{-7} Torr of H_2 at 760 K, the LEED pattern changed to $p(1\times n)$ although the ratio of $Rh3d_{5/2}/Pt4d_{3/2}$ changed little. This phenomenon suggests that hydrogen promotes a reconstruction of the Pt/Rh(110) surface. On the other hand, when this Pt/Rh(110) surface was heated in 1×10^{-7} Torr of O_2, the $Rh3d_{5/2}/Pt4d_{3/2}$ ratio increased with time from 1.1 to finally

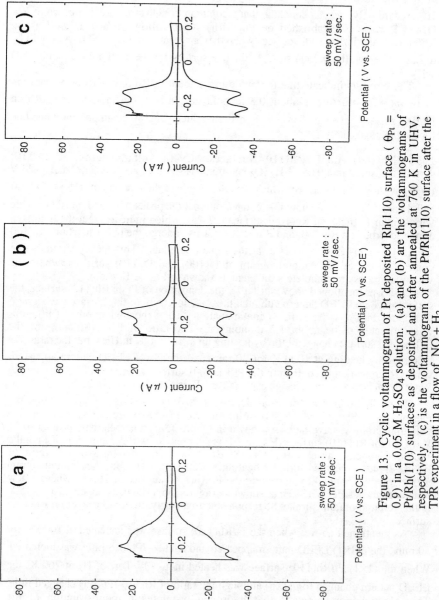

Figure 13. Cyclic voltammogram of Pt deposited Rh(110) surface (θ_{Pt} = 0.9) in a 0.05 M H_2SO_4 solution. (a) and (b) are the voltammograms of Pt/Rh(110) surfaces as deposited and after annealed at 760 K in UHV, respectively. (c) is the voltammogram of the Pt/Rh(110) surface after the TPR experiment in a flow of $NO + H_2$.

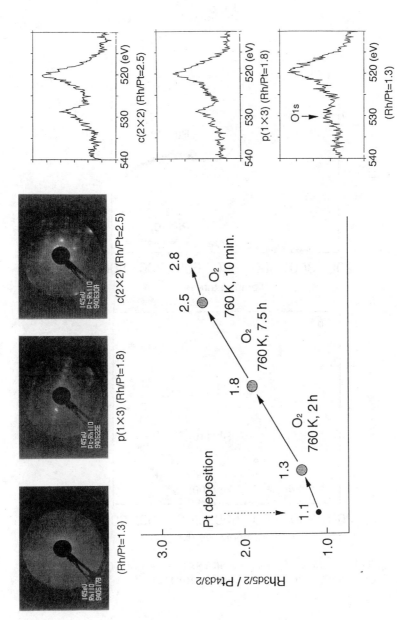

Figure 14. By heating the (1×n) Pt/Rh(110) surface obtained by annealing in H_2 at 760 K in 1×10^{-7} Torr of O_2, Rh atoms are segregated and the LEED pattern changed to (1×3) and finally to c(2×2) structure.

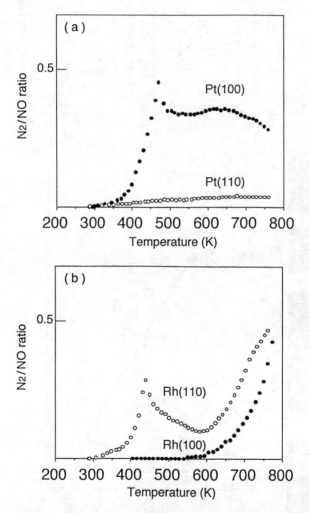

Figure 15. TPR in a flow of a mixture of 5.8×10^{-9} Torr of NO + 1.6×10^{-8} Torr of H_2. (a) Pt(100), Pt(110), (b) Rh(100), Rh(110).

2.8 even at 760 K as show in Figure 14, and the LEED pattern concomitantly changed from $p(1\times3)$ to a $c(2\times2)$ pattern at Rh3d$_{5/2}$/Pt4d$_{3/2}$ = 2.5 .

It should be pointed out that the voltammogram of the Pt/Rh(110) surface is markedly changed by annealing at 760 K although the LEED pattern does not give (1×1) pattern. Figure 13b shows a voltammogram of the Pt/Rh(110) surface annealed at 760 K, where a peak at -0.15 V (SCE) may be responsible to the Pt-layer. In contrast, when the Pt/Rh(110) surface is subjected to the TPR experiment in a flow of NO + H$_2$, the surface is reconstructed and the voltammogram gives a new peak at -0.22 V (SCE) as shown in Figure 13c, which may be responsible to the new surface made by segregated Rh atom by reacting with oxygen.

Catalytic Reaction of NO + H$_2$ on Pt(100), Pt(110), Rh/Pt(100), Rh/Pt(110), Rh(100), Rh(110), Pt/Rh(100), Pt/Rh(110) Surfaces. Temperature programmed reaction (TPR) was performed on Pt(100), Pt(110), Rh(100) and Rh(110) surfaces by raising the temperature in a flow of a mixture of 5.8×10^{-9} Torr of NO and 1.6×10^{-8} Torr of H$_2$. As shown in Figure 15a, Pt(100) surface is quite active for the reaction of NO + H$_2$ → 1/2 N$_2$ + H$_2$O but Pt(110) surface is surprisingly inactive for this reaction (*52*). On the other hand, the TPR of Rh(110) and Rh(100) surfaces in a flow of NO + H$_2$ gives an activity sequence of Rh(100) < Rh(110) as shown in Figure 15b (Sasahara, A.; Tanaka, K., submitted to publication.). Desorption peak of N$_2$ appeared at 450 K - 480 K in Figure 15 is observed on Pt(100) and Rh(110) surfaces only on the way of raising temperature. This N$_2$ desorption does not reflect the catalytic reaction but is a kind of desorption mediated reaction of NO. Therefore, it does not appear when the temperature is lowered in a mixture of NO + H$_2$ because no NO is adsorbed on the surface at high temperature. Following increase of N$_2$ pressure in the TPR is reversible with respect to the temperature, which is given by the catalytic reaction of NO + H$_2$.

If we compare the catalytic activity of a Rh deposited Pt(110) surface to Pt(110) surface for the reaction of NO + H$_2$, as shown in Figure 16a, both the Pt(110) surface and the Pt(110) surface covered with more than several layer of Rh are inactive. Contrary to this, the Pt(110) surface with less than one monolayer of Rh is quite active for this reaction. For example, a (1×2) Rh/Pt(110) surface (θ_{Rh} = ca. 0.4) prepared by treating with O$_2$ and H$_2$ has almost equal catalytic activity to the $p(3\times1)$ Rh/Pt(100) surface for the reaction of NO with H$_2$ (see Figure 16d). Similar enhancement was observed on Pt/Rh(110) surface as shown in Figure 16b. Pt/Rh(110) surface annealed at 1000 K is not so active but the Rh extracted $p(1\times n)$ Pt/Rh(110) surface prepared by heating in O$_2$ and H$_2$ is so high as that on the $p(3\times1)$ Rh/Pt(100) surface (see Figure 16d). Such reconstructive activation of the bimetallic surfaces was more dramatically demonstrated on a Pt deposited Rh(100) surface (θ_{Rh} = 1.1) (*21-22*). As it was mentioned above, Pt/Rh(100) surface is stable in vacuum so that the Pt fraction changes little by annealing at 1000 K. Catalytic activity of the Pt/Rh(100) surface annealed at 1000 K is not so high as shown in Figure 16c, which is almost equal to that on Rh(100) surface. If this surface is heated in O$_2$ or NO at 600 K, the surface undergoes reconstruction and the $p(3\times1)$ structure is established. As shown in Figure 16c, this reconstructed $p(3\times1)$ surface gives surprisingly high catalytic activity for the reaction of NO + H$_2$.

From these results on the alloy and bimetallic surfaces, it could be deduced that a common active structure is established on Rh/Pt(100), Rh/Pt(110), Pt/Rh(110), and Pt/Rh(100) surfaces by combining Rh-O compound and Pt, which is responsible for

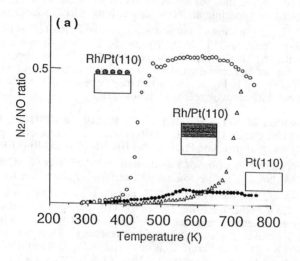

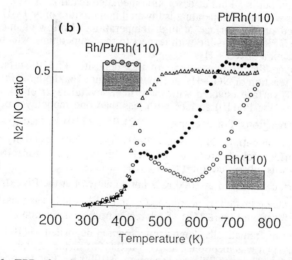

Figure 16. TPR of various bimetallic surfaces in a flow of 5.8×10⁻⁹ Torr of NO + 1.6×10⁻⁸ Torr of H_2. (a) As deposited Rh/Pt(110) and (1×2) Rh/Pt(110) surfaces. (b) As deposited Pt/Rh(110) and (1×n) Pt/Rh(110). (c) Annealed Pt/Rh(100) and p(3×1) Pt/Rh(100). (Adapted from ref. 22.) (d) p(3×1) Rh/Pt(100) surface. Copyright © 1995 Elsevier Science—NL

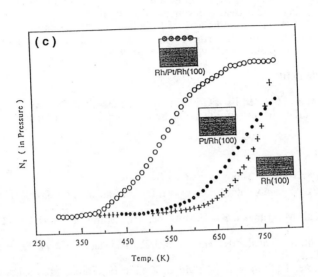

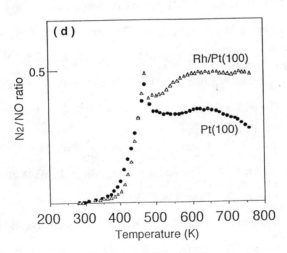

Figure 16. *Continued*

the prominent catalytic activity of the three way catalyst (Sasahara, A.; Tamura, H.; Tanaka, K., submitted to publication.).

Consequently, the roles of Rh atoms in the three way catalyst are summarized as i) NO + H_2 reaction is structure sensitive on Pt and Rh surfaces but is structure insensitive on the Pt deposited Rh as well as on Rh deposited Pt surfaces, that is, the bimetallic surfaces are activated during catalysis by the adsorption of NO and/or O_2, and ii) the active surface composed of Rh-O and Pt atoms may require about 0.3 monolayer of Rh on the Pt surface. iii) Voltammogram is a valuable method to diagnose the surface during catalysis or by adsorption.

Acknowledgments

This work was supported by a Grant-in-Aid for Scientific Research (08232220 and 08834002) of the Ministry of Education, Science and Culture of Japan.

References

1. Ertl,G. *Ang. Chem.* **1990,** *102,* 1258.
2. Eierdal, L.; Besenbacher, F.; Lægsgaard, E.; Stensgaard, I. *Surf. Sci.* **1994,** *312,* 31.
3. Taniguchi, M.; Tanaka, K.; Hashizume, T.; Sakurai, T. *Surf. Sci.* **1992,** *262,* L123.
4. Taniguchi, M.; Tanaka, K.; Hashizume, T.; Sakurai, T. *Chem. Phys. Lett.* **1992,** *192,* 117.
5. Hashizume, T.;Taniguchi, M.; Motai, K.: Lu, H.; Tanaka, K.; Sakurai, T. *J. J. Appl. Phys.* **1991,** *30,* L1529.
6. Nielsen, L. P.; Besenbacher, F.; Lægsgaard, E.; Stensgaard, I. *Phys. Rev. B* **1991,** *44,* 13256.
7. Jensen, F.; Besenbacher, F.; Lægsgaard, E.; Stensgaard, I. *Phys. Rev. B* **1990,** *41,* 10233.
8. Jackman, T. E.; Davice, J. A.; Norton, P. R.; Unertl, W. N.; Griffiths, K. *Surf. Sci.* **1980,** *141,* L313.
9. Christman, K.; Penka, V.; Behm, R. J.; Chehab, F.; Ertl, G. *Solid State Comm.* **1984,** *51,* 487.
10. Kleinle, G.; Penka, V.; Behm, R. J.; Chehab, F.; Ertl, G. *Phys. Rev. Lett.* **1987,** *58,* 148.
11. Tanaka, K. *J. J. Appl. Phys.* **1993,** *32,* 1389.
12. Tsong, T. T.; Ren, D. M.; Ahmad, M. *Phys. Rev. B* **1988,** *38,* 7428.
13. Ren, D. M.; Tsong, T. T. *Surf. Sci.* **1985,** *184,* L439.
14. van Delft, F. C. M. J. M.; Nieuwenhuys, B. E.; Siera, J.; Wolf, R. M. *ISIJ Int.* **1989,** *29,* 550; *Surf. Sci.* **1992,** *264,* 435.
15. Hirano, H.; Yamada, T.; Tanaka, K.; Siera, J.; Nieuwenhuys, B. E. *Surf. Sci.* **1989,** *222,* L804.
16. Hirano, H.; Yamada, T.; Tanaka, K.; Siera, J.; Nieuwenhuys, B. E. *Vacuum* **1990,** *41,* 134.
17. Hirano, H.; Yamada, T.; Tanaka, K.; Siera, J.; Nieuwenhuys, B. E. *Surf. Sci.* **1990,** *226,* 1.
18. Taniguchi, M.; Kuzembaev, E. K.; Tanaka, K. *Surf. Sci.* **1993,** *290,* L711.
19. Tanaka, K. Taniguchi, M. *Topics in Catalysis* **1994,** *1,* 95.
20. Tamura, H.; Tanaka, K. *Langmuir* **1994,** *10,* 4530.
21. Tamura, H.; Sasahara, A.; Tanaka, K. *Surf. Sci.* **1994,** *303,* L379.
22. Tamura, H.; Sasahara, A.; Tanaka, K. *J. Electroanal. Chem.* **1995,** *381,* 95.

23. Gomez, R.; Rodes, A.; Perez, J. M.; Feliu, J. M.; Aldaz, A. *Surf. Sci.* **1995**, *327,* 202.
24. Armand, D.; Clavilier, J. *J. Electroanal. Chem.* **1987**, *233,* 251.
25. Armand, D.; Clavilier, J. *J. Electroanal. Chem.* **1989**, *263,* 109.
26. Yamamoto, K.; Kolb, D.; Kotz, R.; Lehmpfuhl, G. *J. Electroanal. Chem.* **1986**, *214,* 555.
27. Michaelis, R.; Kolb, D. M. *J. Electroanal. Chem.* **1992**, *328,* 341.
28. Kita, H.; Ye, S.; Aramata, A.; Furuya, N. *J. Electroanal. Chem.* **1990**, *295,* 317.
29. Yamamoto, K.; Kolb, D.; Kotz, R.; Lehmpfuhl, G. *J. Electroanal. Chem.* **1979**, *96,* 233.
30. Scortichini, C. L.; Reilley, C. N. *J. Electroanal. Chem.* **1982**, *139,* 233.
31. Zei, M. S.; Batina, N.; Kolb, D. M. *Surf. Sci.* **1994**, *306,* L519.
32. Clavilier, J.; Durand, R.; Guinet, G.; Faure, R. *J. Electroanal. Chem.* **1981**, *127,* 281.
33. Wasberg, M.; Hourani, M.; Wieckowski, A. *J. Electroanal. Chem.* **1990**, *278,* 425.
34. Rhee, C. K.; Wasberg, M.; Horanyi, G.; Wieckowski, A. *J. Electroanal. Chem.* **1990**, *291,* 281.
35. Schmid, M.; Stadler, H.; Varga, P. *Phys. Rev. Lett.* **1993**, *70,* 1441.
36. Schmid, M.; Biedermann, A.; Slama, C.; Stadler, H.; Weigand, P.; Varga, P. *Nucl. Instr. and Meth. in Phys. Res., B* **1993**, *82,* 259.
37. Niehus, H.; Achete, C. *Surf. Sci.* **1993**, *289,* 19.
38. Nagl, C.; Haller, O.; Platzgummer, E.; Schmid, M.; Varga, P. *Surf. Sci.* **1994**, *321,* 237.
39. Matsumoto, Y.; Okawa, Y.; Fujita, T.; Tanaka, K. *Surf. Sci.* **1996**, *355,* 109.
40. Jensen, F.; Besenbacher, F.; Lægsgaard, E.; Stensgaard, I. *Phys. Rev. B* **1990**, *42,* 9206.
41. Wool, Ch.; Wilson, R. J.; Chiang, S.; Zeng, H. C.; Mitchell, K. A. R. *Phys. Rev. B* **1990**, *42,* 11926.
42. Leibsle, F. M. *Surf. Sci.* **1995**, *337,* 51.
43. Chua, F.M.; Kuk, Y.; Silverman, P. J. *Phys. Rev. Lett.* **1989**, *63,* 386.
44. Coulman, D. J.; Wintterlin, J.; Behm, R. J.; Ertl, G. *Phys. Rev. Lett.* **1990**, *64,* 1761.
45. Kuk, Y.; Chua, F. M.; Silverman, P. J.; Meyer, J. A. *Phys. Rev. B* **1990**, *41,* 12393.
46. Matsumoto, Y.; Okawa, Y.; Tanaka, K. *Surf. Sci.* **1995**, *325,* L435.
47. Matsumoto, Y.; Okawa, Y.; Tanaka, K. *Surf. Sci.* **1995**, *336,* L762.
48. Okawa, Y.; Tanaka, K. *Surf. Sci.* **1995**, *344,* L1207.
49. Eierdal, L.; Besenbacher, F.; Lægsgaard, E.; Stensgaard, I. *Ultramicroscopy* **1992**, *42 - 44,* 505.
50. Leibsle, F. M.; Murray, P. W.; Francis, S. M.; Thornton, G.; Bowker, M. *Nature* **1993**, *363,* 706.
51. Dhanak, V. R.; Prince, K. C.; Rosei, R.; Murray, P. W.; Leibsle, F. M.; Bowker, M.; Thornton, G. *Phys. Rev. B* **1994**, *49,* 5585.
52. Sasahara, A.; Tamura, H.; Tanaka, K. *Catal. Lett.* **1994**, *28,* 161.

Chapter 19

Electron Spectroscopy and Electrochemical Scanning Tunneling Microscopy of the Solid–Liquid Interface: Iodine-Catalyzed Dissolution of Pd(110)

Manuel P. Soriaga[1], W. F. Temesghen[1], J. B. Abreu[1], K. Sashikata[2], and K. Itaya[2]

[1]Department of Chemistry, Texas A&M University, College Station, TX 77843
[2]Department of Applied Chemistry, Faculty of Engineering, Tohoku University, Sendai 980–77, Japan

This article showcases the unique capabilities afforded by the combination of electron spectroscopy, scanning tunneling microscopy, and electrochemistry in the study of complex processes at the electrode-electrolyte interface. The interfacial reaction investigated was the anodic dissolution of a Pd(110) single-crystal surface catalyzed by a single chemisorbed layer of zerovalent iodine atoms. Previous work had shown that dissolution of the two other low-index planes, Pd(100) and Pd(111), altered neither the adlattice structures nor the interfacial coverages. In comparison, while the anodic dissolution of the Pd(110)-I surface did not lead to changes in the iodine coverage, it resulted in a disordered surface. The *in situ* STM work provided valuable insight as to why the initial and post-dissolution structures were not identical.

Major advances in the study of the electrode-electrolyte interface at the atomic level were ushered in two decades ago by the integration of traditional electrochemical methods with modern surface-sensitive analytical techniques [1]; such a strategy, now commonly referred to as the ultrahigh vacuum-electrochemistry (UHV-EC) approach, made possible critical correlations between interfacial structure, interfacial composition, and interfacial reactivity [2]. The subsequent development, less than a decade later, of procedures for the fabrication, regeneration, and verification of single-crystal electrode surfaces without recourse to expensive surface science equipment [3] also had a profound influence in the progress of electrochemical surface science. The recent resurgence in research in surface electrochemistry has been instigated by the evolution of *in situ* methods [4] that allow the simultaneous implementation of electrochemical and surface-analytical experiments; the most prominent of these techniques is scanning tunneling microscopy (STM) [5].

While the impressive strides in electrochemical surface science brought about by the combination of STM and electrochemistry (STM-EC) may prompt the abandonment of other surface characterization methods, a lesson learned by the pioneers of modern surface science need only be recalled: the sheer complexity of heterogeneous processes cannot be unraveled by just one technique [6]. The singular strength of STM-EC lies in its integration with other methods such as those that make possible the determination of interfacial energetics, composition, and electronic structures; the combination of STM-EC with UHV-EC serves as an example.

The primary motivation of this paper is to illustrate the use and to showcase the power of UHV-STM-EC, even in a less-than-ideal scenario in which the UHV and STM experiments were performed separately in two different laboratories. The case studied involves the dissolution of Pd(110) that occurs only when a monolayer of iodine is present on the surface [7,8]. The structural and compositional features of the halogen-metal interface that characterize the adsorbate-catalyzed corrosion were first explored with low-energy electron diffraction (LEED) and Auger electron spectroscopy (AES); initial- and final-state measurements were quite easily obtained. Certain questions, however, such as the mechanism for dissolution, necessitated an *in situ* technique for answers; it was in this regard that the STM-EC work was prompted.

Experimental

The experimental procedures specific to the UHV-EC [7] and STM-EC [5c,8] investigations have been described in detail elsewhere. UHV-EC work, carried out at Texas A&M University (College Station, TX) employed a commercially oriented and metallographically polished, 99.9999%-pure Pd(110) single-crystal electrode. STM-EC studies, undertaken at Tohoku University (Sendai, Japan) were done with Pd(110) single-crystal surfaces prepared from 99.995%-pure polycrystalline Pd wires by the method of Clavilier [3] modified to compensate for the unique chemical properties of Pd metal. Commercial instruments were used in the UHV-EC (Perkin-Elmer, Eden Prairie, MN) and STM-EC (Digital Instruments, Santa Barbara, CA) experiments.

Results and Discussion

Electron Spectroscopy. Figure 1 shows current-potential curves for an untreated (clean) and an iodine-coated Pd(110) facet, formed on a single-crystal bead, in halide-free 0.05 M H_2SO_4. As was demonstrated previously [6], the exceedingly large anodic peak at about 1.1 V (vs. RHE) represents the $Pd^0_{(s)}$-to-$Pd^{2+}_{(aq)}$ anodic stripping that occurs only when interfacial iodine is present; in the absence of iodine, only the passivating oxide layer is formed, as indicated by the solid curve. If the potential applied to the I-coated surface is held below the peak potential, the current, as expected from a material-limited dissolution process, does not decay but remains essentially constant. The data in Figure 1 bear strong similarity to those obtained for Pd(111)-($\sqrt{3} \times \sqrt{3}$)R30°-I [7a] and Pd(100)-c(2×2)-I [7b], cases for which layer-by-layer dissolution has been demonstrated [8]. Such mode of dissolution has been exploited for electrochemical "digital etching" in which disordered Pd(111) and Pd(100) surfaces were rendered atomically smooth by dissolution of the (unstable) disordered domains [9]. It may be mentioned that the reactivity of the I-coated Pd(110) towards Pd dissolution is considerably higher than those of the I-treated Pd(111) and Pd(100) surfaces [10]. Indicated in Figure 1 are the potentials at which the UHV-EC surface characterization and the STM-EC experiments were performed.

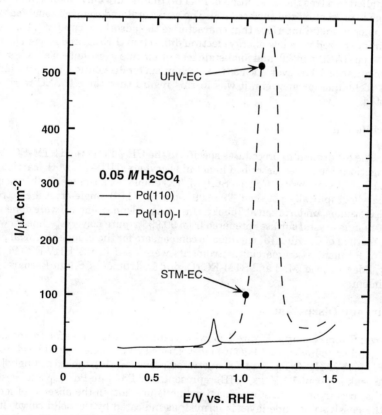

Figure 1. Current density-vs-potential curves in the surface-oxidation region in 0.05 M H_2SO_4 for a Pd(110) facet on a single-crystal bead, clean (solid curve) and I-coated (dashed curve). Potential sweep rate, r = 10 mV s^{-1}.

Photographs are shown in Figure 2 of LEED patterns for UHV-prepared Pd(110) before (A) and after (B) immersion in 0.05 M H_2SO_4 containing 1 mM KI. Figure 2(A) is the familiar (1 × 1) LEED pattern for the clean, unreconstructed Pd(110) surface. The LEED pattern in Figure 2(B) for the KI-immersed Pd(110) surface has been referred to as a distorted-hexagonal or pseudohexagonal structure that results when the adsorbate coverage Θ_I ($\equiv \Gamma_I/\Gamma_{Pd}$, where Γ is the surface packing density expressed in units of mole cm^{-2}) exceeds 0.5 [11]. When the latter adlattice is heated above 700 K, a fraction of the interfacial iodine is desorbed and a stable Pd(110)-c(2 × 2)-I adlattice of coverage Θ_I = 0.5 is formed; as detailed elsewhere [12], the Pd(110)-c(2 × 2)-I structure can also be produced directly, under electrochemical conditions, without the *ex situ* heat treatment. For the present manuscript, only the anodic dissolution of the easier-to-form Pd(110)-pseudohexagonal-I adlattice was investigated. Figure 3 shows a real-space model of a (high-coverage) Pd(110)-pseudohexagonal-I interface; the actual coverage depends upon how distorted the overlayer structure is from perfect hexagonal symmetry [11].

The LEED pattern obtained after the removal of the equivalent of 10 monolayers (ML) of Pd, each monolayer represented by the Faradaic charge (0.30 mC) required for the 2-electron anodic dissolution of 9.36 x 10^{14} Pd(110) atoms cm^{-2}, is shown in Figure 2(C). The degradation of the post-dissolution LEED pattern is obvious. Thus, in contrast to the cases of Pd(100)-c(2 × 2)-I and Pd(111)-($\sqrt{3}$ × $\sqrt{3}$)R30°-I [6], the iodine-catalyzed corrosion of Pd(110) results in a disordered adlayer.

To ascertain whether or not the disorder is due to removal of the chemisorbed iodine during the dissolution process, Auger electron spectra were obtained prior to and after the 10-ML dissolution. The results are shown in Figure 4 where it can be seen that, identical to those obtained for the Pd(100)-c(2 × 2)-I and Pd(111)-($\sqrt{3}$ × $\sqrt{3}$)R30°-I adlattices [6], the iodine coverage has remained essentially unchanged. In other words, the final-state disorder of the Pd(110)-I interface is not due to unwanted side reactions that may have led to iodine removal.

Electrochemical Scanning Tunneling Microscopy. The results from the *in situ* STM experiments are summarized in terms of the (low-resolution) images shown in Figure 5; the top image [Figure 5(A)] was acquired at the start of the dissolution reaction (at the potential shown in Figure 1), whereas Figure 5(B) was obtained after four minutes of dissolution. The STM image of the Pd(110)-I surface prior to the dissolution experiments was similar to that given in Figure 5(A) except that the pits were not present. The pre-dissolution image was characterized by the *absence* of wide terraces, despite metallographic polishing and thermal annealing, and the predominance of steps that run parallel to the {110} direction.

Upon initiation of the iodine-catalyzed dissolution, circular-shaped pits several nm in diameter and about 0.5 nm in depth immediately appeared as shown in Figure 5(A). After four minutes of further dissolution, dramatic changes resulted in the STM image, as can be viewed in Figure 5(B). The notable changes are as follows: (a) The pits, originally minuscule, have, individually or collectively, increased considerably in size. (b) The shapes of the pits have been transformed from circular to rectangular with straight edges (steps) that run parallel to the {110} and {100} directions. (c) The rectangular terraces are preferentially elongated towards the {100} direction. (d) New (smaller) pits are formed at the bottom of the rectangular pits; consequently, the surface becomes progressively roughened, as was initially indicated by the LEED data [Figure 2(C)]. The changes in the STM images in Figures 5(A) and 5(B), in particular the formation of preferentially-elongated rectangular pits, point to a selective dissolution mechanism.

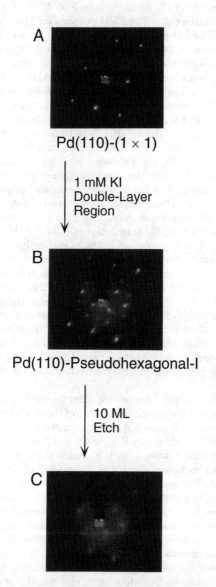

Figure 2. Low-energy electron diffraction (LEED) patterns for clean (iodine-free) Pd(110) (A) and Pd(110)-pseudohexagonal-I before (B) and after (C) anodic dissolution of *approximately* 10 monolayers of Pd(110) surface atoms. Beam energy = 60 eV; beam current = 2mA.

Pd(110)-Pseudohexagonal-I

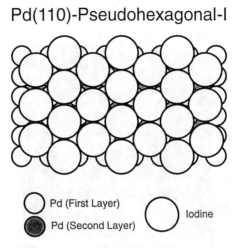

Figure 3. Real-space model for a (high-coverage) Pd(110)-pseudo-hexagonal-I adlattice; the actual coverage depends upon how different the overlayer structure is from perfect hexagonal symmetry [11].

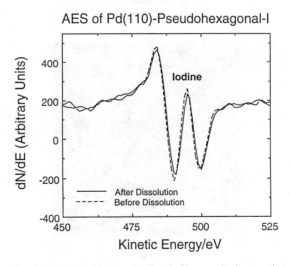

Figure 4. Auger electron spectrum, in the iodine emission region, of Pd(110)-pseudohexagonal-I before (dashed curve) and after (solid curve) removal of approximately 10 monolayers of Pd(110) surface atoms. These spectra correspond to the LEED patterns (B) and (C), respectively, in Figure 3. Incident beam energy = 2 keV; beam current = 1 μA.

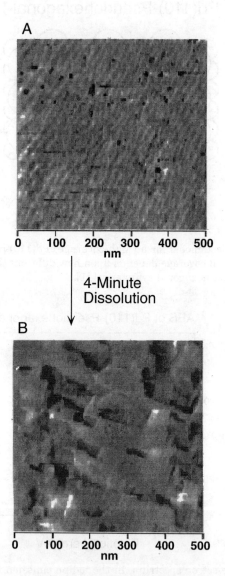

Figure 5. Scanning tunneling microscopy images of I-pretreated Pd(110) at the start (A) and after four minutes (B) of anodic dissolution in 0.05 M H_2SO_4.

Postulated Mode of Dissolution of Pd(110)-I

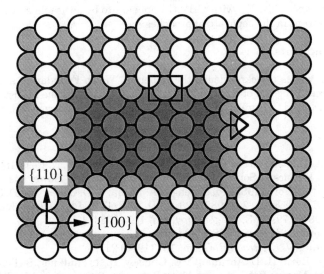

Figure 6. Proposed model for the dissolution of an I-pretreated Pd(110) surface that contains a rectangular pit two atomic layers in depth. The rectangle represents the (100) plane exposed at the step along the {100} direction, whereas the triangle depicts the (111) plane at the edge along the {110} direction.

To help rationalize the formation of rectangular pits brought about by the dissolution process, a schematic model is presented in Figure 6 of a Pd(110) surface that contains a rectangular pit two atomic layers in depth. It can be seen from the model that the step-edge along the {110} direction consists of the (111) plane (represented by the triangle), whereas the step-edge in the {100} direction is constituted by the (100) plane (depicted by the rectangle). It is not difficult to envisage how rectangular pits are formed if it is postulated that selective dissolution occurs along the {100} and the {110} directions. Corrosion in other directions would result in kinks, highly unstable defect sites that would only be dissolved immediately [9] to regenerate the (100) and (111) step-edges. The propensity of the rectangular pits towards elongation along the {100} direction indicates that the (111) step is preferentially (albeit slightly) corroded over the (100) edge. Such preferential dissolution is not inconsistent with the fact that the iodine-catalyzed dissolution of Pd(100) occurs at (slightly) more positive potentials than that of Pd(111) [10].

Conclusions

Two of the many advantages of STM in surface electrochemical studies have been illustrated in the present manuscript: (i) its adaptability for measurements *under reaction conditions*, and (ii) its ability to resolve nanometer-scale structural features. Such advantages have helped vault STM into the forefront in experimental surface science. On the other hand, STM is unable to provide other critical information such as surface energetics, composition, and electronic structure; hence, additional surface spectroscopic techniques would always be required if a more complete understanding of complex heterogeneous processes is contemplated. A strategy that combines STM-EC with UHV-EC appears to be desirable since it bridges the gaps inherent in the separate approaches.

Acknowledgments

MPS wishes to acknowledge the Robert A. Welch Foundation and the Energy Resources Program of Texas A&M University. KI thanks the Ministry of Education, Science, and Culture and the ERATO-Itaya Electrochemiscopy Project, JRDC.

References

1.   (a) A. T. Hubbard. *Accts. Chem. Res.* 13 (1980) 177. (b) P. N. Ross. In "Chemistry and Physics of Solid Surfaces." R. Vanselow and R. Howe, eds. Springer-Verlag, New York (1982). (c) E. Yeager, A. Homa, B. D. Cahan and D. Scherson. *J. Vac. Sci. Technol.*, 20 (1982) 628. (d) D. M. Kolb. *Z. Phys. Chem. Neue Folge.*, 154 (1987) 179. (e) E. Yeager. *Surf. Sci.* 101 (1980) 1.

2.   M. P. Soriaga. *Prog. Surf. Sci.* 39 (1992) 525.

3.   J. Clavilier. *J. Electroanal. Chem.* 107 (1980) 211.

4.   (a) A. J. Bard, H. D. Abruña, C. E. Chidsey, L. R. Faulkner, S. Feldberg, K. Itaya, O. Melroy, R. W. Murray, M. D. Porter, M. P. Soriaga and H. S. White. *J. Phys. Chem.* 97 (1993) 7147. (b) H. Abruña (Ed.) "Electrochemical Interfaces: Modern Techniques for In-Situ Surface Characterization." VCH: New York (1991).

5.   (a) G. Binnig and H. Rohrer. Surf. Sci. 157 (1985) L373. (b) M. M. Dvek, M. J. Heben, N. S. Lewis, R. M. Penner and C. F. Quate. In M. P. Soriaga, Ed. "Electrochemical Surface Science." ACS Books: Washington, D.C. 1988. (c) K. Itaya. In A. T. Hubbard, Ed. "The Handbook of Surface Imaging and Visualization." CRC Press: Boca Raton, FL. 1995.

6.   (a) G.A. Somorjai. "Introduction to Surface Chemistry and Catalysis." Wiley-Interscience: New York, NY. 1995. (b) A. T. Hubbard, Ed. "The Handbook of Surface Imaging and Visualization." CRC Press: Boca Raton, FL. 1995.

7.   (a) J. R. McBride, J. A. Schimpf and M. P. Soriaga. *J. Electroanal. Chem.* 350 (1993) 317. (b) J. A. Schimpf, J. R. McBride and M. P. Soriaga. *J. Phys. Chem.* 97 (1993) 10518.

8.   M. P. Soriaga, J. A. Schimpf, J. B. Abreu, A. Carrasquillo, W. Temesghen, R. J. Barriga, J.-J. Jeng, K. Sashikata and K. Itaya. *Surf. Sci.* 335 (1995) 273.

9.   (a) J. B. Abreu, R. J. Barriga, W. Temesghen, J. A. Schimpf and M. P. Soriaga. *J. Electroanal. Chem.* 381 (1995) 239. (b) J. A. Schimpf, A. Carrasquillo and M. P. Soriaga. *Electrochim. Acta.* 40 (1995) 1203.

10.   Y. Matsui, Y. Imamura, H. Hosoya, K. Sashikata, K. Itaya and M. P. Soriaga. *Langmuir.* Submitted.

11.   U. Bardi and G. Rovida. *Surf. Sci.* 128 (1993) 145.

12.   W. F. Temesghen. Ph. D. Dissertation. Texas A&M University. College Station, TX. 1996.contains a rectangular pit two atomic layers in depth. The rectangle represents the (100) plane exposed at the step along the {100} direction, whereas the triangle depicts the (111) plane at the edge along the {110} direction.

Chapter 20

Electrocatalysis of Formic Acid and Carbon Monoxide with Probe Adlayers of Carbon and Ethylidyne on Pt(111)

D. E. Sauer, R. L. Borup, and E. M. Stuve[1]

Department of Chemical Engineering, University of Washington, P.O. Box 351750, Seattle, WA 98195–1750

Mechanistic details of CO and HCOOH electrocatalysis were examined with the aid of carbon and ethylidyne-derived adlayers on Pt(111) in perchloric acid electrolyte. H, CO, and HCOOH all adsorb or react on the electrode even with more than one monolayer of carbon present. Islands of graphitic carbon enhance CO oxidation, while a full carbon overlayer inhibits the reaction with respect to the carbon-free surface. The peak turnover rate for HCOOH oxidation on carbon-free sites of the electrode is constant at 3 s^{-1} over a range of low carbon coverages, but increases sharply and passes through a maximum of 18 s^{-1} for a carbon coverage of 80% of the surface area. These results indicate site blocking at low carbon coverage and formation, at high carbon coverages, of platinum ensembles that optimize complete oxidation of HCOOH at the expense of formation of the surface poison. Reaction on an ethylidyne (CCH_3) covered surface was faster than on the clean surface, and the results provide an estimate of 2-3 platinum atoms for the optimum ensemble size.

Modification of electrode surfaces is one of the most common means of improving electrocatalytic activity. For example, formic acid oxidation on platinum electrodes is enhanced by a variety of metallic adsorbates [1] including lead [2], bismuth [3-5], antimony [6], selenium [7], and ruthenium [8,9]. It is well known that this reaction suffers from self-poisoning by carbon monoxide [1,10-12] and a hydrogen containing intermediate, either COH or HCO [2,13]. To enhance the reaction the surface modifier must reduce or eliminate the poisoning species and perhaps also offer some other form of enhancement beyond control of the poisoning species. Mechanisms advanced to explain control of the poisoning species include [1]: a "third-body" effect, in which steric hindrance by the modifier atom controls poisoning [7,8,14,15];

---

[1]Corresponding author

suppression of hydrogen adsorption, which is thought to suppress the hydrogen containing poisoning species [16]; and a bifunctional mechanism in which the poison is oxidatively removed by enhanced oxygen transport to the surface nominally through redox behavior of the modifier [8,17,18]. In addition, lead, bismuth, and ruthenium enhance the reaction beyond control of the poisoning species.

A given modifier can therefore offer electrocatalytic enhancement in a number of ways. The interactions among modifier, substrate, reactants, and electrolyte are doubtlessly complex in cases such as this, so it is of interest to see if some particular aspect of the surface reaction, for example ensemble requirements, can be isolated and studied in detail. This is possible through the use of a *probe adlayer*, that is, an adlayer specifically designed to isolate and probe a given mechanistic aspect of the reaction. In this paper we present our first attempts to study carbon monoxide and formic acid electrooxidation in the presence of probe adlayers of carbon or derived from ethylidyne.

To examine the ensemble requirements of a reaction, a suitable probe adlayer should restrict access to the surface by the reacting species in a known and controllable way. As will be shown, both the carbon and ethylidyne adlayers are inert in the sense that they are not consumed by the reaction. The two adlayers differ in their distribution on the surface, however: carbon adsorbs as graphitic islands, whereas ethylidyne adsorbs as isolated molecules in a p(2x2) structure. Information about ensemble requirements can then be obtained through studies of the respective reactions as a function of probe adlayer coverage. Of course, these adlayers cannot isolate ensemble requirements perfectly, as other effects, such as ion adsorption, may occur to some degree. Nonetheless, they represent a simplified electrocatalytic substrate with which definitive information about the surface reaction mechanism can be determined.

A more direct motivation for studies of carbonaceous adlayers comes from the recent finding that formic acid oxidation can occur *through* a hydrocarbon residue formed by formic acid itself on platinum [2]. One implication from this study is that carbonaceous residues could be present in other electrocatalytic reactions, perhaps influencing them in some way, but otherwise remaining quite difficult to detect.

This paper focuses on the effects of the carbon and ethylidyne-derived adlayers in electrooxidation. We present brief accounts of how the adlayers were characterized, both electrochemically and in vacuum, and will describe adlayer characterization in more detail in a separate publication [19]. To our knowledge, the effects of carbon adlayers on single crystal electrodes have not been studied, although the situation is somewhat analogous to platinum particles deposited on graphite electrodes, about which quite a bit is known [20,21]. The ethylidyne adlayer is derived from adsorption of ethylene in vacuum [22,23]. Electrochemical adsorption of ethylene and electroreduction to ethane have also been previously studied [24-28].

Electrodes containing the desired probe adlayer were transferred from vacuum directly to an attached electrochemical cell. Carbon monoxide was adsorbed either by solution or gas phase dosing, while formic acid was supplied only through solution. Vacuum measurements included thermal desorption, Auger spectroscopy, and low energy electron diffraction and were primarily intended to establish coverages and the state of the probe adlayer. In this paper we focus on the electrochemical measurements, as these highlight the influence of the probe adlayer on the reaction.

Experimental Procedure

The experiments were conducted in a combined ultrahigh vacuum chamber (ultimate pressure of 10^{-8} Pa) and electrochemical cell that has been described previously [29,30]. It was equipped with a quadrupole mass spectrometer for thermal desorption spectroscopy, conventional four-grid optics for low energy electron diffraction, and a single-pass cylindrical mirror energy analyzer for Auger spectroscopy. The electrochemical cell was directly attached to the vacuum system, but isolated from it through dual, differentially pumped Teflon seals. The cell was constructed of Teflon and had an electrolyte volume of approximately 0.2 cm^3. The Pt(111) electrode, with a superficial area of 0.4 cm^2, was cut to within 0.5° and polished to mirror smoothness by standard metallographic techniques. Residual surface contaminants were removed by potential cycling in the electrochemical cell followed by annealing in vacuum to restore surface order. Voltammograms were measured with an EG&G 362 potentiostat and Hewlett-Packard 7090A plotter.

Carbon and ethylidyne adlayers were deposited on the sample by adsorption of ethylene (Matheson, 99.5%) in the vacuum chamber. A saturated coverage of ethylidyne (0.25 per Pt atom) was formed by room temperature dosing of ethylene at 7×10^{-6} Pa for 10 s [22,23]. The p(2x2) structure for this adlayer was confirmed by electron diffraction. Ethylidyne prepared by gas phase adsorption of ethylene has been shown to be stable upon immersion into ethylene-free electrolyte at potentials between 0.15 and 0.7 V [26]. Lower potentials promote electroreduction [19] and higher potentials some form of oxidation [19,26]. The oxidation is reversible so long as the upper potential limit does not exceed 0.9 V [19], and formic acid voltammograms in the presence of an ethylidyne-derived adlayer (e.g., Figure 5) could be repeated many times without change.

A carbon adlayer was produced by subsequent annealing of the ethylidyne covered surface to 873 K. A single ethylene dose and anneal cycle resulted in a carbon coverage of 0.2 monolayers. Higher carbon coverages were obtained by repeating the dosing and annealing procedure as often as required; the maximum coverage obtainable was 1.3 monolayers. Following electrochemical measurements the carbon and ethylidyne adlayers were removed by argon sputtering and annealing to 1410 K. Any residual carbon was removed by cycling the sample between room temperature and 873 K in 10^{-5} Pa of oxygen until the sample was clean as judged by Auger spectroscopy.

Carbon coverages were measured by Auger spectroscopy according to the method previously published [31]. The carbon coverage is in terms of a saturated monolayer, that is, $\theta_C = 1$ corresponds to a surface fully covered with carbon. For the graphitic form (basal plane) of carbon this corresponds to 2.57 carbon atoms per platinum surface atom [32]. The Auger calibration was confirmed by titration of the empty surface sites with carbon monoxide in vacuum [33]; these independent methods agreed to within 10%.

At high coverages the carbon adlayer was graphitic in nature, as determined by the characteristic line shape in Auger spectroscopy [34,35] and rings in the electron diffraction pattern that appeared for coverages of 0.4 monolayers or greater, in agreement with previous results [36,37]. At lower coverages the state of carbon, graphitic vs. carbidic, was difficult to ascertain because of the overlap of the carbon

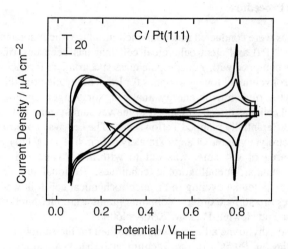

Figure 1. Cyclic voltammograms of Pt(111) in 0.1 mol/dm^3 HClO$_4$ with a carbon coverage of (in the direction of the arrow): 0.00, 0.28, 0.70, and 1.2 monolayer.

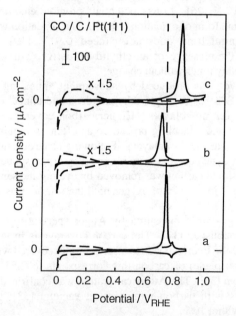

Figure 2. Cyclic voltammograms of carbon-modified Pt(111) in 0.1 mol/dm^3 HClO$_4$ without CO (dashed lines) and with a saturation coverage of solution-dosed CO (solid lines). The electrode had carbon coverages of: (a) 0.0, (b) 0.79, and (c) 1.3 monolayer.

and platinum peaks in the Auger spectrum. In view of the annealing procedure used in deposition, we assume that the carbon adlayer is graphitic at all coverages.

Electrolytes were prepared with water of 18 MΩ cm conductivity purified by a Barnstead Nanopure system and doubly distilled perchloric acid (70 wt%, G. F. Smith Chemicals). Carbon monoxide was dosed from solution by a 1 min exposure of the sample to 0.1 mol/dm^3 perchloric acid saturated with carbon monoxide at 1 atm while maintaining the potential at 0.3 V. The carbon monoxide solution was then removed and the sample rinsed once with blank electrolyte prior to electrochemical measurements. Electrode potentials were measured with respect to a gold/gold oxide reference electrode and converted to the reversible hydrogen electrode (RHE) scale. All cyclic voltammograms were recorded at room temperature with a sweep rate of 50 mV/s. To avoid electrooxidation of the carbon adlayers the sample potential was not allowed to exceed 0.95 V.

Special note about coverages: Because of the nature of the different reactants and adlayers, a common basis for reporting surface coverage is not practical. This creates confusion as some coverages (carbon) are reported as a fraction of a saturated monolayer of adsorbate, while others (carbon monoxide and ethylidyne) are reported as a fraction of the substrate surface atom density. We therefore adopt IUPAC terminology [38,39] and use the term *monolayer* to indicate the maximum amount of adsorbate for which each adsorbed atom or molecule is in direct contact with the substrate. A full monolayer corresponds to $\theta = 1$, as previously noted for carbon. Coverages of carbon monoxide and ethylidyne will be reported as ratios of adsorbate species per substrate surface atom. The reader should note that this definition of monolayer differs from our previous usage as well as that of some other researchers.

Results

The voltammetric behavior of Pt(111) is shown as a function of carbon coverage in Figure 1. The outermost voltammogram for the carbon-free surface is characteristic of clean Pt(111), as noted by the plateau in the hydrogen region below 0.3 V and the sharp spike, known as the butterfly peak, at 0.78 V. The total charge enveloped by the voltammogram is 660 μC/cm^2, in good agreement with the recommended value of 640 μC/cm^2 [40]. With the addition of carbon hydrogen adsorption/desorption shifts to lower potential, but retains 85% of its area even for a fully carbon covered surface. In contrast, the butterfly peak gradually disappears with increasing carbon coverage.

Electrooxidation of solution dosed carbon monoxide is shown in Figure 2 (solid lines) as a function of carbon coverage. The voltammogram for the carbon-free surface is in excellent agreement with earlier studies of carbon monoxide oxidation [30,41]. With increasing carbon coverage the amount of oxidized CO decreases gradually; for the coverage in curve (b) the area under the oxidation peak is about half that of the carbon-free surface, but the peak overpotential is decreased by 35 mV. When the carbon coverage exceeds 1 monolayer the amount of oxidized CO increases as does the overpotential, the latter by 100 mV with respect to the carbon-free surface.

The maximum coverages of solution dosed and vapor dosed carbon monoxide, as determined by the area under the oxidation peak, are compared as a function of carbon coverage in Figure 3. In agreement with previous studies [30,42,43] about 40% more carbon monoxide adsorbs by solution dosing than by vapor dosing. For

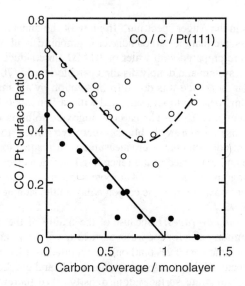

Figure 3. Amount of electrooxidized solution dosed (open circles) and vapor dosed (solid circles) carbon monoxide as a function of carbon coverage on Pt(111). The solid line represents a site blocking model for vapor dosed carbon monoxide. The dashed line is a guide to the eye.

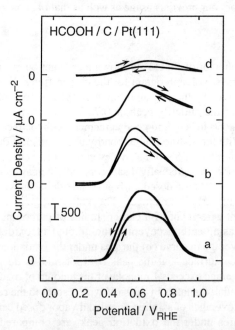

Figure 4. Cyclic voltammograms of carbon-covered Pt(111) in 0.1 mol/dm^3 HClO$_4$ + 0.1 mol/dm^3 HCOOH for carbon coverages of: (a) 0, (b) 0.24, (c) 0.61, and (d) 1.12 monolayer.

both dosing methods the adsorbed amount decreases approximately linearly with increasing carbon coverage. The amount of vapor dosed CO decreases to essentially a trace amount, less than 0.1 per platinum atom, for $\theta_C \geq 1$. However, the amount of solution dosed CO passes through a minimum at $\theta_C \approx 0.8$ and increases to nearly the carbon-free value at the highest carbon coverages studied. Thus, the presence of solution during carbon monoxide adsorption has a large influence on the extent of adsorption and subsequent reaction.

Cyclic voltammograms of HCOOH electrooxidation as a function of carbon coverage are shown in Figure 4. The voltammogram of the clean surface is in good agreement with previous studies [11,44-46] and illustrates the pronounced hysteresis characteristic of this system; the peak oxidation current on the positive potential sweep is approximately 35% less than that on the negative sweep. As the carbon coverage increases the extent of oxidation decreases, again gradually, while the nature of the hysteresis changes considerably. At low carbon coverage the maximum current occurs on the negative potential sweep, but at high coverage it occurs on the positive sweep. For 0.61 monolayer of carbon the voltammogram shows nearly no hysteresis at all. Also noteworthy is that, like carbon monoxide, reaction occurs even with carbon coverages in excess of 1 monolayer.

A probe adlayer of ethylidyne (CCH_3) has a similar effect on the hysteresis of HCOOH oxidation, but with a subtle difference in reaction rate. Figure 5 shows that the hysteresis is greatly reduced in the presence of ethylidyne, with the current of the positive sweep only 13% less than that of the negative sweep. While the oxidation peak is more narrow in the presence of ethylidyne, the maximum current of the negative sweep exceeds that of the clean surface by 9% and occurs at 30 mV lower overpotential. The adlayer composed of 0.25 ethylidyne per platinum surface atom thus maintains a slightly greater reaction rate than does the clean surface. Since ethylidyne does not participate in the reaction the reaction rate per bare platinum atom is significantly enhanced with respect the clean surface (see below).

Discussion

Influence of the Carbon Adlayer in H and CO Adsorption. Small, but appreciable, carbon coverages have little influence on the basic electrochemistry of platinum, as may be seen in Figure 1. The voltammogram changes only modestly from a surface with no carbon to one with 0.28 monolayers. Voltammograms recorded at intermediate coverages showed a steady progression between the two curves. Cyclic voltammetry, therefore, is relatively insensitive to carbon contamination. Indeed, early voltammograms of Pt(111) [47] resemble those we obtained for $\theta_C = 0.05$–0.1, from which we conclude that they were most likely influenced by carbon contamination.

Carbon adlayers exhibit at least two effects on electrochemical adsorption and reaction. Low carbon coverages give rise to simple site blocking. For example, in Figure 3 the uptake of vapor dosed carbon monoxide decreases linearly with carbon coverage, a result clearly due to site blocking. For solution dosed carbon monoxide the uptake first decreases, reaches a minimum, and then increases. The slope of the decreasing portion is the same as that for vapor dosing, indicating a similar site blocking effect. The mechanism for site blocking is easily explained by graphitic carbon islands distributed about the surface. This effect is also consistent with the 35 mV reduction in overpotential at the minimum, a reduction that corresponds to a decrease in activation energy of 65 kJ/mol [48]. Submonolayer coverages of carbon monoxide oxidize with similarly reduced overpotentials on carbon-free platinum [41,42], an effect attributed to enhanced reactivity of the edges of carbon monoxide islands. By analogy, the reduced overpotential with carbon adlayers follows from disruption of tightly packed carbon monoxide by carbon islands. Site blocking is also evident for the butterfly peak in Figure 1, as the area of this peak decreases linearly with carbon coverage.

Carbon coverages in excess of 0.6 monolayers show a more unusual behavior, most strikingly evident by the minimum and rise in CO uptake from solution shown in Figure 3. Even a full monolayer of carbon allows adsorption of hydrogen, carbon monoxide, and formic acid, as well as reaction of the latter two. We now examine some possible explanations of this effect to see if the results reported here can be used to elicit a mechanism.

The change in behavior of the carbon adlayer at higher coverages can be explained by two limiting cases: (1) restructuring of the carbon adlayer in a manner to make free platinum sites available for adsorption and reaction, or (2) a permeable adlayer which allows species to adsorb and react underneath itself. The feasibility of the restructuring model can be examined as follows. First, we assume that platinum sites made free by restructuring behave in the same fashion as the carbon-free surface, and that x (x = H, CO) adsorbs on these sites to the same maximum coverage locally as occurs on the clean surface. Since Auger spectroscopy showed no change in carbon coverage after any of the electrochemical experiments, the effective thickness (in monolayers) of the restructured carbon adlayer $\theta_{C,res}$ can be determined by a carbon balance

$$\theta_{C,res} = \theta_C / (1 - \theta_{x/Pt}/\theta^s_{x,Pt}) \tag{1}$$

where $\theta_{x/Pt}$ is the measured amount of adsorbed x and $\theta^s_{x/Pt}$ is the maximum ratio of x to Pt atoms on the clean surface: 2/3 for H [40] and 0.7 for CO (Figure 3). For the surface with $\theta_C = 1.2$ the measured amount of adsorbed hydrogen is $\theta_{H/Pt}/\theta^s_{H/Pt}$ = 0.85 (Figure 1) so the restructured adlayer would have an effective thickness of 8.0 monolayers. The measured amount of adsorbed CO on the surface with $\theta_C = 1.3$ was $\theta_{CO/Pt} = 0.5$ (Figure 3), which gives an effective thickness of 4.5 monolayers for the restructured adlayer. While the value of 4.5 seems at least plausible, the value of 8.0 does not. Furthermore, the restructuring model cannot explain the 100 mV increase in CO oxidation overpotential (Figure 2c), since carbon monoxide on the free platinum sites should have the same reactivity as it does on the carbon-free surface.

For the permeable adlayer model, adsorption may occur on the entire platinum surface up to the saturation limits that hold for the carbon-free surface. The slightly reduced uptakes of hydrogen and carbon monoxide on the carbon-covered surface are indeed consistent with the carbon-free limits. In addition, the permeable adlayer model can explain the 100 mV increase in overpotential as an effect due either to steric hindrance of the reaction by the adlayer, or diffusional resistance of H_2O (the reactant) or CO_2 (the product) through the adlayer. Thus, we attribute the increase in

CO adsorption at near monolayer carbon coverages to permeability of the carbon adlayer. While the cause of the permeable adlayer is still unknown, we note that permeation through carbon layers has been seen before [49]. Further details of this effect require additional study, especially by scanning tunneling microscopy.

Formic Acid Oxidation with Carbon: Hysteresis Reversal. The hysteresis in the voltammogram of the clean surface, Figure 4a, is one manifestation of the well known tendency of this reaction to self-poison [1]. Reaction begins on the positive potential sweep once a sufficient overpotential is reached, but goes through a maximum and decreases at high positive potentials. This decrease is generally considered to result from blocking of reaction sites by adsorption of oxygen or hydroxide [15,50]. Reaction again increases on the negative sweep because of oxygen or hydroxide desorption, but with a significantly greater maximum current. The difference between the two maxima is caused by the poison, which forms at potentials below 0.4 V and is oxidized off the surface at the positive potential limit [1,16]; hence, the negative sweep has the higher maximum.

With increasing carbon coverage the hysteresis between positive and negative sweeps decreases and eventually reverses, with the positive sweep showing the greater maximum at high carbon coverage (Figure 4). For $\theta_C = 0.61$, somewhat close to the carbon coverage of the minimum in Figure 3, there is essentially no hysteresis at all. This response can result from (1) a decrease in the amount of poison forming at potentials less positive than the oxidation peak, (2) an increase in a different poison that predominates at potentials more positive than the oxidation peak, or both (1) and (2). The first explanation alone does not explain the results of Figure 4, since the condition of no hysteresis would occur once the poison is eliminated, thereby precluding reversal of hysteresis. Such a case occurs for selenium enhancement of formic acid oxidation on Pt(111) [7]. Explanation (2) alone *can* account for the results if it occurs at all carbon coverages, including zero. In this case the poison predominant at positive potentials would be some type of oxygen containing species, and the amount of hysteresis would consequently be sensitive to the positive potential limit of the voltammogram, which we have not yet investigated thoroughly.

The most likely scenario to explain the results of Figure 4 is a combination of both (1) and (2). The presence of carbon monoxide and a hydrogen containing intermediate at the lower potentials is well documented for the carbon-free surface [1,2,10-13]. The influence of surface carbon is to reduce the amount of poison formed at lower potentials, and hence increase (relatively) the voltammetric maximum on the positive sweep. Verification of (1) would require quantitative measurements of poison formation (for example, by chronocoulometry), which we have not yet performed. For (2) carbon-induced adsorption of an oxygen containing species at high potentials would decrease reactivity until the potential decreases on the negative sweep to a value sufficient to remove that species. The effect on the voltammogram is a reduction of the current maximum on the negative sweep as a function of increasing carbon coverage, in agreement with Figure 4.

Formic Acid Oxidation with Carbon and Ethylidyne Probe Adlayers. We now turn to a more quantitative treatment of formic acid oxidation in terms of carbon's role as a probe adlayer and comparison with an ethylidyne-derived probe adlayer. To

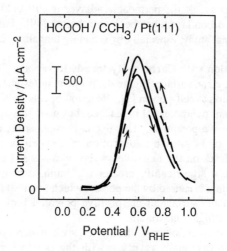

Figure 5. Cyclic voltammogram of electrooxidation of 0.1 mol/dm³ HCOOH in 0.1 mol/dm³ HClO₄ on clean Pt(111) (dashed line) and with 0.25 ethylidyne molecules per platinum atom (solid line).

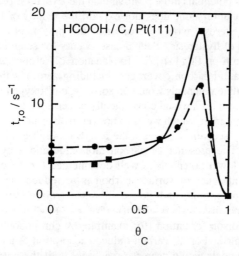

Figure 6: Peak turnover rates (reactions per bare platinum atom per second) for formic acid oxidation on Pt(111) as a function of carbon coverage. The data were obtained from cyclic voltammograms (squares - positive sweep; circles - negative sweep) at 50 mV/s in 0.1 mol/dm³ HCOOH plus 0.1 mol/dm³ HClO₄. The lines are guides to the eye.

understand the role of a probe adlayer, its surface structure must be known. As previously discussed, carbon adsorbs as graphitic islands in vacuum. While we do not have the means for checking the carbon adlayer in situ, the available data suggest that graphitic islands are present in the electrochemical situation as well: the voltammograms in blank electrolyte (Figure 1) show no distinct features attributable to a strong interaction with, say, an atomic carbon species, and the reduced overpotential of carbon monoxide oxidation (Figure 2) is consistent with islands of carbon monoxide brought about by graphitic islands

As was true for both hydrogen and carbon monoxide, reaction of formic acid is possible even for carbon coverages greater than 1 monolayer. The reaction rate at the highest carbon coverage is only about 10% of that for the clean surface, however. In principle, reaction could occur at defects in the carbon adlayer which appear either during growth in vacuum or as a response to the solution (the restructuring model). In any event, a significant distinction cannot be made between the restructuring and permeability models for this relatively small effect. Noteworthy, however, is a previous finding of formic acid oxidation on polycrystalline platinum fully covered with a carbonaceous residue [2]. The residue was derived from irreversibly adsorbed $H^{13}COOH$, and subsequent reaction in a solution containing $H^{12}COOH$ yielded exclusively $^{12}CO_2$ at normal reaction potentials (below 0.7 V), as measured by differential electrochemical mass spectrometry.

With carbon growth in the form of islands, progressively higher carbon coverages result in smaller patches of bare platinum atoms and correspondingly smaller ensembles of atoms to support surface reactions. The ensemble requirement of formic acid oxidation is thus revealed through the rate of reaction as a function of carbon coverage, as illustrated in Figure 4. Since reaction occurs both on the clean and carbon-covered surfaces, it is necessary to remove the contribution of reaction on the full carbon adlayer from the measured reaction rate to obtain a value for reaction on the bare platinum surface. The total (measured) reaction rate r_t is then a linear combination of reaction on the bare r_o and carbon-covered surfaces r_C,

$$r_t(\theta_c) = \left(1 - \theta_c\right) r_o(\theta_c) + \theta_c r_c, \qquad (2)$$

where both r_t and r_o depend on the carbon coverage, but r_C is assumed to be a constant given by the reaction rate on a fully carbon covered surface. The reaction rates in Eq. (2) may be equivalently expressed as turnover rates t_r with units of reaction events per platinum atom per second. The turnover rate for the bare platinum surface is then given by

$$t_{r_o}(\theta_c) = \left[t_{r,t}(\theta_c) - \theta_c\, t_{r,c}\right] / \left(1 - \theta_c\right), \qquad (3)$$

where the other subscripts have the same meanings as as those in Eq. (2).

Figure 6 shows the turnover rate corrected for the bare platinum surface for the peak reaction rates of the positive (squares) and negative (circles) potential sweeps of Figure 4. The value of $t_{r,C}$ for the positive sweep data was taken from the peak reaction rate of the positive sweep in Figure 4d, and similarly, $t_{r,C}$ for the negative sweep data was obtained from the peak in the negative sweep. (The trends in Figure 6 are independent of the value chosen for $t_{r,C}$ for any reasonable value, including zero.) Low to intermediate carbon coverages have no influence on the peak turnover rates of the bare surface; this is precisely the effect expected for site blocking. At higher coverages the bare surface turnover rate increases dramatically, reaching values of about 18 and 12 s^{-1} for the positive and negative sweeps, respectively. The turnover rate then falls off dramatically at higher coverages.

The intrinsic reactivity of bare platinum is greatly enhanced by high carbon coverages. For the positive potential sweep this enhancement is nearly five-fold, which is similar to the enhancement derived from a 50% Pt-Ru alloy vs. the pure platinum surface [9]. The maximum in turnover rate shows that an optimum carbon coverage, and hence an optimum ensemble size, exists for this reaction. The enhanced turnover rates occur in concert with the reversal in hysteresis, so the optimum ensemble size is associated with removal of the poison formed at low potentials. The optimum ensemble size thus controls the poison either by eliminating its production, through some form of steric hinderance, or by assisting its removal, through the high local concentration of other adsorbates. An ensemble requirement of three surface atoms has been reported for complete oxidation of formic acid vs. a requirement of four atoms for formation of the poison [1]. This effect highlighted by the carbon probe adlayer corresponds to the third-body effect [1,7,8,14,15] often cited for metallic reaction modifiers, but without the complications of possible surface redox behavior or other unique chemistry of the modifier.

The different surface arrangement of ethylidyne as compared with graphitic carbon, provides an additional test of the ensemble requirements of reaction. Ethylidyne adsorbs in a p(2x2) structure in vacuum with a CCH_3/Pt ratio of 0.25. In this structure one platinum atom of the unit cell has no direct interaction with ethylidyne and the three others each form one bond with the alpha carbon. Since three of the four platinum atoms of the unit cell interact with ethylidyne, the equivalent ensemble size must be less than four, with a reasonable value being two or three. If the p(2x2) unit cell were maintained in the electrochemical environment, then formic acid oxidation ought to be enhanced by ethylidyne. A glance at Figure 5 shows that this is so; the hysteresis associated with the poison is nearly eliminated and the maximum current on the negative sweep is 9% greater, and at 30 mV less overpotential, than in the absence of ethylidyne. The turnover rate based on the superficial electrode area thus increases by nearly 10%, whereas the turnover rate per unoccupied platinum atom increases by more than 35% from 3.8 to 5.2 s^{-1} on the positive sweep. This latter figure is based on the assumption that ethylidyne occupies just one platinum atom; for occupation of more than one platinum atom the reported turnover rate would be even greater.

Conclusions

Mechanistic details of electrocatalytic reactions can be probed through selectively modifying the surface with probe adlayers. Carbon adsorbs in graphitic islands at low coverage, which restrict available surface sites for adsorption and reaction. As carbon coverage increases the amount of adsorbed CO decreases, consistent with site-blocking by carbon, while the reactivity of the bare sites to HCOOH increases. This last effect is attributed to formation, by the carbon adlayer, of smaller ensembles of bare platinum atoms which promote complete oxidation of HCOOH over formation of the poison. As inferred by HCOOH reaction on an ethylidyne adlayer, the optimum ensemble size is 2-3 Pt atoms.

A full carbon monolayer is permeable to reactants and products of surface reactions, as significant quantities of H, CO, and HCOOH adsorb or react through the adlayer. One consequence of this finding for applied electrocatalysis is that a typical platinum particle catalyst on a carbon support may maintain activity in the event that the particle becomes fully covered with carbon. This effect will thus help in interpreting the behavior of aged or sintered catalysts.

Acknowledgements

We gratefully acknowledge support of this work by the National Science Foundation (CTS-9103543; CTS-9502971) and the donors to the Petroleum Research Fund administered by the American Chemical Society.

References

1. Parsons, R.; VanderNoot T., *J. Electroanal. Chem.* **1988**, *257*, 9.
2. Xia, X.; Iwasita, T., *J. Electrochem. Soc.* **1993**, *140*, 2559.
3. Clavilier, J.; Fernández-Vega, A.; Feliu, J. M.; Aldaz, A., *J. Electroanal. Chem.* **1989**, *258*, 89.
4. Chang, S.-C.; Ho, Y.; Weaver, M. J., *Surface Sci.* **1992**, *265*, 81.
5. Herrero, E.; Feliu, J. M.; Aldaz, A., *J. Electroanal. Chem.* **1993**, *350*, 73.
6. Kizhakevariam, N.; Weaver, M. J., *Surface Sci.* **1994**, *310*, 183.
7. Llorca, M. J.; Herrero, E.; Feliu, J. M.; Aldaz, A., *J. Electroanal. Chem.* **1994**, *373*, 217.
8. Gasteiger, H. A.; Markovic, N.; Ross, P. N., Jr.; Cairns, E. J., *Electrochim. Acta* **1994**, *39*, 1825.
9. Markovic, N. M.; Gasteiger, H. A.; Ross, P. N., Jr.; Jiang, X.; Villegas, I.; Weaver, M. J., *Electrochim. Acta* **1995**, *40*, 91.
10. Bewick, A., *J. Electroanal. Chem.* **1983**, *150*, 481.
11. Chang, S.-C.; Leung, L.-W. H.; Weaver, M. J., *J. Phys. Chem.* **1990**, *94*, 6013.
12. Iwasita, T.; Nart, F. C.; Lopez, B.; Vielstich, W., *Electrochim. Acta* **1992**, *37*, 2361.
13. Wolter, O.; Willsau, J.; Heitbaum, J., *J. Electrochem. Soc.* **1985**, *132*, 1635.
14. Angerstein-Kozlowska, H.; McDougall, D.; Conway, B. E., *J. Electrochem. Soc.* **1973**, *120*, 756.
15. Hartung, T.; Willsau, J.; Heitbaum, J., *J. Electroanal. Chem.* **1986**, *205*, 135.
16. Sun, S. G.; Clavilier, J., *J. Electroanal. Chem.* **1988**, *240*, 147.
17. Motoo, S.; Watanabe, M., *J. Electroanal. Chem.* **1976**, *69*, 429.
18. Beden, B.; Kadirgan, F.; Lamy, C.; Leger, J. M., *J. Electroanal. Chem.* **1981**, *127*, 75.
19. Sauer, D. E.; Ph.D. Thesis, University of Washington **1994**; Sauer, D. E.; Borup, R. L; Stuve, E. M., in preparation.
20. Antonucci, P. L.; Alderucci, V.; Giordano, N.; Cocke, D. L.; Kim, H. *J. Appl. Electrochem.* **1994**, *24*, 58.

21. Frelink, T.; Visscher, W.; van Veen, J. A. R., *J. Electroanal. Chem.* **1995,** *382,* 65.

22. Steininger, H.; Ibach, H.; Lehwald, S., *Surface Sci.* **1982,** *117,* 685.

23. Koestner, R. J.; Frost, J. C.; Stair, P. C.; Van Hove, M. A.; Somorjai, G., *Surface Sci.* **1982,** *116,* 85.

24. Hubbard, A T.; Young, M. A.; Schoeffel, J. A., *J. Electroanal. Chem.* **1980,** *114,* 273.

25. Semrau, G.; Heitbaum, J., in *The Chemistry and Physics of Electrocatalysis,* McIntyre, J. D. E.; Weaver, M. J.; Yeager, E. B., The Electrochemical Society: Pennington, NJ, 1984, PV 84-12; pp 639.

26. Wieckowski, A.; Rosasco, S. R.; Salaita, G. N.; Hubbard, A. T.; Bent, B. E.; Zaera, F.; Godbey, D.; Somorjai, G. A., *J. Am. Chem. Soc.* **1985,** *107,* 5910.

27. Hourani, M.; Wieckowski, A., *Langmuir* **1990,** *6,* 379.

28. Schmiemann, U.; Müller U.; Baltruschat, H., *Electrochim. Acta* **1995,** *40,* 99.

29. Borup, R. L.; Sauer, D. E.; Stuve, E. M., *Surface Sci.* **1993,** *293,* 10.

30. Borup, R. L.; Sauer, D. E.; Stuve, E. M., in preparation.

31. Biberian, J. P.; Somorjai, G. A., *Appl. Surface Sci.* **1979,** *2,* 352.

32. Davis, S. M.; Gordon, B. E.; Press M.; Somorjai, G. A., *J. Vac. Sci. Tech.* **1981,** *19,* 231.

33. Borup, R. L.; Sauer, D. E.; Stuve, E. M., *J. Electroanal. Chem.* **1994,** *374,* 235.

34. Goad, J. P.; Riviere, J. C., *Surface Sci.* **1971,** *25,* 609.

35. Chesters, M. A.; Hopkins, B. J.; Winton, R. I., *Surface Sci.* **1976,** *59,* 46.

36. Lang, B., *Surface Sci.* **1975,** *53,* 317.

37. Weinberg, W. H.; Deans, H. A.; Merrill, R. P., *Surface Sci.* **1974,** *41,* 312.

38. Burwell, R. L., *Pure Appl. Chem.* **1976,** *46,* 71.

39. Mills, I.; Cvitas, T.; Homann, K.; Kallay, N.; Kuchitsu, K., *Quantities, Units and Symbols in Physical Chemistry,* 2nd ed., Blackwell Scientific Publications: Oxford, 1993.

40. Clavilier, J.; Rodes, A.; El Achi, K.; Zamakchari, M. A., *J. Chim. Phys. C* **1991,** *88,* 1291.

41. Zurawski, D.; Chan, K. Wieckowski, A. *J. Electroanal. Chem.* **1987,** *230,* 205.

42. Zurawski, D.; Wasberg, M.; Wieckowski, A., *J. Phys. Chem.* **1990,** *94,* 2076.

43. Dalbeck, R.; Buschmann, H. W.; Vielstich, W., *J. Electroanal. Chem.* **1994,** *372,* 251.

44. Adzic, R. R.; O'Grady, W. E.; Srinivasan, S., *Surface Sci.* **1980,** *94,* L191.

45. Sun, S. G.; Clavilier, J.; Bewick, A., *J. Electroanal. Chem.* **1988,** *240,* 147.

46. Fernández-Vega, A.; Feliu, J. M.; Aldaz, A.; Clavilier, J., *J. Electroanal. Chem.* **1991,** *305,* 229.

47. Clavilier, J., *J. Electroanal. Chem.* **1980,** *107,* 211.

48. Brett, C. M. A.; Oliveira-Brett, A. M., *Electrochemistry,* Oxford University Press: Oxford, 1993, p. 73. The transfer coefficient is assumed to be 1/2.

49. Kinoshita, K., *Carbon: Electrochemical and Physicochemical Properties,* Wiley: New York, 1988.

50. Gasteiger, H. A.; Markovic, N.; Ross, P. N., Jr.; Cairns, E. J.., *J. Phys. Chem.* **1994,** *98,* 617.

Chapter 21

Photooxidation Reaction of Water on an n-TiO$_2$ Electrode

Investigation of a Previously Proposed New Mechanism by Addition of Alcohols to the Electrolyte

Y. Magari, H. Ochi, S. Yae, and Y. Nakato[1]

Department of Chemistry, Faculty of Engineering Science, and Research Center for Photoenergetics of Organic Materials, Osaka University, 1–3 Machikaneyama, Toyonaka, Osaka 560, Japan

The mechanism of the photooxidation reaction of water on an n-TiO$_2$ (rutile) electrode in acidic electrolyte has been studied by in situ photoluminescence as well as photocurrent measurements. Addition of alcohols to the electrolyte caused an appearance of a shoulder in the photocurent-potential curve and an increase in the illumination intensity limited photocurrent, together with a decrease in the photoluminescence intensity. Detailed analyses of the results have given strong support to our previously proposed new mechanism that surface Ti-OH group is not oxidized by photogenerated holes but that Ti-OH group or OH$^-$ ion present in bulk defects near the electrode surface is oxidized, and thus the photooxidation reaction of water on n-TiO$_2$ (rutile) in acidic solutions proceeds mainly through the oxidation of such bulk species by the holes.

The electrolysis of water into hydrogen and oxygen is an important reaction in energy conversion technologies. Though hydrogen evolution proceeds efficiently on some metal electrodes, oxygen evolution has a fairly high overvoltage, resulting in a significant loss in the energy conversion efficiency (1). For finding efficient electrode materials for the oxygen evolution, it seems inevitably necessary to clarify its reaction mechanism in detail on a molecular level.

The oxygen evolution reaction occurs at large positive potentials where most metal electrodes are covered with thin (or rather thick) native oxide layers. This implies that studies of the reaction mechanism on metal oxide electrodes are very important. In this respect, the photooxidation reaction of water on an n-TiO$_2$ electrode (2) is very interesting. A number of studies have been made on the mechanism of this reaction (3-11), but the details still remain unclear because the water oxidation is a four-electron process and will proceed via many reaction intermediates at the electrode surface.

We found previously that the n-TiO$_2$ (rutile) electrode showed a photoluminescence (PL) band peaked at 840 nm which could be explained as arising from a surface reaction intermediate or a species closely related to it (12). This enabled us to use in situ

[1]Corresponding author

PL measurements for the mechanistic studies (12-15). From detailed studies on n-TiO$_2$ and that doped with transition metal ions such as Cr^{3+}, we proposed a new mechanism that the reaction in acidic solutions is initiated by the oxidation of bulk OH species, not surface OH species (14). However, we later found a paper (16) reporting that Cr^{3+} ions contained in TiO$_2$ as an impurity emitted a PL band accidentally at the same wavelength (840 nm) as that observed by us till then. Also, a criticism to our conclusion was reported (17) probably due to confusion of the interpretation of the PL band. Thus it had become necessary to re-investigate our mechanism in detail, and we started systematic studies (18,19). The purpose of the present paper is to give confirmative experimental evidence to our new mechanism.

Previously Proposed New Mechanism

First we explain our new mechanism for the photooxidation reaction of water on the n-TiO$_2$ (rutile) electrode. It has been assumed in many studies (3-7) that the reaction is initiated by the oxidation of surface Ti-OH group with photogenerated holes, producing surface $\cdot$ OH radicals. The radicals will combine with each other, giving hydrogen peroxide, which is easily oxidized by the holes, thus resulting in oxygen evolution. On the other hand, we proposed (14) that the surface Ti-OH group cannot be oxidized by photogenerated holes and thus the reaction is, in acidic solutions, initiated by the oxidation of Ti-OH group or OH$^-$ ion present in bulk defects (such as dislocations) near (and exposed to) the electrode surface (cf. Figure 2 discussed later). The resultant bulk radicals, HO $\cdot$ $_{in}$, come out to the surface either by diffusion through the defect or by electron transfer between adjacent Ti-OH group in the defect, followed by the processes to oxygen evolution similar to the above-mentioned conventional mechanism. The PL band was assigned to an electronic transition from the conduction band to the vacant 2p-level of the HO $\cdot$ $_{in}$ radicals.

It is to be noted here that the above new mechanism does not exclude the possibility of the oxidation of surface species by the holes in neutral and alkaline solutions because much easily oxidized surface species such as Ti-O$^-$ will be formed in these solutions (the point of zero charge for TiO$_2$ is about 5.5). The oxidation of surface species may also occur for anatase-type TiO$_2$ because it has a wider band-gap than rutile-type TiO$_2$.

The key of our new mechanism lies in that the rate constant for the oxidation of Ti-OH (or OH$^-$ ion) in the bulk defect by photogenerated holes is much higher than that for the oxidation of surface Ti-OH group, as theoretically verified (15). In our model, the bulk defect is assumed to be a kind of crystal dislocations or narrow atomic gaps and therefore the Ti-OH or OH$^-$ ion in the bulk defect is present in the "naked" form, i.e., not hydrated by water molecules, contrary to the surface Ti-OH group which is strongly hydrated by water molecules in the solution. In other words, the Ti-OH or OH$^-$ ion in the bulk defect is mainly stabilized by the electronic polarization of the TiO$_2$ crystal, whereas the surface Ti-OH group is largely stabilized by the orientational polarization of water molecules. Thus, the reorganization energy (or the activation energy) for electron transfer in Marcus theory (20) is much less in the oxidation of the bulk species than in the surface species. In this respect, our new mechanism is of much interest, opening a possibility of a new active path for electrode reactions.

Our recent experiments have shown that the PL-emitting species (luminescent species) decays with time according to the square-root law of it (19), indicating that the luminescent species disappears by diffusion, namely, it lies in the TiO$_2$ bulk near the surface, in agreement with our mechanism.

Experimental

Single crystal TiO$_2$ (rutile) wafers, 10×10 mm^2 in area and 1.0 mm thick, cut perpendicular to the c-axis and having a purity of 99.99%, were obtained from Earth Jewelry Co. Ltd. They were polished with alumina powder of diameters 3.0, 1.0, 0.3 and 0.1 μ m successively (sometimes 0.06-μ m powder was used), etched in concentrated sulfuric acid containing 33 wt% (NH$_4$)$_2$SO$_4$ at 200$\sim$250℃ for 30 min, washed, annealed in air at 1300℃ for 3$\sim$4 h, and then slightly reduced by heating at 550 to 700℃ for 30 min under a stream of hydrogen for getting n-type semiconductivity. The resistivity (ρ) of the n-TiO$_2$ wafer was measured by painting indium-gallium alloy on both faces of the wafer, and it ranged from 0.2 to 2.0 Ω cm. After the alloy was removed by immersing in an HCl solution, ohmic contact for electrode preparation was obtained again on one face of the wafer with indium-gallium alloy.

Photocurrent (j) vs. potential (U) curves for the n-TiO$_2$ electrodes were obtained with a commercial potentiostat and potential programmer, using a Pt plate as the counter electrode and an Ag/AgCl electrode as the reference electrode. Illumination was performed by a 365-nm band from a 500-W high-pressure mercury lamp, taken out with a Shimadzu-Bausch-Lomb monochromator and appropriate glass filters. Photoluminescence spectra were measured with a Jobin-Yvon H20 monochromator combined with a lens and glass filters and a Hamamatsu Photonics R316 or R712 photomultiplier cooled at -20℃. PL intensity vs. potential curves were measured simultaneously with the j-U curve measurements. Investigations of effects of addition of alcohols were performed by introducing alcohols to the solution by use of a syringe so as not to cause any change in the optical path and the positions of the electrochemical cell and the electrode.

The electrolyte solutions were prepared by use of water purified from deionized water with a Milli-Q Water Purification System. Chemicals (H$_2$SO$_4$ and alcohols) were of reagent grade and used without further purification. The electrolyte solution in the cell was in most cases stirred during measurements. The concentration unit, mol/dm^3, is abbreviated as M in the present work.

Results

We reported in a recent paper (*18*) that a "fresh" n-TiO$_2$ electrode (i.e., the electrode just prepared by the procedures of polishing, etching, annealing and hydrogen reduction at an elevated temperature as described in the experimental section) showed only a weak photocurrent and no PL in the first potential scan, but both the photocurrent and the PL intensity increased very much with time nearly in parallel to each other and finally reached maxima while the the potential scans were repeated under illumination in 0.05M H$_2$SO$_4$. Such initial changes were in most cases accompanied by the formation of a lot of micropores at the electrode surface.

The above initial changes clearly indicate that our observing PL band is not arising from Cr^{3+} ions which may be present in TiO$_2$ as an impurity, but from a pecies produced by interfacial electrochemical reactions. The above results also indicate that the n-TiO$_2$ electrode is activated by the flow of anodic photocurrent in 0.05M H$_2$SO$_4$ probably

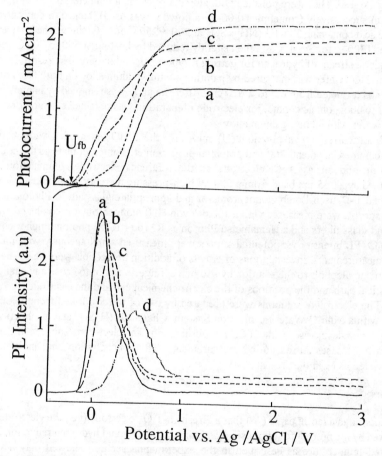

Figure 1. Photocurrent vs. potential and PL intensity vs. potential for activated n -TiO$_2$ in 0.05M H$_2$SO$_4$ (pH 1.20) containing methanol of a 0.0 M, b 0.1 M, c 1.0 M, and d 6.0 M. U$_{fb}$: flat-band potential.

through photoetching, which occurs slightly and competitively with the photooxidation reaction of water. We concluded previously (*18*) that such activation of the electrodes is due to the increase in the specific rate (or cross section) of the photooxidation reaction of water, not due to the increase in the area of the electrode surface. It is likely that the "bulk defect" mentioned in the preceding section, acting as active sites for the reaction, is produced in the course of the photoetching. The details will be reported in a separate paper (*19*).

It should be mentioned here that similar activation of the n-TiO$_2$ electrodes occurs by chemical etching in (hot) H$_2$SO$_4$ aqueous solutions (*18*), which is often used for cleaning of the TiO$_2$ surface in the literature. Besides, once-activated n-TiO$_2$ electrodes keep their activated state for a long time, at least, more than several weeks (*18*), unless they are polished, annealed and reduced in a hydrogen atmosphere as described in the experimental section. From these features, it is probable that many studies reported thus far on TiO$_2$ have been made unconsciously using such activated electrodes.

In the present work, the "intentionally activated" n-TiO$_2$ electrodes were used in all cases. Figure 1 shows photocurrent (j)-potential (U) and PL intensity-U curves for an n-TiO$_2$ electrode in 0.05M H$_2$SO$_4$ with and without methanol. No current is observed in the dark in potentials more positive than ca. -0.3 V vs. Ag/AgCl. Curve a is for no methanol, where the photocurrent above 0.2 V is mainly due to the photooxidation reaction of water. When methanol is added, the j-U curve shows a shoulder at -0.3 to 0.2 V, together with an increase in the illumination intensity limited photocurrent above ca. 0.8 V. The appearance of the shoulder is attributed to the oxidation of adsorbed methanol (Ti-OCH$_3$) as discussed later. The increase in the illumination intensity limited photocurrent is explained by the well-known current doubling mechanism (*21*) in which intermediate radicals, produced by the oxidation of methanol by photogenerated holes, inject electrons into the conduction band of n-TiO$_2$, thus two electrons flowing per one absorbed photon at maximum.

The above results were analyzed on the basis of our previously proposed new mechanism with addition of some processes arising from the presence of methanol. The main processes are shown in Figure 2. When photogenerated holes (h) migrate to the electrode surface, they will cause processes 1), 2) or 3) with the probabilities α, β and γ respectively ($\alpha + \beta + \gamma = 1$). Process 1) is the oxidation of surface Ti-OH group usually assumed in the conventional mechanism. In process 2), the holes react with adsorbed methanol (Ti-OCH$_3$), producing adsorbed radicals (CH$_3$O · $_{ad}$), which cause the electron injection into the conduction band of n-TiO$_2$ (current doubling). The probability of the electron injection can be taken to be neary one because the radicals are adsorbed on the n-TiO$_2$ surface. In process 3), the holes react with Ti-OH group or OH$^-$ ion in the bulk defect near (and exposed to) the electrode surface, expressed as Ti-OH$_{in}$ or HO$^-_{in}$ in Figure 2. The Ti-OH$_{in}$ may be formed by splitting of Ti-O-Ti bonds in the bulk defect, followed by dissociative adsorption of water molecules.

$$\text{Ti-O-Ti} + \text{H}_2\text{O} \rightarrow \text{Ti-OH} + \text{HO-Ti} \tag{1}$$

The resultant HO · $_{in}$ radicals in the bulk come out to the surface either by diffusion or by electron transfer between adjacent Ti-OH$_{in}$, finally producing surface radicals, HO · $_s$, similar to process 1). Some of the surface radicals combine with each other, forming hydrogen peroxide and finally leading to oxygen evolution, whereas the other radicals

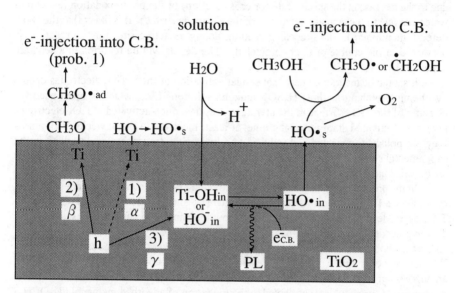

Figure 2. Various photooxidation processes at the activated n-TiO$_2$ electrode in acidic electrolyte containing methanol. See text for details.

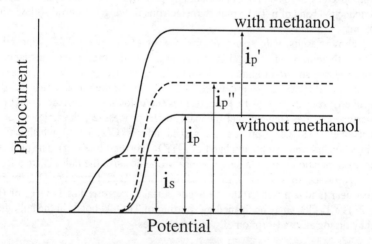

Figure 3. Division of the observed photocurrent-potential curve for calculating a contribution of methanol oxidation. — observed curve and ··· estimated curve.

react with methanol in the solution, producing $CH_3O \cdot$ or $CH_2OH \cdot$ radicals in the solution. The radicals will inject electrons into the conduction band of n-TiO$_2$ with the probability less than unity because the radicals are in the solution in this case.

The α, β and γ values were determined as follows: The observed illumination intensity limited photocurrent in the absence of methanol, i_p in Figure 3, indicates the flux of the holes coming to the electrode surface with no recombination with the electrons in the conduction band under large band bending. The observed j-U curve in the presence of methanol, represented by i_p' in Figure 3, can be divided into two curves represented by i_s and i_p'' ($i_p' = i_s + i_p''$), where the curve represented by i_p'' is obtained simply by expanding the curve represented by i_p in the direction of the ordinate axis. The i_s (subscript s means shoulder) is attributed to the oxidation of adsorbed methanol (process 2) as explained later, and therefore expressed, by taking account of the current doubling, as follows (cf. Figure 2):

$$i_s = i_p \cdot \beta \times 2 \qquad (2)$$

The i_p'' arises from various oxidation processes originating from processes 1) and 3) in Figure 2, and is rather difficult to analyse quantitatively. Here we assume that $\alpha \doteqdot 0$ for simplifying the analysis. This assumption is verified experimentally as described later. We also pay attention to the potential region between the flat-band potential (U$_{fb}$) and the onset potential (U$_{on}$) of the photocurrent in the absence of methanol (lying at +0.2 V in Figure 1). In this potential region, a sufficient number of photogenetrated holes come to the electrode surface, as indicated by the appearance of the high photocurrents i_s in the high methanol concentrations, but no photocurrent is observed in the absence of methanol. The latter fact implies that the HO $\cdot_{in}$ radicals produced by process 3) recombine efficiently with the electrons in the conduction band, not causing the production of the surface HO $\cdot_s$ radicals and the subsequent processes. The occurrence of the efficient methanol oxidation in this potential region is attributed to the fact that the $CH_3O \cdot_{ad}$ radicals produced are quickly converted to H_2CO by the electron injection into the conduction band (current doubling) and removed from the electrode surface, thus resulting in little electron-hole recombination. The production of the surface HO $\cdot_s$ radicals and the subsequent processes occur in potentials more positive than U$_{on}$ where the electron-hole recombination is suppressed by the presence of large band bending.

Based on the above argument and assumption, the γ value can be determined directly from the decrease of the PL intensity by the addition of methanol (Figure 1). The PL is emitted by the recombination of HO $\cdot_{in}$ and the electrons in the conduction band as mentioned before (Figure 2). Therefore, the PL intensity is proportional to the concentration of HO $\cdot_{in}$, which is in turn proportional to γ. The γ value is thus given by the ratio of the PL intensity in the presence of methanol against that in the absence of it.

$$\gamma = I_{PL}(\text{alcohol}) / I_{PL}(\text{no alcohol}) \qquad (3)$$

where I_{PL} is the PL intensity.

The β and γ values thus determined are shown in Figure 4 as a funcion of the methanol concentration in the solution. The $(1 - \beta)$ value, which is plotted in Figure 4 instead of β, determined from the photocurrent i_s, agrees well with the γ value

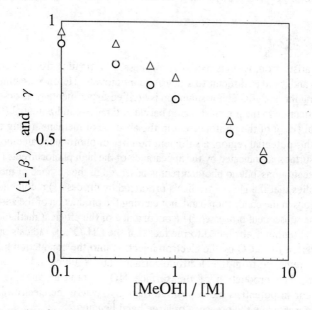

Figure 4. $(1 - \beta)$ and γ values vs. the concentration of methanol, [MeOH].
○: $(1 - \beta)$ and △: γ.

determined from the PL intensity at every methanol concentration, i.e., $1 - \beta \fallingdotseq \gamma$ or $\beta + \gamma \fallingdotseq 1$. Since $\alpha + \beta + \gamma = 1$, we can conclude that $\alpha \fallingdotseq 0$ as assumed above. Although this conclusion is obtained by the analysis of the electrode behavior between U_{fb} and U_{on}, this conclusion should hold similarly in potentials above U_{on} because the holes at the electrode surface lie at the top of the valence band and have the same energy (or reactivity), irrespective of the amount of band bending. Further details on this conclusion will be discussed in the next section.

Similar experiments were done for ethanol and *iso*-propanol. The $(1 - \beta)$ and γ values obtained are shown in Figures 5 and 6. The relation, $1 - \beta \fallingdotseq \gamma$, holds also for these alcohols, supporting the above conclusion. The $(1 - \beta)$ or γ values for *iso*-propanol are larger than those for methanol in the high alcohol concentration region. This is most probably due to the weaker adsorptivity of *iso*-propanol to the TiO$_2$ surface due to larger steric hindrance.

We have assumed in Figure 2 that adsorbed methanol (Ti-OCH$_3$) is oxidized by the holes. In order to confirm this assumption, we examined the dependence of i_s on the methanol concentration. If methanol is dissociatively adsorbed as follows,

$$CH_3OH = CH_3O_{ad} + H_{ad} \tag{4}$$

and the adsorption is of a Langmuir type, the surface coverage of CH$_3$O$_{ad}$, θ, is expressed in the following

$$\theta = (Kc)^{1/2}/\{1 + (Kc)^{1/2}\} \tag{5}$$

where c is the concentration of methanol in the solution and K is a constant. As i_s should be proportional to θ, we can expect that $1/i_s$ changes linearly with $c^{-1/2}$. This relation was really observed as shown in Figure 7, indicating that our assumption is correct. Similar adsorption behavior was reported by Herrmann et al. for photocatalytic decomposition of alcohols by TiO$_2$ powder (22).

Discussion

As has been mentioned in the preceding section, the $(1 - \beta)$ and γ values agree with each other for various alcohols in their wide concentration range. This clearly indicates that $\alpha \fallingdotseq 0$, i.e., a contribution of process 1) in Figure 2 is negligibly small, and strongly supports our previously proposed new mechanism. This conclusion is unambiguous because both the $(1 - \beta)$ and γ values are determined directly from the experimentally measured quantities, the photocurrent i_s and the PL intensity respectively. If $\alpha \neq 0$, we cannot expect any agreement between the $(1 - \beta)$ and γ values, because the photocurrent i_s is solely determined by process 2) (cf. Figure 7) and the PL intensity is solely determined by process 3) as far as the PL-emitting species is considered to be a bulk species as shown in Figure 2.

One might expect that the experimental results in the present work can be explained within the framework of the conventional mechanism, without introducing process 3) in Figure 2, if a certain surface species such as the surface HO $\cdot_s$ radical is assumed to be the PL-emitting species. However, this assumption is not acceptable because our experiments have shown that the PL-emitting species is a bulk species (19) as mentioned

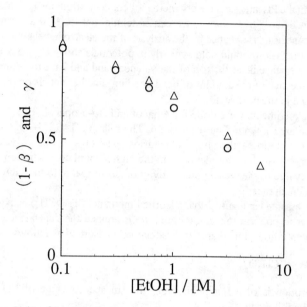

Figure 5. $(1 - \beta)$ and γ values vs. the concentration of ethanol, [EtOH].
○: $(1 - \beta)$ and △: γ.

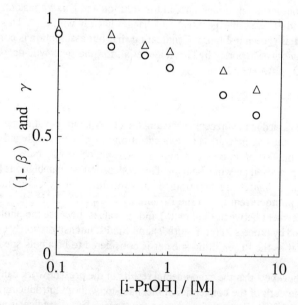

Figure 6. $(1 - \beta)$ and γ values vs. the concentration of *iso*-propanol, [*i*-PrOH].
○: $(1 - \beta)$ and △: γ.

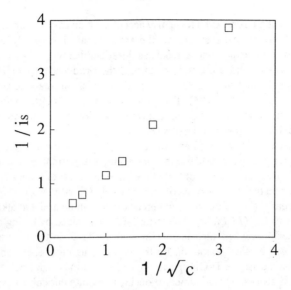

Figure 7. $1/i_s$ vs. $1/c^{1/2}$, both in an arbitrary unit.

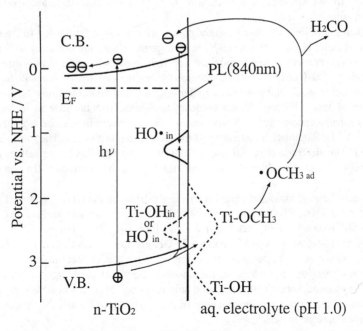

Figure 8. Energy level diagram for n-TiO$_2$ in acidic electrolyte with methanol. C.B. conduction band, V.B. valence band, E$_F$ Fermi level. Energy levels depicted by broken lines are only roughly estimated.

in the foregoing section. Also, even if we may accept such an assumption, by taking a step backward, the conventional mechanism (the assumption of $\alpha \neq 0$) leads to serious contradictions with the experimental resuls as described below.

First, if $\alpha \neq 0$ and process 1) occurred, the produced surface HO $\cdot_s$ radicals should react with alcohols in the solution in case where alcohols are added, especially in their high concentrations (1 to 6 M). This process should contribute to the photocurrent i_s, and such a contribution should cause a severe deviation of the $1/i_s$ vs. $c^{-1/2}$ plot from the linearity, contrary to the experiments (Figure 7).

Next, if process 1) occurred and the surface HO $\cdot_s$ radicals were produced, the radicals should give rise to a considerable amount of photocurrent even in the absence of alcohols. In this context, it is to be noted that the onset potential of the photocurrent (U_{on}) in acidic solutions (in the absence of alcohols) is largely (about 0.5 V) more positive than the flat-band potential (U_{fb}) (Figure 1), whereas the U_{on} in neutral and alkaline solutions lies much closer to U_{fb} (*12,14,18*). The latter fact is explained by taking account of the formation of easily oxidized surface species such as Ti-O⁻ in the neutral and alkaline solutions, followed by their oxidation by the holes. This implies, if considered in the opposite way, that the surface Ti-OH species in acidic solutions is difficult to be oxidized. In other words, if process 1) is assumed to occur, it is quite difficult to explain the large pH-dependence of the deviation of U_{on} from U_{fb}.

Finally, it is worth noting that Ohtani et al. reported (*23*), from studies on the photocatalytic activity of TiO_2 powder, that a reaction site for the photooxidation of water is different from that for the photooxidation of organic compounds, in accordance with our conclusion.

The result of Figure 7 shows that the photocurrent, i_s, is in proportion to the surface coverage of Ti-OCH$_3$, θ, determined by the Langmuir adsorption isotherm. This implies that an adsorption equilibrium is maintained approximately even under illumination where the oxidation of Ti-OCH$_3$ proceeds continuously. Such an approximation will, however, hold only under weak illumination or under high alcohol concentrations. Actually, the experimental data in Figure 7 have a tendency to deviate from the linearity in the region of small methanol concentrations. Thus, the β (and γ) values have to be regarded to be a function of the illumination intensity and the alcohol concentration. They will also be a function of the electrode potential which determines the hole flux to the surface as well as the probability of recombination between HO $\cdot_{in}$ and the electrons in the conduction band.

Figure 8 shows various electronic processes relating to the photooxidation reaction of water on n-TiO_2. The present work has shown that the oxygen 2p-level for Ti-OH group or OH⁻ ions in the TiO_2 bulk and that for surface Ti-OCH$_3$ group are above the top of the valence band, whereas that for surface Ti-OH group is below the top of the valence band, though the accurate positions are unknown. This implies that the oxygen 2p-level has various energies, depending on the chemical bonds of the oxygen atom and the surrounding atmosphere. Further detailed studies along this line will be helpful for the clarification of the atomic-level mechanism of the water oxidation and the exploration of new active electrode materials.

Literature Cited

1. Trasatti, S.; O'Grady, W. E. *Adv. Electrochem. Electrochem. Eng.* **1981**, *12*, 177.

2. Honda, K.; Fujishima, A. *Nature (London)* **1972**, *238*, 37.
3. Augustynski, J. *Structure and Bonding* **1988**, *69*, 1 and papers cited herein.
4. Salvador, P. *New J. Chem.* **1988**, *12*, 35.
5. Kiwiet, N. J.; Fox, M. A. *J. Electrochem. Soc.* **1990**, *137*, 561.
6. Garcia Gonzalez, M. L.; Salvador, P. *J. Electroanal. Chem.* **1992**, *325*, 369.
7. Nogami, G.; Sei, H.; Aoki, A.; Ohkubo, S. *J. Electrochem. Soc.* **1994**, *141*, 3410.
8. Salama, S. B.; Natarajan, C.; Nogami, G.; Kennedy, J. H. *J. Electrochem. Soc.* **1995**, *142*, 806.
9. Kratochvilova, K.; Hoskovcova, I.; Jirkovsky, J.; Klima, J.; Ludvik, J. *Electrochim. Acta* **1995**, *40*, 2603.
10. Shaw, K.; Christensen, P.; Hamnett, A. *Electrochim. Acta* **1996**, *41*, 719.
11. Nosaka, Y.; Koenuma, K.; Ushida, K.; Kira, A. *Langmuir* in press.
12. Nakato, Y.; Tsumura, A.; Tsubomura, H. *J. Phys. Chem.*, **1983**, *87*, 2402.
13. Nakato, Y.; Tsumura, A.; Tsubomura, H. *Chem. Phys. Lett.* **1982**, *85*, 387.
14. Nakato, Y.; Ogawa, H.; Morita, K.; Tsubomura, H. *J. Phys. Chem.* **1986**, *90*, 6210.
15. Nakato, Y.; Tsubomura, H. *Denki Kagaku* **1989**, *57*, 1108 (in Englsih).
16. Grabner, L; Stokowski, S. E.; Brower, Jr., W. S. *Rhys. Rev. B* **1970**, *2*, 590.
17. Smandek, B.; Gerischer, H. *Electrochim. Acta* **1989**, *34*, 1411.
18. Nakato, Y.; Akanuma, H.; Shimizu, J. -I.; Magari, Y. *J. Electroanal. Chem.* **1995**, *396*, 35.
19. Nakato, Y.; Akanuma, H.; Shimizu, J. -I.; Magari, Y. *J. Phys. Chem.* to be submitted.
20. Marcus, R. A. *Ann. Rev. Phys. Chem.* **1964**, *15*, 155.
21. Morrison, S. R.; Freund, T. *J. Chem. Phys.* **1967**, *47*, 1563.
22. Herrmann, J. M. a private communication.
23. Ohtani, B.; Nishimoto, S. *J. Phys. Chem.* **1993**, *97*, 920.

Chapter 22

The Behavior of Pyrazine and Monoprotonated Pyrazine Adsorbed on Silver Electrodes

A Surface-Enhanced Raman Scattering Study

A. G. Brolo and D. E. Irish

Guelph-Waterloo Centre for Graduate Work in Chemistry, Department of Chemistry, University of Waterloo, Waterloo, Ontario N2L 3G1, Canada

Electrochemical processes involving pyrazine (pz) and monoprotonated pyrazine cation (pzH$^+$) adsorbed on a silver surface have been investigated by Surface-Enhanced Raman Scattering (SERS). The investigation of the faradaic and non-faradaic behaviour of pz adsorbed on silver by SERS is a good example of the application of this technique to the study of solid-liquid interfaces. Electrochemical SERS can be used to infer the orientation of the adsorbed molecule. The results presented in this work indicate that pz adsorbs end-on via the lone pair electrons on the nitrogen, and this orientation is not potential dependent. On the other hand, the pzH$^+$ adsorption mode does change with potential. A flat adsorbed cation is predominant at potentials more negative than -300 mV. The end-on adsorbed pzH$^+$ dominates the SERS spectrum at potentials more positive than -170 mV. Electrochemical SERS is also useful for the *in situ* study of faradaic processes. The electroreduction of pz was observed at potentials more negative than -900 mV, and the reduction product was identified as the 1,4-dihydropyrazine cation (DHPz$^+$).

Surface-Enhance Raman Scattering (SERS) has been widely used to monitor electrode processes (1). The advantage of SERS over other spectroelectrochemical methods is that it makes possible the *in situ* vibrational characterization of the electrode/aqueous solution interface. Being so, SERS is used in the study of electrochemical solid-liquid interfaces to infer the orientation of adsorbed molecules and to characterize electrochemical products. In this work, the adsorption of pyrazine and some related species on silver electrodes was investigated.

A typical SERS spectroelectrochemical cell is shown in Figure 1. The working electrode (mainly Cu, Ag, or Au) must be activated. The activation procedure consists of application of oxidation-reduction cycles (ORC's) which create the required roughness. This activation procedure is distinct for different working electrodes. In this work a silver working electrode in a 1 M KBr medium was used; the activation procedure consisted of ORC's from -800 mV to -50 mV (versus SCE) at 5 mV/s under laser illumination (514.5 nm; 100 mW). Figure 2 shows a typical voltammogram obtained in the absence and in the presence of pz. The presence of the organic molecule affects the electrochemical behaviour of silver in bromide medium. Activating in the presence of pz leads to a better SERS signal-to-noise ratio.

The roughness features produced in the activation procedure -- ranging from 30 nm to 100 nm -- support the excitation of the metal's surface plasmons by both the incident and the scattered radiation. The coupling, involving the photon and the metallic resonance, increases the local electric field, thus leading to an enhancement in the Raman signal (2). This contribution to the SERS phenomenon is called the *Electromagnetic (EM)* mechanism. Another contribution to the overall effect is known as the *Chemical or Charge Transfer (CT)* mechanism; it comes from the atomic scale roughness (adatoms and/or small clusters), also called "active sites". In this case, enhancement is attributed to a resonance involving the CT band of the active-site/adsorbed-molecule complex and both the incident and the scattered light; this is a resonance Raman-like mechanism (3).

The SERS selection rules can be used to infer molecular orientation. For a situation where only the EM mechanism is important, a set of SERS selection rules can be derived. The expression for the effective polarizability when an enhanced local electric field of a rough surface bathes the adsorbed molecule is given by equation 1, below (4):

$$\alpha_{eff} = \frac{(\chi + 1)^2}{(\varepsilon_i + \chi)(\varepsilon_s + \chi)} \begin{bmatrix} \alpha_{XX} & \alpha_{XY} & \kappa_s\alpha_{XZ} \\ \alpha_{YX} & \alpha_{YY} & \kappa_s\alpha_{YZ} \\ \kappa_i\alpha_{ZX} & \kappa_i\alpha_{ZY} & \kappa_i\kappa_s\alpha_{ZZ} \end{bmatrix} \quad (1)$$

The subscripts i and s stand for incident and scattered radiation. The term χ corresponds to the polarizability of the metal aggregate of the rough surface, ε_i and ε_s are the metal dielectric constants (relative to the surrounding medium) at the incident and scattered wavelengths, κ is related to the metal dielectric constant, and Z is the axis perpendicular to the electrode surface. The frequency of the incident photon is generally close to the frequency of the scattered light; hence, as a good approximation, one can consider: $\kappa_i \approx \kappa_s \approx \kappa$. Assuming that the direction of the vibration determines the contribution from the molecular polarizability tensor, an enhancement given by the relation $1:|\kappa|^2: |\kappa|^4$ is expected for modes transforming as (α_{XX}, α_{XY}, α_{YX}, α_{YY}), (α_{XZ}, α_{YZ}, α_{ZX}, α_{ZY}), and α_{ZZ}, respectively. In other words, the vibrational modes with polarizability changes preferentially perpendicular to the electrode surface are expected to be more intense than the parallel ones. Additionally, the CT process would enhance the symmetrical vibrational modes (4).

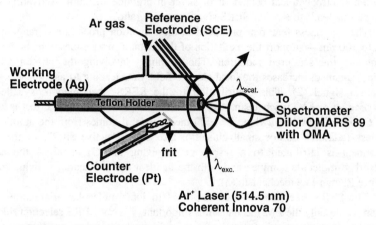

Figure 1. Spectroelectrochemical cell for *in situ* SERS experiments.

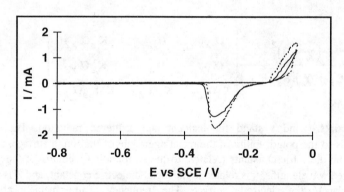

Figure 2. Activation procedure for a silver electrode in 1.0M KBr in the presence (solid line - [pz] = 0.01 M) and absence (dotted line) of pyrazine. ORC from -800 mV to -50 mV at 5 mV/s.

Application of Electrochemical SERS to the Determination of the Orientation of Molecules adsorbed on an Electrode Surface.

Adsorption of Pyrazine on Silver Electrodes. Pyrazine can interact with the metallic surface by either the lone pair electrons of nitrogen or via the π cloud of the ring. These interactions induce the two modes of adsorption presented in Figure 3. A comparison involving the normal Raman spectrum (NRS) of a 1 M pyrazine solution and the SERS spectrum (E_{app} = -600 mV vs. SCE) is presented in Figure 4, for the region between 500 cm^{-1} to 800 cm^{-1}. A band at ca. 630 cm^{-1} (in Figure 4) is very weak for the solution but shows up with fair intensity on the electrode surface. On the other hand, the band at 697 cm^{-1} is stronger for pz in solution than for the adsorbed molecule. The forms of the two normal vibrational modes, obtained from ref. 5, are presented below:

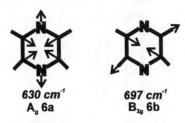

630 cm^{-1}
A$_g$ 6a

697 cm^{-1}
B$_{2g}$ 6b

The polarizability change for the 630 cm^{-1} vibration is mainly along the N-N axis of the pz molecule. In contrast, the polarizability change for the 697 cm^{-1} mode is mainly perpendicular to the N-N molecular axis. The SERS selection rules predict that modes with polarizability changes perpendicular to the electrode surface should be most enhanced. Hence, the spectrum presented in Figure 4 suggests that pz is adsorbed on the electrode surface via its N lone pair, as illustrated in Figure 3a. In fact, this conclusion cannot be drawn based solely on these two bands; the entire spectrum must be considered. A summary of the observed pz SERS bands is presented in Table I. A detailed analysis of the pz SERS spectrum (6, 7) showed that the A$_g$ modes (which contain the α_{ZZ} term of the molecular polarizability, considering an end-on adsorption) dominate the spectrum, followed by the B$_{2g}$ and B$_{3g}$ modes (which contain the α_{XZ} and α_{YZ} components, respectively). The B$_{1g}$ vibrational modes (containing the α_{XY} term) give the weakest signals. These results confirm the end on adsorption. Additional bands due to normally forbidden pz modes ("u - type") are also present. These NRS-forbidden modes were activated by the large electric field gradient near the electrode surface (8). When an inhomogeneous electromagnetic field bathes the molecule adsorbed on the metallic surface, the induced dipole moment (μ) is given by the expression (8):

$$\mu = \alpha E + \frac{1}{3} A \nabla E \qquad (2)$$

Here A is a third rank tensor, called the quadrupole polarizability, operating over the field gradient. This tensor transforms as the hyperpolarizability. α is the

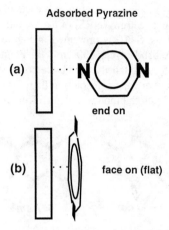

Figure 3. Possible orientations for adsorption of pyrazine on a metallic surface.

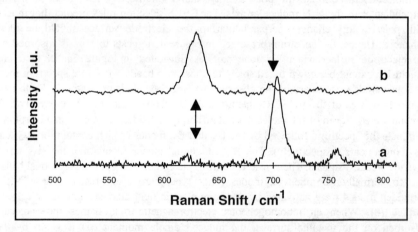

Figure 4. Comparison of pz spectra obtained in different situations. (a) NRS of pz from a 1 M pz solution. (b) SER spectra of pz adsorbed on a silver electrode from a 0.1 M pz + 1 M KBr solution.

polarizability, E is the electric field and ∇E is the field gradient. The vector's and tensor's Cartesian components were omitted for simplification. The second term in the induced dipole expression is very important for rough surfaces. The tensor A transforms as a product of three translations (8). Considering an image-charge approach, the translations in the X and Y directions should yield $-X$ and $-Y$ images inside the metal (Figure 5). The Z direction translation vector, however, would have an image with the same sign and intensity yielding a positive resultant as indicated in Figure 5. As the quadrupole polarizability can be thought of as a product of three translations, the products that do not change sign on the image-charge transformation would be active (8). For instance, the A_{ZZZ}, A_{XYZ}, A_{YYZ} and A_{XXZ} should be surface active and the A_{XXX}, A_{YYY}, A_{YZZ}, A_{XYY}, A_{YXX} and A_{XZZ} produce surface inactive vibrations. It is also important to point out that the assumption that X and Y components are completely screened by the surface is not fully valid when one is working in the visible range of the electromagnetic spectrum (8).

Figure 6 shows the SER spectrum of pz adsorbed on silver at ca -700 mV. The 360 cm^{-1} band is an A_u out of plane band. It is practically impossible to detect this band in the IR of pz because the A_u mode does not contain any component from the dipole moment. However, this mode contains the A_{XYZ} component of the quadrupole polarizability, and hence it is expected to be surface active by the field gradient model (8), and it is observed (Table I and Figure 6).

Moreover, following the field gradient mechanism, the bands that contain the A_{ZZZ} components should also be enhanced. For pz under D_{2h} symmetry, B_{1u} modes span A_{ZZZ}. Figure 6 shows that, in fact, the B_{1u} mode bands are observed with very weak intensity at 1118 and 1040 cm^{-1}.

Some B_{2u} out of plane bands also show up with weak intensity, but are undoubtedly present at 430 and 797 cm^{-1} (Figure 6) The B_{2u} modes span the A_{YYY}, A_{XXY} and A_{YZZ} terms of the quadrupole polarizability; hence, these bands are expected to be surface inactive for a pz molecule adsorbed perpendicular to the surface by the field gradient model. The presence of unexpected "surface inactive" bands for the end on adsorption may reflect the limitations of the field gradient model. In order to derive the activity of "u" modes in the SER spectrum, Sass et al (8) considered an image-charge approach, but it is well known that the molecular dipole screening from the electrons in the metal in the X and Y direction is not as effective in the visible region as in the IR. Hence, for Raman spectroscopy, the X and Y contributions to the polarizability change may become important.

Figure 7 shows the dependence of the ring breathing mode (at ca. 1000 cm^{-1}) SER intensity on the applied potential (potential profile). The SER intensity increases as the potential becomes negative, and reaches a maximum positive to the potential-of-zero-charge, pzc ~ -912 mV (SO_4^{-2} medium (9)). The "bell shaped" curve may reflect the change of surface concentration with applied potential (10); however, additional charge-transfer contributions cannot be discarded (11). All pz SERS bands presented the "bell shaped" potential profile as shown in Figure 7; however, their relative intensities did not alter with change of the applied potential, which suggests that the pz orientation does not change with alteration of the applied potential (7).

Table I: SERS bands of pyrazine arranged according to their species type:

Species	Frequency / cm^{-1}	Number	components
A_g	3030 (s)	2	$\alpha_{xx}, \alpha_{yy}, \alpha_{zz}$
	1590 (vs)	8a	
	1215 (m)	11+16b	
	1235 (s)	9a	
	1020 (vs)	1	
	630 (m)	6a	
B_{1g}	916 (vw)	10a	α_{xy}
B_{2g}	3020 (m)	7b	α_{xz}
	1514 (m)	8b	
	1324 (w)	3	
	700 (m)	6b	
B_{3g}	741 (w)	4	α_{yz}
A_u	360 (w)	16a	A_{xyz}
B_{1u}	1118 (vw)	18a	$A_{zzz}, A_{yyz}, A_{zzx}$
	1040 (vw)	12	
B_{2u}	797 (w)	11	$A_{yyy}, A_{xxy}, A_{yzz}$
	430 (w)	16b	

The SER intensities are given in parentesis.
s - strong, m - medium, w - weak, v - very

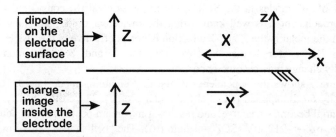

Figure 5. Electrical dipoles oriented on the electrode surface and their images inside the metal.

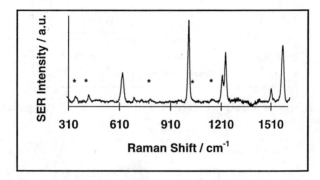

Figure 6. SER spectrum of pz adsorbed on silver electrode. E = -700 mV. The Ag electrode was activated in 1.0 M KBr solution in the presence of 0.01 M pz. * designates the NRS forbidden bands discussed in text.

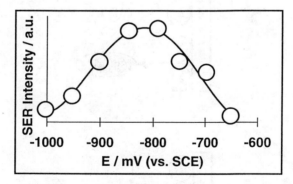

Figure 7. Dependence of the pz's ring breathing mode (ca. 1000 cm⁻¹) SER intensity with the applied potential (potential profile). Pz adsorbed on silver electrode from a 0.1 M pz + 1 M KBr solution.

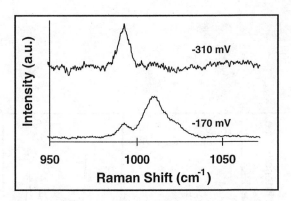

Figure 8. Spectra of pzH⁺ adsorbed on a silver electrode at two potentials.

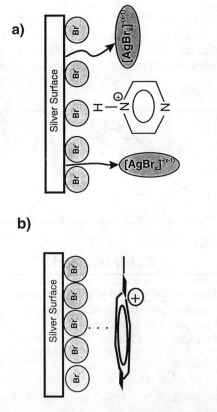

Figure 9. Two pzH⁺ orientations suggested from the SER results.

Adsorption of Monoprotonated Pyrazine Cation on Silver Electrodes. Pyrazine is a very weak base ($pK_{a1} = 0.65$ (12)), and hence the monoprotonated pz spectrum was obtained in a 1 M $HClO_4$ + 1 M KBr solution. Figure 8 shows spectra of pzH^+ adsorbed on a silver electrode at two different potentials. One band at ca. 990 cm^{-1} is observed for an applied potential of -310 mV vs. SCE. A broad envelope containing two bands (one at ca. 1015 cm^{-1} and another at ca. 1030 cm^{-1}, corresponding to the modes number 1 and 12 in Wilson's notation, respectively) appears as the applied potential becomes more positive (see the spectrum at -170 mV presented in Figure 8). The forms of these two normal vibrational modes are presented below (13):

1015 cm⁻¹   **1030 cm⁻¹**

It is important to point out that a silver oxidation wave is observed in the voltammogram at -170 mV, and the spectral intensity has some time-dependence due to that electrochemical reaction. The changes of the spectral features with potential are consistent with a potential-induced reorientation of the pzH^+ cation. By analogy to pz two modes of adsorption are expected for pzH^+ (figure 9). However, the monoprotonated cation is not expected to interact directly with a positively charged surface. The adsorption of cations on specifically adsorbed halides has been observed by SERS (14). The NRS of pzH^+ in solution contains two bands in this region (6): one at ca. 1020 cm^{-1} (ring breathing mode) and another at 1030 cm^{-1} (ring extension). According to the SERS selection rules, the ring breathing mode is expected to be active for both adsorbed orientations (either end-on or flat). The 1030 cm^{-1} band should be active only for end-on adsorbed pzH^+. Moreover, the interaction involving the π system of the ring with the electrode surface is expected to shift the ring breathing mode to a lower wavenumber; this, in fact, occurs for the SER spectrum of flat adsorbed benzene (15). Therefore, we can conclude that pzH^+ is predominantly adsorbed flat on the specifically adsorbed halide layer at potentials more negative than -300 mV (Figure 9b). The adsorbed cation stands up as the applied potential is switched to positive values. The formation of silver oxidation products in this halide medium may stabilize an end-on adsorbed cation. The envelope containing two bands is due to this new orientation (Figure 9a).

Electrochemical reduction of Pyrazine in 1.0 M KBr Medium

Figure 10 shows SER spectra of pyrazine adsorbed on a silver electrode at applied potentials more negative than -900 mV vs. SCE. The pz features vanish as the potential becomes negative, and new bands due to a pz electrochemical reduction product can be observed. Figure 10 is a good example of the use of the SERS technique as a tool to monitor *in situ* faradaic processes. The reduction of pz is a well characterized process.

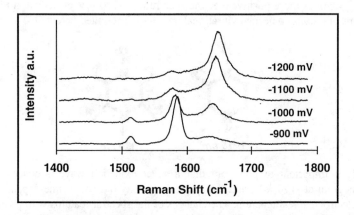

Figure 10. SER spectra of pzH⁺ adsorbed on silver electrodes at several potentials.

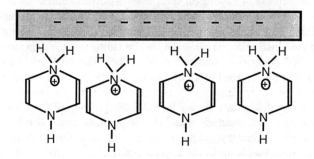

Figure 11. The electrochemically-produced 1,4 dihydropyrazine cation adsorbed on a silver electrode at a potential more negative than the pzc.

Several authors agree that the main product is the 1,4-dihydropyrazine cation (DHPz$^+$) (16). The vibrational bands observed in the SER spectrum of pz at potentials more negative than -900 mV are consistent with the formation of that ion (7). The DHPz$^+$ may interact directly with the electrode surface at potentials more negative than the pzc, as presented in Figure 11. However, the re-oxidation of the DHPz$^+$ to pz is not complete; hence, some of the bands due to this electroreduced pz were also observed at potentials more positive than the pzc (after the DHPz$^+$/pz oxidation current was observed). The adsorption of the cation on a positively charged electrode may occur through the halide layer in a mechanism similar to the adsorption of the pzH$^+$ cation on a halide-coated silver electrode. The presence of DHPz$^+$ trapped on the electrode surface, even at potentials more positive than the pzc, can lead to misassignments of the pz SERS bands (7). This result suggests that care must be taken during the activation procedure in SERS to avoid electrochemically generated trapped species.

Conclusions

The SERS spectra of pz and pzH$^+$ adsorbed on silver electrodes were measured. Orientations of these molecules on the electrode surface were obtained from the analysis of the relative intensities of the SERS bands, and subsequent application of the SERS selection rules. Pz adsorbs end-on, via the N lone pair. Several pz bands which are normally forbidden in the NRS appeared in the SER spectrum with fair intensity. The presence of these bands was explained, based on the field gradient mechanism; nevertheless, this mechanism does not explain the origin of all forbidden pz bands. The pz SER spectrum is potential dependent; however, there is no evidence of reorientation in the potential range studied. pzH$^+$ is predominantly adsorbed flat on the halide-coated silver electrode at potentials more negative than -300 mV. An end-on adsorbed cation is observed at potentials more positive than -170 mV. The pz molecule was also electroreduced in halide medium at potentials more negative than -900 mV. Bands due to the reduction product were observed. The reduced pz was identified as the 1,4-dihydropyrazine cation. This cation may be trapped on the electrode surface, and thus its bands can be observed even at potentials more positive than -900 mV.

Acknowledgment:

This work was supported by grants from the Natural Sciences and Engineering Research Council of Canada.

Literature Cited

1 Pettinger, B., in *Adsorption of Molecules at Metal Electrodes*, Lipkowski, J., and

Ross, P. N., Eds., VCH, New York, (1992), Ch. 6, p. 285.

2 Moskovits, M., *Rev. Mod. Phys.*, **1985**, *57*, 783.

3   Otto, A., Mrozek, I., Grabhorn, H. and Akeman, W, *J. Phys. Condens. Matter*,

1992, *4*, 1143.

4   Craighton, J. A., in *Spectroscopy of Surfaces*; Clark, R. J. H. and Hester, R. E.,

Eds., Advances in Spectroscopy; Wiley, Chichester, 1988, vol. 16, Ch. 2, p. 37.

5   Lord, R. C., Marston, A. L. and Miller, F. A., *Spectrochim. Acta*, **1976**, *32A*, 605.

6   Brolo, A. G. and Irish, D. E., *Z. Naturforsh., part A*, **1995**, *50a*, 274.

7   Brolo, A. G. and Irish, D. E., *J. Electroanal. Chem.*, **1996**, in press.

8   Sass, J. K., Neff, H., Moskovits, M. and Holloway, S., *J. Phys. Chem.*, **1981**, 85,

621.

9 *Encyclopedia of Electrochemistry of the Elements*; Bard, A. J.; ed.; Dekker, New

York, 1973, Vol. VIII.

10   Guidelli, R., in *Adsorption of Molecules at Metal Electrodes*, Lipkowski, J., and

Ross, P. N., Eds., VCH, New York, (1992), Ch. 1, p. 1.

11   Lombardi, J. R., Birke, R. L., Lu, T. and Xu, J., *J. Chem. Phys.*, **1986**, *84*, 4174

12   Chia, A. S-C. and Trimble Jr., R. F., *J. Phys. Chem.*, **1961**, *65*, 863.

13   Lin-Vien, D.; Colthup, N. B.; Fatley, W. G.; Grasselli, J. G.; *The Handbook of

Infrared and Raman Characteristic Frequencies of Organic Molecules*; Academic,

Boston, 1991.

14   Deng, Z. and Irish, D. E., *Langmuir*, **1994**, *10*, 586.

15   Moskovits, M. and DiLella, D. P., *J. Chem. Phys.*, **1980**, *73*, 6068.

16   Swartz, J. and Anson, F. C., *J. Electroanal. Chem.*, **1980**, *114*, 117.

Chapter 23

Temperature Dependence of Growth of Surface Oxide Films on Rhodium Electrodes

Francis Villiard and Gregory Jerkiewicz[1]

Département de chimie, Université de Sherbrooke,
Sherbrooke, Québec J1K 2R1, Canada

Surface oxides on Rh electrodes were formed by anodic polarization at potentials, E_P, between 0.70 and 1.40 V, RHE, with an interval of 0.05 V for polarization times, t_P, up to 10 000 s and at temperatures, T, between 278 and 348 K. This procedure results in thin films having their charge density, q_{OX}, of less than 1 260 $\mu C\,cm^{-2}$, thus their thickness, X, of up to 2 ML of $Rh(OH)_3$. Cyclic-voltammetry, CV, reveals one states, OC1, in the oxide reduction profiles. Increase of T leads to augmentation of the oxide thickness but it does not influence its surface state; thermodynamics of their reduction are not affected by T variation. Plots of q_{OX} versus $\log t_P$ or $1/q_{OX}$ versus $\log t_P$ for a wide range of T and E_P allow one to discriminate between the logarithmic and the inverse-logarithmic oxide growth kinetics. Two kinetic regions are observed in the oxide formation plots, each one giving rise to a distinct growth mechanism. Oxides having $X \leq 1\,ML$ of RhOH are formed according to the logarithmic kinetics and the process is limited by the rate of the place exchange between the Rh surface atoms and the electroadsorbed OH groups. Formation of oxides having X between 1 ML of RhOH and 2 ML of $Rh(OH)_3$ follows the inverse-logarithmic kinetics and the process is limited by the rate of escape of Rh^{3+} from the metal into the oxide. Theoretical treatment of the data in the region corresponding to $X \rangle 1\,ML$ of RhOH leads to determination of the potential drop across the film and the electric field within the oxide layer, the latter being of 10^9 V m^{-1}.

Surface oxide films on various transition metals can be formed by application of electrochemical techniques or by exposure to an oxidizing atmosphere and the extent of surface oxidation, thus the oxide thickness, is affected by the oxidation conditions.

[1]Corresponding author

In the case of electroformation of oxides, the extend of surface oxidation depends on the nature of the metals, the polarization conditions (polarization potential, current density or time, E_P, i_P and t_P, respectively) and the electrolyte composition and pH (*1-20*). The oxide formed on an electrode surface markedly affects anodic Faradaic electrode processes at the double-layer by: (i) affecting the reaction energetics; (ii) changing electronic properties of the metal electrode; (iii) imposing a barrier to the charge transfer; (iv) affecting the adsorption properties of the reaction intermediates and products (*14-17*). Knowledge of the chemical and electronic state of the surface oxide is of major importance in electrocatalysis since it determines the electrocatalytic properties of the surface at which various anodic Faradaic reactions take place.

At noble metals, the growth of submonolayer and monolayer oxides can be studied in detail by application of electrochemical techniques such as cyclic-voltammetry, CV (*11-20*) and such measurements allow precise determination of the oxide reduction charge densities. Complementary X-Ray photoelectron spectroscopy (XPS), Auger electron spectroscopy (AES), infra-red (IR) or ellipsommetry experiments lead to elucidation of the oxidation state of the metal cation within the oxide and estimation of the thickness of one oxide monolayer (*12,21-23*). Coupling of electrochemical and surface-science techniques results in meaningful characterization of the electrified solid/liquid interface and in assessment of the relation between the mechanism and kinetics of the anodic process under scrutiny and the chemical and electronic structure of the electrode's surface (*21-23*).

Rhodium, Rh, like other noble metals, forms surface oxides upon anodic polarization even in the region of water stability, thus below the thermodynamic reversible potential of the oxygen evolution reaction, $E_{OER}^{\circ} = 1.23$ V, SHE (*1-20*). In aqueous H_2SO_4 solution, the oxide growth on Rh commences at 0.55 V, RHE (reversible hydrogen electrode), and up to ca. 1.40 V, RHE, a complete monolayer (ML) of $Rh(OH)_3$ is formed as revealed by coupled CV and XPS measurements (*21*). Polarization at potentials between 3.00 and 4.00 V, RHE, leads to formation of oxides comprising two various species, namely $Rh(OH)_3$ and $RhO(OH)$. Surface oxides on Rh were formed in aqueous H_2SO_4 solution by application of potentiostatic polarization at E_P up to 2.4 V, RHE, for polarization times up to 10^4 s. The data demonstrated that these conditions result in thin films having their thickness of less than 4 equivalent ML of $Rh(OH)_3$. Subsequently, kinetics of the oxygen evolution reaction, OER, on pre-oxidized Rh electrodes were evaluated and a relation between the oxide thickness and the kinetic parameters of the OER was established (*24,25*). Some of the most recent data on electrochemical processes refer to Rh single-crystal electrodes and the experiments were focused on the adsorption of hydrogen, small inorganic/organic species, and interaction of anions with the substrates; they showed that a great deal of information may be related to surface specific parameters (*26-31*).

In this paper, the authors demonstrate the first data on the temperature dependence of formation of monolayer oxides at Rh in aqueous H_2SO_4 solution. The growth of Rh surface oxides was accomplished by potentiostatic polarization at $0.70 \le E_P \le 1.40$ V, RHE, for $t_P \le 10^4$ s. Extensive studies at $278 \le T \le 348$ K followed by a theoretical treatment result in unambiguous distinction between two different growth mechanisms.

Experimental

Electrode Preparation. The Rh electrode was of 99.99% purity; the preparation methodology, which was important in order to ensure excellent reproducibility of the experimental results, is described elsewhere *(24,25,32,33)*. The shape of the CV profiles recorded in aqueous H_2SO_4 solution indicated that both the electrode and the electrolyte were free of impurities *(32,33)*. It was discussed elsewhere *(24,25)* that formation of thin oxides does not affect the electrode surface which sustains its surface as it is revealed from CV profiles for the under-potential deposition of H (UPD H). The real surface area of the Rh electrodes was determined by accepting the charge of $210 \ \mu C \ cm^{-2}$ as the charge necessary to form a monolayer of H_{UPD} *(1,24,25)*, allowing for the double-layer charging; it was found to be $0.26 \pm 0.01 \ cm^2$. No anodic dissolution of the anodically formed films was observed during the course of research. This conclusion is based on the observation that the q_{ox} versus $\log t_p$ or the $1/q_{ox}$ versus $\log t_p$ relations do not deviate from linearity *(17)*.

Solution and Electrochemical Cell. A high-purity solution was prepared from BDH Aristar grade H_2SO_4 and Nanopure water. The cleanliness of the solution was verified by recording CV profiles which were found in agreement with those in the literature *(1-3,18,19,24,25)*. The electrochemical cell employed was a standard, all-glass, three-compartment one with sleeved stopcocks. The glassware was pre-cleaned according to the well-established procedure *(32,33)*. During the experiments, pre-cleaned and pre-saturated with water vapor H_2 gas was bubbled through the reference electrode, RE, compartment in which a Pt/Pt black electrode was immersed.

Temperature Measurements. The electrochemical cell was immersed in a water bath, Haake W13, and the temperature was controlled to within $\pm 0.5 \ K$ by means of a thermostat, Haake D1; the water level in the bath was maintained above the electrolyte in the cell. The temperature in the water bath and the electrochemical cell were controlled by means of thermometers, $\pm 0.5 \ K$, and a K-type thermocouple, 80 TK Fluke, and the data were found to agree to within $\pm 0.5 \ K$.

Instrumentation. The electrochemical instrumentation included: (a) EG&G Model 273 potentiostat-galvanostat, (b) 80486 computer, and (c) EG&G M270 Electrochemical Software and programs developed in this Laboratory. All potentials quoted in the paper were measured with respect to the reversible hydrogen electrode, RHE; upon anodic polarization the potentials were held constant to within 1 mV and were monitored on a Fluke 45 digital multimeter.

Results and Discussion

General Remarks. Polarization of Rh electrodes at E_p between 0.70 and 1.40 V, RHE, with an interval of 0.05 V, for t_p up to 10 000 s leads to formation of monolayer oxide films (Figures 1 and 2). The oxide-formation CV profiles reveal one peak and a wide wave overlapping each other in the 0.55 and 1.40 V, RHE, range.

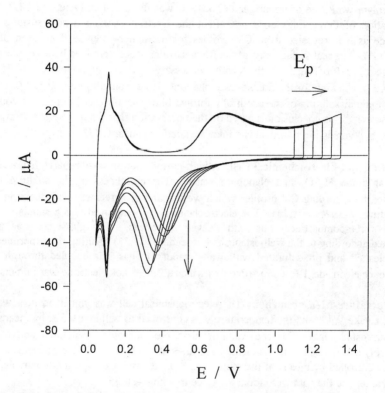

Figure 1. Series of CV profiles for Rh in 0.5 M aq H_2SO_4 solution taken at the sweep rate $s = 50$ mV s^{-1}; the electrode real surface area equals 0.26 ± 0.01 cm^2. The profiles demonstrate samples of the effect of increase of the polarization potential, E_P, from 1.10 to 1.35 V, RHE, on the oxide thickness while sustaining T constant at 298 K and the polarization time, t_P, at 1000 s. The CV oxide-reduction profiles reveal only one oxide state, OC1, which shifts towards less-positive potentials.

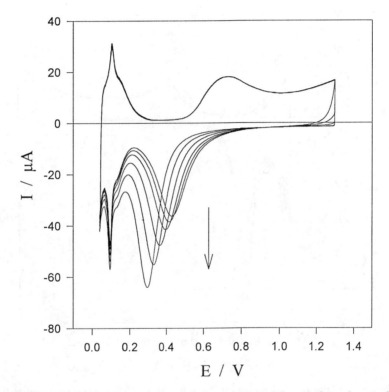

Figure 2. Series of CV profiles for Rh in 0.5 M aq H_2SO_4 solution taken at the sweep rate $s = 50$ mV s^{-1}; the electrode real surface area equals 0.26 ± 0.01 cm^2. The profiles demonstrate samples of the effect of increase of the polarization time, t_p, from 0 to 10 000 s, on the oxide thickness while sustaining T constant at 308 K and the polarization potential, E_p, at 1.30 V, RHE. The CV oxide-reduction profiles reveal only one oxide state, OC1, which shifts towards less-positive potentials.

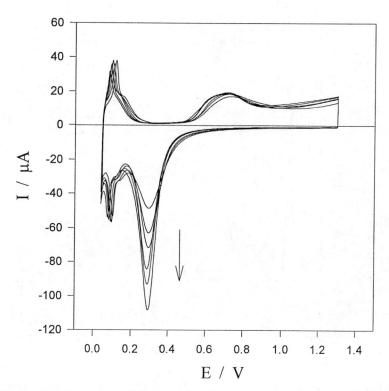

Figure 3. Series of CV profiles for Rh in 0.5 M aq H_2SO_4 solution taken at the sweep rate $s = 50$ mV s^{-1}; the electrode real surface area equals 0.26 ± 0.01 cm^2. The profiles demonstrate samples of the effect of increase of T from 288 K to 348 K, on the oxide thickness while sustaining the polarization time, t_P, constant at 10 000 s and the polarization potential, E_P, at 1.30 V, RHE. The CV oxide-reduction profiles reveal only one oxide state, OC1, which does not change its peak potential.

The CV oxide-reduction profiles reveal only one feature designated the OC1 peak (*24,25*); it increases its overall reduction charge density and it shifts towards less-positive potentials upon extension E_p or t_p. In the case of the increase of E_p (for $t_p = const$ and $T = const$, Figure 1), the oxide-reduction peak shifts from 0.43 to 0.32 V, RHE. In the case of the increase of t_p (for $E_p = const$ and $T = const$, Figure 2), the oxide-reduction profile shifts from 0.43 to 0.29 V, RHE. Similar behavior is observed for other values of E_p and t_p than those referred to in Figures 1 and 2. An increase of E_p or t_p always results in enhanced oxide reduction charge densities, thus in thicker oxide films of the same surface-chemical nature (*21,24,25*). This perception is based on the observation that the same oxide state is distinguishable in the CV reduction profiles. It should be added that extension of E_p or t_p not only increases the oxide thickness but it also affects the thermodynamic of its reduction. Clearly, lower potentials are required to reduce the same but thicker oxide film when E_p or t_p is increased, thus it is reasonable to conclude that thicker oxide films are thermodynamically more stable than the thin ones.

It is informative to deliberate on the surface processes which occur during the oxide formation and reduction. Elsewhere (*21*), it was demonstrated based upon electrochemical and XPS measurements that the electroformation of Rh surface oxides proceeds in two steps. The first step involves electrodeposition of OH and the process occurs in the 0.55 – 0.75 V, RHE, potential range (see equation 1) whereas the second step leads to initial formation of $Rh(OH)_3$ (see equation 2). Although the mechanism of the very initial formation of Rh oxide films is not well understood yet and it awaits further investigation, it seems reasonable to assume that the very initial development of RhOH is followed by the place-exchange between the Rh surface atoms and the electroadsorbed OH groups. Evidence for the place exchange between Rh surface atoms and adsorbed OH groups is based on theoretical treatment of the oxide growth mechanism and X-Ray scattering data as is presented in refs. *34* and *35*.

The oxide reduction behavior is more straightforward and the existence of one peak in the reduction profiles (OC1) indicates that this is an one-step process.

Formation
$$Rh + H_2O \ \rightarrow \ RhOH + H^+ + e \qquad\qquad 0.55 - 0.75\ V \qquad\qquad (1)$$

$$RhOH + 2\,H_2O \ \rightarrow \ Rh(OH)_3 + 2\,H^+ + 2\,e \quad 0.75 - 1.40\ V \qquad (2)$$

Reduction
$$Rh(OH)_3 + 3\,H^+ + 3\,e \ \rightarrow \ Rh + 3\,H_2O \qquad\qquad\qquad (3)$$

A new behavior is presented in Figure 3 and it shows an impact of T increase on the oxide growth. The experimental data, which is the main effect presented in the current paper, show that the oxide films increase their thickness when the temperature is raised to higher values. However, the oxide-reduction peak does not shift towards less-positive potentials when the oxide thickness increases (compare it with the behavior reported above on increase of E_p or t_p). Lack of shift of the oxide reduction peak potential indicates that the thermodynamics of the oxide reduction are

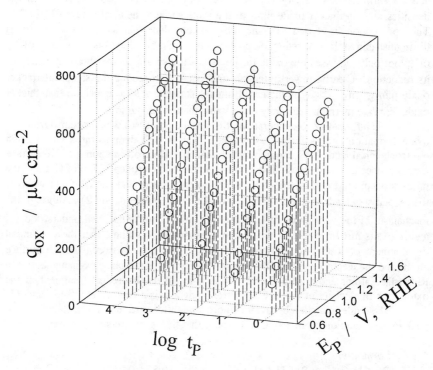

Figure 4A. Rh oxide growth plots expressed as the reduction charge density, q_{OX}, versus the polarization potential, E_P, and log of the polarization time, $\log t_P$, for $T = 278$ K; E_P is between 0.70 and 1.40 V, RHE, and $\log t_P$ is between 0 and 4. The data demonstrate linear relations between q_{ox} and $\log t_P$ for every $E_P = $ const.

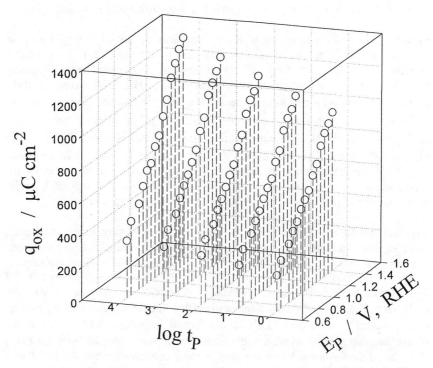

Figure 4B. Rh oxide growth plots expressed as the reduction charge density, q_{ox}, versus the polarization potential, E_P, and log of the polarization time, $\log t_P$, for $T = 348$ K; E_P is between 0.70 and 1.40 V, RHE, and $\log t_P$ is between 0 and 4. The data demonstrate linear relations between q_{ox} and $\log t_P$ for every $E_P = $ const.

not affected by temperature and the oxides have the same Gibbs free energy of reduction even though the films becomes thicker when the temperature is raised.

The anodic polarization of Rh electrodes at $0.70 \leq E_p \leq 1.40$ V, RHE, for $1 \leq t_p \leq 10^4$ s and at $278 \leq T \leq 348$ K results in oxides whose reduction charge density is between 210 and 1260 $\mu C \, cm^{-2}$, thus in films whose thickness is between 1 ML of RhOH and 2 ML of $Rh(OH)_3$. It is worthwhile mentioning that the electroformation of oxide films on Rh proceeds slower than that on Pt under equivalent polarization conditions (15-17). Thus one may conclude that the oxide formation-reduction behavior at Rh differs not only quantitatively but also qualitatively from that at Pt.

Surface-Chemical Composition of Rh Oxide Films by XPS. The XPS data on electrochemically formed Rh oxide films are limited to one paper (21). They indicate that the initial electro-oxidation of Rh involves an electroadsorbed OH group, $Rh - OH_{ads}$, and the process takes place beyond 0.55 V, RHE. Upon extension of the polarization potential a new surface species, $Rh(OH)_3$, is formed; one monolayer of $Rh(OH)_3$ is grown upon reaching 1.40 V, RHE. At potential higher than 1.40 V, RHE, formation of RhO(OH) commences on top of 3 ML of $Rh(OH)_3$ (25). The existing XPS data indicate that extension of the polarization potential beyond some 0.8 - 0.9 V, RHE, results in oxide films with Rh in the +3 oxidation state. Finally, the experimental data and results presented in ref. 21 indicate that the OC1 oxide-reduction peak corresponds to the surface process shown in equation 3.

Temperature Effect. The influence of T variation on the electrochemical formation of Rh surface oxides has never been investigated before and in this respect the paper represents a new contribution. In the course of research the authors investigated changes of the oxide growth behavior brought about by T variation between 278 and 348 K for all the E_p and t_p values cited above. A representative sample of the results is shown in Figure 3 and the data demonstrate that upon T increase (for E_p = const and t_p = const) the amount of the oxide formed increases. One of the major observation that arises from these studies is that the T increase results in thicker oxide films than those formed at low T but the CV oxide-reduction profiles do not change their characteristics and show the same OC1 peak. On the basis of this observation and the above cited XPS data, one may conclude that the same oxide state, namely $Rh(OH)_3$, is formed when T is raised from 278 to 348 K (for $0.70 \leq E_p \leq 1.40$ V, RHE, and $1 \leq t_p \leq 10^4$ s). This is an important observation with respect to studies of the kinetics and mechanism of electroformation of Rh oxides.

The authors evaluated the oxide-reduction charge densities for all the E_p, t_p and T values discussed in Experimental. Some representative sample of the results are shown in Figure 4 in the form of 3D plots of q_{ox} versus $(\log t_p, E_p)$. It should be added that the formation of very thin (monolayer) oxide films often proceeds by either the logarithmic or the inverse-logarithmic growth kinetics and it is illustrative to represent the oxide charge densities as a function of log of E_p and t_p, $\log t_p$. This representation is advantageous in a subsequent analysis of the oxide growth kinetics and mechanism (see Oxide Growth Kinetics and Mechanism).

The results shown in Figure 4 show that at the lowest T, T = 278 K, the oxide reduction charge density, q_{ox}, increase from 100 to 790 μC cm^{-2} upon extension of E_p and t_p; the lowest q_{ox} corresponds to some 0.5 ML of RhOH whereas the largest one to some 1.25 ML of $Rh(OH)_3$. In the case of polarization at T = 348 K, q_{ox} increases from 220 to 1350 μC cm^{-2} upon extension of E_p and t_p; the lowest q_{ox} corresponds to 1.05 ML of $Rh(OH)_3$ and the largest one to 2.14 ML of $Rh(OH)_3$. The thickness of the oxides formed at the highest T and E_p and the longest t_p only slightly exceeds 2 equivalent ML of $Rh(OH)_3$. Finally, it is important to emphasize that an increase of T by 70 degrees extends the oxide film by not more than 1 ML of $Rh(OH)_3$, thus the oxide growth is slow in time.

Oxide Growth Kinetics and Mechanism.   Formation of oxide films by potentiostatic polarization and their characterization by CV enables distinction of various oxide states as a function of the polarization conditions, here E_p, t_p and T. This method allows precise determination of the thickness of oxide films with accuracy comparable to the most sensitive surface science techniques (*4-7,11-20*). CV may be considered the electrochemical analog of temperature programmed desorption, TPD, and one may refer to it as potential programmed desorption, PPD. Theoretical treatment of such determined oxide reduction charge densities by fitting of the data into oxide formation theories leads to derivation of important kinetic parameters of the process as a function of the polarization conditions. The kinetics of electro-oxidation of Rh at the ambient temperature were studied and some representative results are reported in ref. *24*. The present results are an extension of the previous experiments and they involve temperature dependence studies.

In the very initial (submonolayer) development of the surface oxide, a film of chemisorbed O-containing species (here OH) exists on the metal surface. The O-containing species create a surface dipole moment which induces the place exchange process, thus the O-containing species and the metal surface atoms, M, undergo the place exchange which results in formation of a new 2D surface structure (*10,11,34*). The arrangement of this structure depends on the surface crystallography of the metal substrate (*20,24*). The place exchange mechanism gives rise to the logarithmic kinetics and the process is driven by the electric field associated with the presence of the surface dipole moment and its flip-over due to the place exchange (*34*). During the process the electrons pass from the metal surface atoms to the O-containing adsorbates through tunneling or thermoionic emission. Upon completion of the electron transfer, a strong electric field becomes established across the metal/oxide interface and it is responsible for the subsequent pulling metal cations from the metal surface into the oxide at the metal/solution interphase or for the ion transport through the oxide film. In this region the inverse-logarithmic kinetics predominate (*24,36-38*).

Applicability of various oxide growth mechanisms to any metal-oxide system may be tested by evaluation of the X_O and X_1 parameters (see equation 4), and by plotting oxide thicknesses, X (which may also be expressed as the oxide mass, m, or the oxide reduction charge density, q_{ox}) or their reciprocals, 1/X (1/m or 1/q_{ox}), versus logarithm of the polarization times, log t_p. The X_O and X_1 parameters are defined as follows:

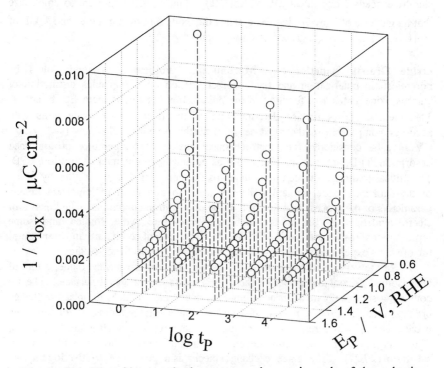

Figure 5A. Rh oxide growth plots expressed as reciprocals of the reduction charge density, $1/q_{ox}$, versus the polarization potential, E_p, and log of the polarization time, $\log t_p$, for $T = 278$ K; E_p is between 0.70 and 1.40 V, RHE, and $\log t_p$ is between 0 and 4. The data demonstrate that the $1/q_{ox}$ versus $\log t_p$ relations (for $E_p = $ const) are linear for oxide films having their thickness, X, more than 1 ML of RhOH.

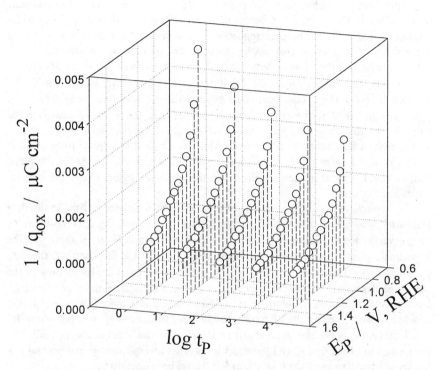

Figure 5B. Rh oxide growth plots expressed as reciprocals of the reduction charge density, $1/q_{ox}$, versus the polarization potential, E_P, and log of the polarization time, $\log t_P$, for $T = 348\,K$; E_P is between 0.70 and 1.40 V, RHE, and $\log t_P$ is between 0 and 4. The data demonstrate that the $1/q_{ox}$ versus $\log t_P$ relations (for $E_P = $ const) are linear for oxide films having their thickness, X, more than 1 ML of RhOH.

$$X_O = \sqrt{\frac{\kappa \, k \, T}{8 \, \pi \, n \, q^2}} \qquad\qquad X_1 = \frac{q \, a \, V}{k \, T} \qquad\qquad (4)$$

where κ is the relative permittivity of the oxide film, n the number of metal ions per unit volume expressed in cm^3, q the metal cation charge, V the potential drop across the oxide, a the distance between metal cations in the film (or between the metal atom and the metal cation at the inner metal/oxide interphase, designated a'), k Boltzmann's constant; these parameters represent critical oxide thicknesses. If the experimentally observed oxide thickness, X, is below or above X_O or X_1, then a specific oxide-growth mechanism might become applicable. However, this is a prerequisite and the applicability of a given mechanism has to be tested by data fitting procedures. A detail discussion of the relations between X, X_O and X_1 is given in refs. *24,36–38*.

Here, the authors emphasize that in the case of monolayer Rh oxides $X \langle X_O$ and $X \langle X_1$ (very thin oxide films, thus beyond the submonolayer limit where the process is governed by the place exchange; see ref. *24*), the electrons penetrate the oxide by the tunneling of thermoionic effect and leave metal cations behind, thus a strong electric field becomes established. This electric field for typical films of 50 Å or less is of the order of 10^9 V m^{-1} and it enforces directional escape of the metal cations from the metal surface into the oxide; this mechanism gives rise to the inverse-logarithmic kinetics.

Applicability of the logarithmic or inverse-logarithmic mechanisms to the data presented above may be established by plotting q_{OX} versus $\log t_P$ (Figure 4) or $1/q_{OX}$ versus $\log t_P$ (Figure 5). Experimental data shown in Figure 4 show that the logarithmic plots are linear over the whole range of E_P. On the contrary, the data shown in Figure 5 indicate that the $1/q_{OX}$ versus $\log t_P$ plots are linear only in the case of oxides whose thickness is more than 1 ML of RhOH; deviations from linearity are observed in the case of submonolayer oxide films. These observations lead to the conclusion that: (i) either the Rh oxide growth is logarithmic in t_P over the whole E_P and T range, (ii) or the process is logarithmic in the E_P and T range corresponding to the formation of the first RhOH layer and it becomes inverse-logarithmic beyond this thickness limit, thus once the first ML of RhOH has been completed.

It should be recognized that during the electroadsorption of the first ML of the OH groups, the changes of the surface dipole moment are significant. However, they become less pronounced upon commencement of the place exchange when a chess-board-like pattern is formed with OH and M occupying adjacent surface sites. Moreover, the deposition of the second, third etc. monolayer of OH results in much smaller changes of the surface dipole moment than the adsorption of the first ML of OH. Thus it is reasonable to conclude that the electroadsorption of the first ML of OH is logarithmic in time whereas the growth of subsequent layers of Rh surface oxides becomes inverse-logarithmic. This conclusion is in agreement with the Mott-Hauffe-Ilschner theorem that formation of the very-first oxide film follows the logarithmic law and a changeover to the inverse-logarithmic one is observed once the initial oxide monolayer has been completed (*38*).

The inverse-logarithmic theory of growth of surface oxides implies that the rate of monolayer oxide formation is controlled by the rate of escape of the metal cation, here Rh^{3+}, from the metal into the oxide and it leads to the following kinetic formula (see Figure 6):

$$\frac{dX}{dt} = N \Omega v \exp\left(-\frac{H}{kT}\right) \exp\left(\frac{q a' V}{X k T}\right) \tag{5}$$

where N is the surface density of atoms, Ω is the volume of oxide per metal atom, v stands for the surface metal atom vibrational frequency (usually of the order of 10^{12} s^{-1}), $H = H_i + U$, H_i represents for the enthalpy of solution of the metal cations in the oxide, U is the activation energy of diffusion, V / X is the electric field E across the oxide, and a' is as in Figure 6. Integration of this relation (24) leads to the following simplified relationship:

$$\frac{1}{X} = \frac{1}{X_1} \ln\left(\frac{X_1 u}{X_L^2}\right) + \frac{1}{X_1} \ln t_P \tag{6}$$

where $X_1 = q a' V / k T$, $u = N \Omega v \exp(-H / k T)$ and X_L is the limiting oxide thickness (36-38).

The experimental results corresponding to oxide films thicker than 1 ML of RhOH (Figure 5) were fitted into the Mott-Cabrera theory by converting the oxide reduction charge densities into oxide thicknesses. This conversion is accomplished by accepting the charge density of formation of 1 ML of $Rh(OH)_3$ as 630 $\mu C \ cm^{-2}$ and its thickness as 4.16 Å (24). The slope of the $1 / q_{ox}$ versus $\log t_P$ plots equals:

$$\frac{1}{X_1} = 2.3026 \frac{k T}{q a' V} \tag{7}$$

where the 2.3026 coefficient originates from transformation of ln to log and a' = 2.105 Å (24). It should be added that the slopes of the $1 / q_{ox}$ versus $\log t_P$ plots depends on the polarization potential, E_P (Figure 5), but this does not affect the electric field across the oxide (see the discussion below). An analysis of equation 7 indicated that the slope of the $1 / q_{ox}$ versus $\log t_P$ plots allows one to determine the potential drop across the oxide film, V. Subsequently, one can determine the electric field, E, across the oxide by dividing the potential drop, V, by the oxide thickness, X. Calculations indicate that the potential drop across the oxides is of the order of $0.15 - 0.80$ V and the respective electric field of the order of 10^9 $V \ m^{-1}$. The kinetic parameters determined according to the Mott-Cabrera treatment and summarized in Table I indicate agreement between the assumptions on which the theory is based and the experimental results. Thus it is concluded that the formation of monolayer oxide films on Rh electrodes having their nominal thickness between 1 ML of RhOH and 2 ML of $Rh(OH)_3$ proceeds according to the inverse-logarithmic kinetics and the process is limited by the rate of escape of the Rh^{3+} cation from the metal into the oxide at the inner metal/oxide interface. An analysis of the results shown in Figures 4

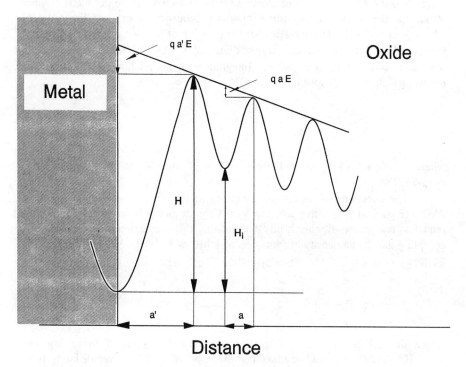

Figure 6. Schematic representation of the potential energy profile for a metal cation escaping from the metal surface into the growing surface oxide film assisted by the strong field established across the interface; a' is half of the distance between the surface metal atom and the metal cation in the oxide film; a is half of the distance between two metal cations in the oxide film (a' $\rangle$ a); $H = H_i + U$, where H_i is the heat of solution of the metal cations, U is the activation energy of diffusion of the metal cation in the oxide; E is the electric field across the oxide and $E = V / X$ (Adopted from ref. *24*).

and 5 indicates that no changes are observable in the q_{ox} versus $(\log t_P, E_P)$ or the $1/q_{ox}$ versus $(\log t_P, E_P)$ relations that could be associated with the temperature increase, thus indicating that the mechanism and the kinetics of the oxide growth are not affected by T under the E_P and t_P conditions presented in the paper.

Table I. Summary of the Kinetic Parameters of the Monolayer Rh Oxide Growth Determined on the Basis of the Mott-Cabrera Theorem

Temperature T / K	Potential drop for $E_P = 0.7 - 1.4$ V V / V	Oxide thickness for $E_P = 0.7 - 1.4$ V X / nm	Electric field across the oxide E / V m^{-1}
278	0.15 - 0.57	0.18 – 0.46	$0.80 - 1.30 \times 10^9$
288	0.16 - 0.65	0.19 – 0.52	$0.85 - 1.30 \times 10^9$
298	0.15 - 0.68	0.19 – 0.56	$0.85 - 1.15 \times 10^9$
308	0.20 - 0.78	0.21 – 0.62	$0.95 - 1.30 \times 10^9$
318	0.24 - 0.65	0.23 – 0.60	$0.95 - 1.15 \times 10^9$
328	0.23 - 0.68	0.25 – 0.66	$0.95 - 1.10 \times 10^9$
338	0.30 - 0.74	0.27 – 0.77	$0.90 - 1.10 \times 10^9$
348	0.27 - 0.79	0.29 – 0.76	$0.85 - 1.05 \times 10^9$

This paper demonstrate new results on the temperature-dependence of growth of monolayer oxides on Rh electrodes on the basis on electrochemical measurements and the existing XPS data. In a subsequent paper, the authors will report on detail XPS and LEES measurements on Rh oxide films using coupled electrochemical and surface-science techniques. Subsequently, they will report on theoretical treatment of the formation of submonolayer oxides based on the logarithmic growth kinetics.

Conclusions

1. Anodic polarization of Rh electrodes at E_P between 0.70 and 1.40 V, RHE, for t_P up to 10 000 s and at T between 278 and 348 K results in thin oxide films which reveal only one state in the LSV reduction profiles and having their reduction charge density of less than $1\,260$ μC cm^{-2}, thus up to 2 ML of Rh$(OH)_3$.

2. Temperature increase leads to enhancement of the oxide thickness but it does not influence the oxide surface-chemical composition. Thermodynamics of the oxide reduction are not affected by the temperature variation; the oxide-reduction peak potential depends on E_P and t_P but do not on T.

3. Plots of q_{ox} versus $\log t_P$ and $1/q_{ox}$ versus $\log t_P$ for a wide range of T allow one to discriminate between the logarithmic and inverse-logarithmic growth kinetics. Two kinetic regions are distinguishable in the oxide growth; the oxides having $X \leq 1$ ML of RhOH are formed according to the logarithmic kinetics and the process is limited by the rate of the place exchange between Rh surface atoms and the electroadsorbed OH groups; the formation of oxide having X between 1 ML of

RhOH and 2 ML of $Rh(OH)_3$ follows the inverse-logarithmic kinetics and the process is limited by the rate of escape of the Rh^{3+} cation from the metal into the oxide.
4. Theoretical treatment of the oxide growth in the X $\rangle$ 1 ML of RhOH region leads to determination of the potential drop across the film and the electric field within the oxide layer, the latter being of the order of 10^9 V m^{-1}.

Acknowledgments

Acknowledgment is made to the NSERC of Canada and le FCAR du Québec for support of this research.

Literature Cited

1. Woods, R. In *Electroanalytical Chemistry*; Bard, J., Ed.; Marcel Dekker: New York, 1977, Vol. 9; pp 27-90.
2. Burke, L. D. In *Electrodes of Conductive Metallic Oxides*, Trasatti, S., Ed.; Elsevier: New York, 1980, Part A; pp 141-181.
3. Burke, L. D. In *Modern Aspects of Electrochemistry*, Bockris, J.O'M., Conway, B.E., Eds.; Plenum Press: New York, 1986, No. 18; pp 169-248.
4. Shibata, S. *Bull. Chem. Soc. Jpn.* **1965**, *38*, 1330-1337; *J. Electroanal. Chem.* **1978**, *89*, 37-58; see also Shibata, S.; Sumino, M. *Electrochim. Acta* **1975**, *20*, 739-746.
5. Damjanovic, A.; Dey, A.; Bockris, J. O'M. *J. Electrochem. Soc.* **1966**, *113*, 739-746.
6. James, S.D. *J. Electrochem. Soc.* **1969**, *116*, 1681-1688.
7. Balej, J.; Spalek, O. *Collect. Czech. Chem. Commun.* **1972**, *37*, 499-512.
8. Burke, L. D.; O'Sullivan, E. J. M. *J. Electroanal. Chem.* **1979**, *93*, 11-18; **1979**, 97, 123-125; **1980**, *112*, 247-252; **1981**, *129*, 133-148.
9. Rand, D.A.J.; Woods, R. *J. Electroanal. Chem.* **1972**, *35*, 209-218.
10. Vetter, K. J.; Schultze, J. W. *J. Electroanal. Chem.* **1972**, *34*, 131-139; 141-158.
11. Angerstein-Kozlowska, H.; Conway, B. E.; Sharp, W. B. A., *J. Electroanal. Chem.* **1973**, *43*, 9-36.
12. Gottesfeld, S. *J. Electrochem. Soc.* **1980**, *127*, 272-277.
13. Angerstein-Kozlowska, H.; Conway, B. E.; Hamelin, A.; Stoicoviciu, L. *Electrochim. Acta* **1986**, *31*, 1051-1061; see also Angerstein-Kozlowska, H.; Conway, B. E.; Tellefsen, K.; Barnett, B. *Electrochim. Acta* **1989**, *34*, 1045-1056.
14. Conway, B. E.; Liu, T. C. *Proc. R. Soc. Lond. A* **1990**, *429*, 375-397.
15. Conway, B. E.; Tremiliosi-Filho, G.; Jerkiewicz, G. *J. Electroanal. Chem.* **1991**, *297*, 435-443.
16. Jerkiewicz, G.; Tremiliosi-Filho, G.; Conway, B. E. *J. Electroanal. Chem.* **1992**, *334*, 359-370.
17. Tremiliosi-Filho, G.; Jerkiewicz, G.; Conway, B. E. *Langmuir* **1992**, *8*, 658-667.
18. Cataldi, Z.; De Tacconi, N. R.; Arvia, A. J. *J. Electroanal. Chem.* **1981**, *122*, 367-372.

19. Pallotta, C.; De Tacconi, N. R.; Arvia, A. J. *Electrochim. Acta* **1981**, *26*, 261-273; see also Cataldi, Z.; Lezna, R. O.; Giordano, M. C.; Arvia, A. J. *J. Electroanal. Chem.* **1988**, *261*, 61-75.

20. Conway, B. E.; Jerkiewicz, G. *J. Electroanal. Chem.* **1992**, *339*, 123-146.

21. Peuckert, M. *Surface Sci.* **1984**, *141*, 500-514.

22. Kim, K.S.; Winograd, N.; Davis, R. E. *J. Am. Chem. Soc.* **1971**, *93*, 6296-6297.

23. Allen, G.C.; Tacker, P.M.; Capon, A.; Parsons, R. *J. Electroanal. Chem.* **1974**, *50*, 335-343; see also Hammond, J.S.; Winograd, N. *J. Electroanal. Chem.* **1977**, *78*, 55-69.

24. Jerkiewicz, G.; Borodzinski, J. J. *Langmuir* **1993**, *9*, 2202-2209.

25. Jerkiewicz, G.; Borodzinski, J. J. *J. Chem. Soc., Faraday Trans.* **1994**, *90*, 3669-3675.

26. Hourani, M.; Wieckowski, A., *J. Electroanal. Chem.* **1988**, *244*, 147-161.

27. Wasberg, M., Hourani, M.; Wieckowski, A. *J. Electroanal. Chem.* **1990**, *278*, 425-432.

28. Zelenay, P., Horanyi, G., Rhee, C. K.; Wieckowski, A. *J. Electroanal. Chem.* **1991**, *300*, 499-519; see also Hourani, G., Wasberg, M., Rhee, C. K.; Wieckowski, A. *Croat. Chem. Acta* **1990**, *63*, 373-399.

29. Krauskopf, E. K.; Wieckowski, A. In *Adsorption of Molecules at Metal Electrodes*; Lipkowski, J. and Ross, P. N., Eds.; VCH Publishers: New York, 1992; pp 119-169.

30. Leung, L.-M. H.; Weaver, M. J. *J. Phys. Chem.* **1989**, *93*, 7218-7226.

31. Leung, L.-M. H., Chang, S.-C.; Weaver, M. J. *J. Chem Phys.* **1989**, *90*, 7426-7435.

32. Conway, B. E.; Sharp, W.B.A.; Angerstein-Kozlowska, H.; Criddle, E. E. *Anal. Chem.* **1973**, *45*, 1331-1336.

33. Angerstein-Kozlowska, H. In *Comprehensive Treatise of Electrochemistry*; Yeager, E.; Bockris, J.O'M.; Conway, B. E.; Sarangapani, S., Eds.; Plenum Press: New York, 1984, Vol. 9; pp 15-59.

34. Conway, B. E.; Barnett, B.; Angerstein-Kozlowska, H.; Tilak, B. V. *J. Chem. Phys.* **1990**, *93*, 8361-8373.

35. You, H., Zurawski, D. J., Nagy, Z.; Yonco, R. M. *J. Chem. Phys.* **1994**, *100*, 4699-4702.

36. Cabrera, N.; Mott, N. F. *Rep. Prog. Phys.* **1948-49**, *12*, 163-184; see also Mott, N. F. *J. Chim. Phys.* **1949**, *44*, 172-180.

37. Kubaschewski, O.; Hopkins, B. E. *Oxidation of Metals and Alloys*; Butterworths: London, 1962.

38. Hauffe, K. *Oxidation of Metals*; Plenum Press: New York, 1965.

INDEXES

Author Index

Abreu, J. B., 274
Batina, N., 171
Borup, R. L., 283
Brisard, Gessie M., 142
Brolo, A. G., 310
Fairbrother, D. Howard, 106
Feliu, Juan M., 156
Forrer, P., 210
Furuya, Mikio, 61
Gómez, Roberto, 156
Gasteiger, Hubert A., 87,142
Glosli, James N., 13
Goodman, D. W., 71
Hara, Y., 202
Hossner, L. R., 71
Irish, D. E., 310
Itaya, K., 171,274
Jerkiewicz, Gregory, 1,45,323
Johnston, H. S., 106
Kim, Y.-G., 171
Koinuma, Michio, 189
Kunitake, M., 171
Lucas, C. A., 87
Magari, Y., 297
Marcus, P., 236
Markovic´, Nenad M., 87,142
Matsue, T., 202
Matsumoto, Y., 245
Maurice, V., 236
Murrell, T. S., 71
Nakato, Y., 297
Nooney, M. G., 71
Ochi, H., 297
Ogaki, K., 171

Okawa, Y., 245
Orts, José M., 156
Philpott, Michael R., 13
Repphun, G., 210
Ross, Philip N., Jr., 87,142
Sasahara, A., 245
Sashikata, K., 274
Sauer, D. E., 283
Schmidt, E., 210
Sekine, Namiki, 189
Shiku, H., 202
Siegenthaler, H., 210
Somorjai, G. A., 106
Sorenson, T. A., 115
Soriaga, Manuel P., 274
Spohr, E., 31
Stickney, J. L., 115
Stuve, E. M., 283
Sung, Y.-E., 126
Takeda, T., 202
Tanaka, K., 245
Temesghen, W. F., 274
Thomas, S., 126
Uchida, I., 202
Uosaki, Kohei, 189
Villiard, Francis, 323
Wan, L.-J., 171
Wieckowski, Andrzej, 126
Wilmer, B. K., 115
Yae, S., 297
Yamada, T., 171
Ye, Shen, 189
Zenati, Entissar, 142
Zolfaghari, Alireza, 45

Affiliation Index

Hokkaido University, 189
IBM Almaden Research Center, 13
Kanagawa Industrial Technology Research
    Institute, 61
Lawrence Livermore National
    Laboratory, 13

Osaka University, 297
Texas A&M University, 71,274
Tohoku University, 171,202,274
Universidad d'Alacant, 156
Universität Bern, 210
Universität Ulm, 31

Université de Sherbrooke, 1,45,142,323
Université Pierre et Marie Curie, 236
University of California—Berkeley,
   87,106,142
University of Georgia, 115

University of Illinois at Urbana-
   Champaign, 126
University of Tokyo, 245
University of Washington, 283
University of Waterloo, 310

Subject Index

A

Acidified water surfaces
   degree of surface sensitivity, 111
   experimental description, 107–109
   stratospheric significance, 107
   sulfuric acid, 109–111
   surface composition, 111–113
   surface vs. bulk composition, 113–114
Adsorption
   H_2 on Ni(110) surface, 245–246
   iodide on single-crystal electrodes, 171
   organic molecules on electrode
      surfaces, 172
Adsorption energy, effect on density,
   computer simulation of structure and
   dynamics of water near metal surfaces,
   34–36
Adsorption-induced reconstruction,
   surface metal atom, 245
Adsorption isotherms, development, 4f,5
Alcohols, role in photooxidation reaction
   of water on n-TiO_2 electrode, 297–308
Alloy single-crystal surface, thin anodic
   oxide overlayers, 236–244
Alloy surfaces, chemical reconstruction,
   246–247
Alumina, growth kinetics of phosphate
   films, 83,84f
Anion(s), role in boundary layer at metal-
   electrolyte solution interface, 126–127
Anion adsorption and charge transfer on
   single-crystal electrodes
   advantages and disadvantages, 169
   anion coverages, 161,162f
   ball models for adlayer structures, 161,163f
   charge correction, 167,169
   defects, influence on adsorption, 165,166f

Anion adsorption and charge transfer on
      single-crystal electrodes—Continued
   experimental procedure, 157–159
   polycrystalline electrode activated by
      electrochemical cycling, 165,167
   Pt(531) electrode, 167,168f
   state of electrode surface using cyclic
      voltammetry, 164,165f
   voltammetric profile, 158f,160
Antigen–antibodies, use in glass surfaces,
   202–209
Atomic force microscopy
   development, 7
   micrometer-scale imaging of native
      oxide on silicon wafers, 61–69
   See also Electrochemical atomic force
      microscopy
Atomic layer epitaxy, 117
Atomic level studies of CdTe(100),
   electrochemical digital etching, 115–123
Auger electron spectrometers, 106
Auger electron spectroscopy
   bisulfate anion adsorption on Au(111),
      Pt(111), and Rh(111) surfaces, 126–138
   to monitor composition of acidified
      water surfaces, 106–114
   underpotential deposition of lead on
      Cu(100) and Cu(111), 142–154
   use for bisulfate anion adsorption, 127

B

Bimetallic surfaces
   catalytic reaction of NO + H_2,
      268f,269–272
   characterization
      Pt–Rh(100) and Pt–Rh(110) surfaces,
         265–267,269

Bimetallic surfaces—*Continued*
characterization—*Continued*
Rh–Pt(100) and Pt–Rh(100) surfaces,
254,256–265
chemical reconstruction, 246–247
experimental procedure, 247
preparation by electrochemical
deposition, 247–253,255
Bisulfate anion adsorption on Au(111),
Pt(111), and Rh(111) surfaces
adsorption threshold against estimated
potentials of zero charge, 136f,137–138
Auger electron spectroscopy, 130f,131,133
core-level electron energy loss
spectroscopy, 132f,133
electrode potential, effect on S2p core
loss, 135
experimental description, 127,129
factors affecting energy decrease, 134–135
low-energy electron diffraction,
132f,133–134
voltammetry, 128f,129,131,133–134
Bond energy between metal substrate
and underpotential deposited species,
determination, 45–57
Bromide adsorption, effects on
underpotential deposition of Cu at
Pt(111)–solution interface
adsorption isotherm of Br$^-$ on Pt(111),
91–93
adsorption isotherm of Br$^-$ on Pt(111)–
Cu surface, 99–101
adsorption isotherm of Cu^{2+} on Pt(111)–
Br$^-$ surface, 95,97–98
Cu–Br structure on Pt(111)
ex situ ultrahigh vacuum measurements,
95,96f
X-ray scattering measurements, 101–103
experimental description, 88–91
model, 104
surface structure of Br$^-$ on Pt(111), 93–95

C

Cadmium–tellurium (100), atomic level
studies using electrochemical digital
etching, 115–123

Cadmium–tellurium single crystals,
etching, 115
Carbon monoxide, electrocatalysis
with probe adlayers of carbon and
ethylidyne on Pt(111), 283–294
Carbon probe adlayers on Pt(111),
electrocatalysis of formic acid and
carbon monoxide, 283–294
Carcinoembryonic antigen, use in glass
surfaces, 202–209
Catalysis, bimetallic surfaces, 245–272
Catalytic activity of electrodes, role
of surface property modifications, 142
Charge displacement technique to
determine charge in adsorption process
on single-crystal electrodes, 159
Charge-transfer mechanism, contribution to
surface-enhanced Raman scattering, 311
Charge transfer on single-crystal electrodes,
See Anion adsorption and charge
transfer on single-crystal electrodes
Computer simulation
electrochemical interfaces, 31
structure and dynamics of water near
metal surfaces
experimental description, 31–32
models, 32–34
structure and dynamics of interfacial
water
density profiles, 34
adsorption energy, 34–36
liquid water–liquid mercury
interface, 36–37
molecular polarizability, 37–38
surface corrugation, 38–39
interfacial polarization, 39–40,41f
residence times, 40–42
Copper, underpotential deposition at
Pt(111) interface, effects of Br$^-$
adsorption, 87–104
Copper (100) and (111) single-crystal
surfaces, underpotential deposition
of lead, 142–154
Core-level electron energy loss
spectroscopy, bisulfate anion adsorption
on Au(111), Pt(111), and Rh(111)
surfaces, 126–138

D

Diaphorase, use in glass surfaces, 202–209
Diffuse double layer, model of
  electrochemical interface, 1,3f,5
Digital etching
  description, 115,117
  See also Electrochemical digital etching
Dry etching, 115

E

Effective polarizability, determination, 311
Effective thickness of restructured carbon
  adlayer, determination, 290
Electric double-layer screening, molecular
  dynamics simulation of interfacial
  electrochemical processes, 13–27
Electric potential across double layer,
  molecular dynamics simulation of
  interfacial electrochemical processes,
  14–15
Electrified solid–liquid interface
  contemporary model, 6f,7
  first visual representation, 1,2f
Electroactive polymers
  electropolymerization reaction scheme
    polyhydroxyphenazine, 211–212
    polythiophene, 211
  electropolymerization routines and
    cyclic voltammetry
    polyhydroxyphenazine, 215,217–219
    polythiophene, 214f,215,216f
  equilibrium charges, 220–223
  experimental description,
    211–213,219,223
  instrumentation, 213
  morphology
    polyhydroxyphenazine, 213,225,227,233
    polythiophene, 213,223–226,232
  voltammetric behavior, 229,232
  voltammetric capacities, 220–223
Electrocatalysis of formic acid and CO
  with probe adlayers of carbon and
  ethylidyne on Pt(111)
  electrooxidation of solution-dosed CO
    vs. carbon coverage, 286f,287

Electrocatalysis of formic acid and CO
  with probe adlayers of carbon and
  ethylidyne on Pt(111)—*Continued*
  experimental description, 284,285,287
  formic acid oxidation with carbon and
    ethylidyne probe adlayers, 291–294
  H and CO adsorption vs. carbon adlayer,
    289–291
  HCOOH electrooxidation vs. carbon
    coverage, 288f,289
  hysteresis vs. ethylidyne, 289,292f
  maximum coverages of solution-dosed
    and vapor-dosed CO vs. carbon
    coverage, 287–289
  voltammetric behavior vs. carbon
    coverage, 286f,287
Electrocatalytic activity, modification
  of electrode surfaces, 283
Electrochemical activation technique,
  voltammetric profiles, 157
Electrochemical atomic force microscopy,
  surface structures of GaAs(100) surface
  during electrochemical reactions, 189–200
Electrochemical digital etching
  Auger spectra, 118f,119
  charge passed vs. potential used for
    oxidation, 121,123f
  chromoamperogram
    reduction of previously oxidized
      CdTe(100) surface, 121,122f
    ion bombarded annealed CdTe(100)
      surface, 119,122f
  creation of pitted morphology, 121
  experimental description, 117,119
  low-energy electron diffraction pattern
    of CdTe(100) crystal, 118f,119
  principle, 117
  voltammograms on argon ion bombarded
    annealed CdTe(100) surface, 119,120f
Electrochemical interface, models and
  research, 1–8
Electrochemical metal deposition,
  bimetallic surfaces, 245–272
Electrochemical scanning tunneling
  microscopy
  development, 7
  solid–liquid interface, 274–281

Electrochemically active polymers, classes, 210–211

Electrode surfaces, modification for electrocatalytic activity, 283

Electron spectroscopy
acidified water surfaces, 106–114
solid–liquid interface, 274–281

Electrostatics, molecular dynamics simulation of interfacial electrochemical processes, 15

Ensembles, molecular dynamics simulation of interfacial electrochemical processes, 16

Enzymes, use in glass surfaces, 202–209

Etching of CdTe single crystals, 115

Ethanol, role in photooxidation reaction of water on n-TiO$_2$ electrode, 297–308

Ethylidyne probe adlayers on Pt(111), electrocatalysis of formic acid and CO, 283–294

Ex situ low-energy electron diffraction
organic molecules adsorbed on iodine-modified Au(111), Ag(111), and Pt(111) electrodes, 171–186
underpotential deposition of lead on Cu(100) and Cu(111), 142–154

Ex situ scanning tunneling microscopy, thin anodic oxide overlayers on metal and alloy single-crystal surfaces, 241–244

Ex situ ultrahigh vacuum spectroscopy, effect of Br$^-$ adsorption on underpotential deposition of Cu at Pt(111)–solution interface, 87–104

F

Fixed double layer, model of electrochemical interface, 1,3f

Formic acid, electrocatalysis with probe adlayers of carbon and ethylidyne on Pt(111), 283–294

G

GaAs(100) surface during electrochemical reactions, structures, 189–200

Geometrics, computer simulation of structure and dynamics of water near metal surfaces, 2

Glass surfaces patterned with enzymes and antigen–antibodies
alignment of anti-carcinoembryonic antigen-adsorbed polystyrene beads at glass substrate, 205,207f
carcinoembryonic antigen microspotted substrate image, 205,207f
diaphorase immobilized image, 205,206f
experimental description, 202–204
horseradish peroxidase immobilized image, 205,206f
human chorionic gonadotropin image, 208f,209
human placental lactogen image, 208,209f
photofabricated glass substrate used in dual assay, 205,208–209

Gold (111) electrodes, adsorption of organic molecules, 171–186

Gold (111) surface, bisulfate anion adsorption, 126–138

Growth kinetics of phosphate films on metal oxide surfaces
experimental description, 72,74–75
phosphate adsorption
on alumina, 83,84f
on hematite, 75–81
on titania, 81–83
phosphate reactions
with aluminum and titanium oxides, 73–74
with iron oxides, 72–73

Growth of surface oxide films on Rh electrodes
electrochemical formation, 330–333
experimental description, 324,325
instrumentation, 325
kinetic parameters, 338–339
kinetics, 333
measurement techniques, 324
mechanism, 333–336
monolayer oxide formation rate vs. rate of metal-ion escape from metal into oxide, 336–337
oxide reduction behavior, 329

Growth of surface oxide films on Rh
electrodes—*Continued*
oxide thickness, 328*f*,329,332
polarization potential vs. oxide
thickness, 325–326,329
polarization time vs. oxide thickness,
325–326,329
potential energy profile, 337,338*f*
reduction charge density vs. polarization
potential and polarization time,
334–336
surface chemical composition of Rh
oxide films, 332

H

Halide ions, effect of adsorption on
underpotential deposition of Cu at
Pt(111)–solution interface, 88–104
Hematite, growth kinetics of phosphate
films, 75–81
Highly ordered pyrolytic graphite,
adsorption studies, 172
Horseradish peroxidase, use in glass
surfaces, 202–209
Human chorionic gonadotropin, use in
glass surfaces, 202–209
Human placental lactogen, use in glass
surfaces, 202–209
Hydrogen, adsorption on Ni(110) surface,
245–246
Hydrogen, underpotential deposition, 47–54

I

Imaging of native oxide on silicon wafers
using scanning Auger electron spectroscopy, *See* Micrometer-scale imaging
of native oxide on silicon wafers using
scanning Auger electron spectroscopy
In situ scanning probe microscopy,
electroactive polymers, 210–233
In situ scanning tunneling microscopy
organic molecules adsorbed on
iodine-modified Au(111), Ag(111),
and Pt(111) electrodes, 171–186
solid–liquid interface, 274–281

In situ surface X-ray scattering
spectroscopy, effect of Br⁻ adsorption
on underpotential deposition of Cu
at Pt(111)–solution interface, 87–104
Induced dipole moment, determination, 313
Interfacial electrochemical processes,
molecular dynamics simulation, 13–27
Interfacial electron transfer, molecular
dynamics simulation of interfacial
electrochemical processes, 16
Interfacial polarization, computer
simulation of structure and dynamics
of water near metal surfaces, 39–41*f*
Iodine adlayer structures, 171
Iodine-catalyzed dissolution of Pd(110),
electron spectroscopy and
electrochemical scanning tunneling
microscopy, 274–281
Iodine-modified Au(111), Ag(111), and
Pt(111) electrodes, adsorption of
organic molecules, 171–186
Iron–chromium alloy, thin anodic oxide
overlayers, 236–244
Isopropyl alcohol, role in photooxidation
reaction of water on *n*-TiO₂ electrode,
297–308

K

Kinetics of phosphate films on metal
oxide surfaces, *See* Growth kinetics
of phosphate films on metal oxide
surfaces

L

Lead, underpotential deposition on
Cu(100) and Cu(111), spectroscopic
and electrochemical studies, 142–154
Liquid–solid interface, *See* Solid–liquid
interface
Liquid water–liquid mercury interface,
effect on water density, 36–37
Low energy electron diffraction,
bisulfate anion adsorption on Au(111),
Pt(111), and Rh(111) surfaces,
126–138

M

Metal(s), underpotential deposition, 53,55–57
Metal single-crystal surface, thin anodic oxide overlayers, 236–244
Metal surfaces, computer simulation of structure and dynamics of water, 31–42
Metal oxide surfaces, growth kinetics of phosphate films, 75–81
Methanol, role in photooxidation reaction of water on n-TiO$_2$ electrode, 297–308
Microfabrication, glass surfaces patterned with enzymes and antigen–antibodies, 202–209
Micrometer-scale imaging of native oxide on silicon wafers using scanning Auger electron spectroscopy
atomic force microscopy topographical images, 63,64f
compulsive reduction, 63,65,66f
cross-sectional schematic diagram for native oxide film thickness calculation, 67,68f
element map images, 65,68f
experimental description, 62–63
LVV spectrum, 63,66f
native oxide film thickness, 67,69
secondary electron microscopy contrast pattern, 63,64f
Molecular dynamics simulation of interfacial electrochemical processes
1 M NaCl solution, 21–24
2 M NaCl solution, 24–26
3 M NaCl solution, 26–27
experimental description, 13–14
model details, 18–19
path to useful computation of reactions, 16–18
screening of charged metal electrodes in electrolyte, 19–21
status, 14–16
Molecular polarizability, influence on water density, 37–38
Molecule–metal potentials, molecular dynamics simulation of interfacial electrochemical processes, 15–16

Molecule–molecule potentials, molecular dynamics simulation of interfacial electrochemical processes, 15–16
Monoprotonated pyrazine cation adsorbed on Ag electrodes, See Pyrazine and monoprotonated pyrazine cation adsorbed on Ag electrodes

N

Native oxide, micrometer-scale imaging on silicon wafers using scanning Auger electron spectroscopy, 61–69
Native oxide film, characterization on silicon wafers, 61–62
Nickel, thin anodic oxide overlayers, 236–244
Nickel (110) surface, adsorption of H$_2$, 245–246
Nitric oxide, role in bimetallic surfaces, 268f,269–272

O

Organic molecules adsorbed on iodine-modified Au(111), Ag(111), and Pt(111) electrodes
experimental description, 172
other molecules on I–Au(111), 184f,185,186f
structures of iodine adlayers
on Ag(111), 175–177,178f
on Au(111), 173–175,176f
tetrakis$N$-methylpyridinium-4-yl-21H,23H-porphine on I–Ag(111), 179,182,183f
tetrakis(N-methylpyridinium-4-yl-21H,23H-porphine on I–Au(111), 177,179–181f
tetrakis(N-methylpyridinium-4-yl-21H,23H-porphine on I–Pt(111), 182,184–185
Oxide films, See Surface oxide films
Oxygen evolution reaction, mechanistic studies, 297

P

Palladuim (110), iodine-catalyzed
dissolution using electron spectroscopy
and electrochemical scanning tunneling
microscopy, 274–281
Parsons double-layer model, 5
Passive films, *See* Thin anodic oxide
overlayers
Phosphate films, growth kinetics on metal
oxide surfaces, 75–81
Photooxidation reaction of water on
n-TiO$_2$ electrode
electrode activation, 299,301
energy level diagram, 307f,308
ethanol effect, 305,306f
experimental procedure, 299
isopropyl alcohol effect, 305,306f
methanol effect, 301–305
photocurrent vs. concentration,
305,307f,308
photocurrent vs. potential, 300,301f
photoluminescence intensity vs.
potential, 300,301f
previously proposed new mechanism,
298–299
Platinum (111), electrocatalysis of formic
acid and CO with probe adlayers of
carbon and ethylidyne, 283–294
Platinum (111) electrodes, adsorption
of organic molecules, 171–186
Platinum (111)–solution interface, Br⁻
adsorption effects on underpotential
deposition of Cu, 87–104
Platinum (111) surface, bisulfate anion
adsorption, 126–138
Platinum–rhodium (100) surfaces, structure
and catalysis, 245–272
Platinum–rhodium (110) surfaces, structure
and catalysis, 245–272
Platinum single crystals, voltammetric
profiles, 156–157
Polyhydroxyphenazine, electrochemical
and in situ scanning probe microscopy,
210–233
Polymers, electroactive, *See* Electroactive
polymers

Polythiophene, electrochemical and in situ
scanning probe microscopy, 210–233
Potential of zero total charge,
determination, 159
Probe adlayer, carbon and ethylidyne
on Pt, electrocatalysis of formic acid
and CO, 283–294
Pyrazine and monoprotonated pyrazine
cation adsorbed on Ag electrodes
electrochemical reduction of pyrazine
of pyrazine in 1.0 M KBr medium,
319–321
orientation of adsorbed molecules
monoprotonated pyrazine cation
adsorption on Ag electrodes, 318–319
pyrazine adsorption on Ag electrodes,
313–317

R

Residence times, computer simulation
of structure and dynamics of water
near metal surfaces, 40–42
Rhodium, formation of surface oxides, 324
Rhodium electrodes, effect of temperature
on growth of surface oxide films,
323–339
Rhodium (111) surface, bisulfate anion
adsorption, 126–138
Rhodium–platinum (100) surfaces, structure
and catalysis, 245–272
Rhodium–platinum (110) surfaces, structure
and catalysis, 245–272
Rotating ring-disk method, effect of Br⁻
adsorption on underpotential deposition
of Cu at Pt(111)–solution interface,
87–104

S

Scanning Auger electron microscopy,
micrometer-scale imaging of native
oxide on silicon wafers, 61–69
Scanning electrochemical microscopy
advantages and disadvantages, 202

Scanning electrochemical microscopy—
  Continued
  microfabrication and characterization of
    glass surfaces patterned with enzymes
    and antigen–antibodies, 202–209
Scanning probe microscopy, electroactive
  polymers, 210–233
Scanning tunneling microscope, advantages
  and disadvantages, 189–190
Scanning tunneling microscopy
  development, 7
  solid–liquid interface, 274–281
  thin anodic oxide overlayers on metal and
    alloy single-crystal surfaces, 241–244
  See also In situ scanning tunneling
    microscopy
Secondary electron microscopy,
  micrometer-scale imaging of native
  oxide on silicon wafers, 61–69
Semiconductor(s), underpotential
  deposition, 53,55–57
Silicon wafers
  micrometer-scale imaging of native
    oxide using scanning Auger electron
    spectroscopy, 61–69
  surface characterization methods, 61
Silver (111) electrode adsorption
  organic molecules, 171–186
  pyrazine and monoprotonated pyrazine
    cation, 310–321
Single-crystal electrodes
  anion adsorption and charge transfer,
    156–169
  use for study of crystallographic
    dependence of deposition reactions, 142
Single-crystal surfaces, thin anodic oxide
  overlayers, 236–244
Sodium chloride solution, role in molecular
  dynamics simulation of interfacial
  electrochemical processes, 13–27
Solid–liquid electrochemical interface,
  overview of research, 1–8
Solid–liquid interface
  electrochemical scanning tunneling
    microscopy, 277,280–281
  electron spectroscopy, 275–279

Solution–Pt(111) interface, effect of Br⁻
  adsorption on underpotential deposition
  of Cu, 87–104
Stern double layer, model of
  electrochemical interface, 3*f*,5
Sulfuric acid–water surface, stratospheric
  significance, 107
Surface charge, influencing factors, 72
Surface corrugation, effect on density,
  computer simulation of structure and
  dynamics of water near metal surfaces,
  38–39
Surface-enhanced Raman scattering
  behavior of pyrazine and
    monoprotonated pyrazine cation
    adsorbed on Ag electrodes, 310–321
  effect of pyrazine, 311,312*f*
  selection rules, 311
  spectroelectrochemical cell, 310–311,312*f*
Surface metal atom, adsorption-induced
  reconstruction, 245
Surface oxide films, temperature effect
  on growth on Rh electrodes, 323–339
Surface structures of GaAs(100) surface
  during electrochemical reactions
  atomic force microscopy tip-induced
    modification, 198*f*,199–200
  atomically resolved structure, 191–194
  electrodeposition of Cu, 195–199
  experimental description, 190
  p-GaAs(100) electrode
    electrochemical characteristics, 190–192*f*
    tunneling characteristics, 190–192*f*
Surface X-ray scattering spectroscopy,
  effect of Br⁻ adsorption on
  underpotential deposition of Cu at
  Pt(111)–solution interface, 87–104

T

Temperature, role in growth of surface
  oxide films on Rh electrodes, 323–339
Thermodynamics, underpotential
  deposition of hydrogen, semiconductors,
  and metals, 45–57

Thin anodic oxide overlayers
applications, 236
factors affecting growth, stability, and
breakdown, 236
on metal and alloy single-crystal surfaces
ex situ scanning tunneling microscopy,
241–244
X-ray photoelectron spectroscopy,
237–240
n-TiO$_2$ electrode, photooxidation
reaction of water, 297–308
Titania, growth kinetics of phosphate
films, 81–83
Turnover rate for bare Pt surface,
determination, 292f,293

U

Ultrahigh-vacuum-based experimental
techniques, development, 5,7
Ultrahigh-vacuum spectroscopy, effect
of Br$^-$ adsorption on underpotential
deposition of Cu at Pt(111)–solution
interface, 87–104
Underpotential deposition
Cu at Pt(111)–solution interface, 87–104
experimental description, 46–47
hydrogen, 47–54
metals, 53,55–57
Pb on Cu(100) and Cu(111)
Auger electron spectroscopy, 152
comparison of ultrahigh vacuum and
electropolished prepared Cu(hkl)
surface, 144f,145
effect of chloride, 146–149
experimental description, 143,145
low-energy electron diffraction
measurements, 152–154

Underpotential deposition—*Continued*
Pb on Cu(100) and Cu(111)—*Continued*
methods, 142–143
Pb monolayer coverage, 149–152
semiconductors, 53,55–57
Underpotential shift, determination, 46

V

Voltammetry, electroactive polymers,
210–233

W

Water
near metal surfaces, computer
simulation of structure and dynamics,
31–42
photooxidation reaction on n-TiO$_2$
electrode, 297–308
Water density, computer simulation of
structure and dynamics near metal
surfaces, 34–39
Water–sulfuric acid surface, *See* Sulfuric
acid–water surface
Water surfaces, *See* Acidified water
surfaces
Wet etching, description, 115

X

X-ray photoelectron spectroscopy
effect of temperature on growth of
surface oxide films on Rh electrodes,
323–339
thin anodic oxide overlayers on metal
and alloy single-crystal surfaces,
237–240

Highlights from ACS Books

Bestsellers from ACS Books

The ACS Style Guide: A Manual for Authors and Editors
Edited by Janet S. Dodd
264 pp; clothbound ISBN 0–8412–0917–0; paperback ISBN 0–8412–0943–X

Writing the Laboratory Notebook
By Howard M. Kanare
145 pp; clothbound ISBN 0–8412–0906–5; paperback ISBN 0–8412–0933–2

Career Transitions for Chemists
By Dorothy P. Rodmann, Donald D. Bly, Frederick H. Owens, and Anne-Claire Anderson
240 pp; clothbound ISBN 0–8412–3052–8; paperback ISBN 0–8412–3038–2

Chemical Activities (student and teacher editions)
By Christie L. Borgford and Lee R. Summerlin
330 pp; spiralbound ISBN 0–8412–1417–4; teacher edition, ISBN 0–8412–1416–6

Chemical Demonstrations: A Sourcebook for Teachers, Volumes 1 and 2, Second Edition
Volume 1 by Lee R. Summerlin and James L. Ealy, Jr.
198 pp; spiralbound ISBN 0–8412–1481–6
Volume 2 by Lee R. Summerlin, Christie L. Borgford, and Julie B. Ealy
234 pp; spiralbound ISBN 0–8412–1535–9

From Caveman to Chemist
By Hugh W. Salzberg
300 pp; clothbound ISBN 0–8412–1786–6; paperback ISBN 0–8412–1787–4

The Internet: A Guide for Chemists
Edited by Steven M. Bachrach
360 pp; clothbound ISBN 0–8412–3223–7; paperback ISBN 0–8412–3224–5

Laboratory Waste Management: A Guidebook
ACS Task Force on Laboratory Waste Management
250 pp; clothbound ISBN 0–8412–2735–7; paperback ISBN 0–8412–2849–3

Reagent Chemicals, Eighth Edition
700 pp; clothbound ISBN 0–8412–2502–8

Good Laboratory Practice Standards: Applications for Field and Laboratory Studies
Edited by Willa Y. Garner, Maureen S. Barge, and James P. Ussary
571 pp; clothbound ISBN 0–8412–2192–8

For further information contact:

American Chemical Society
1155 Sixteenth Street, NW ◆ Washington, DC 20036
Telephone 800–227–9919 ◆ 202–776–8100 (outside U.S.)
The ACS Publications Catalog is available on the Internet at
http://pubs.acs.org/books